The British Army on Bloomsday

A Military Companion to James Joyce's *Ulysses*
in Two Volumes

Volume I
The British Army Reference for *Ulysses* Scholars

Peter L. Fishback

Copyright 2020 by Peter L. Fishback.
All rights reserved.

Published 2020 by F.F. Simulations, Inc.
Birmingham, Alabama

ISBN: 978-1-7353525-0-3

LOCCN: 2020915512

This book has also been released in eBook (Kindle) and PDF editions.
PDF ISBN: 978-1-7353525-1-0
eBook (Kindle) ISBN: 978-1-7353525-2-7

Companion Website:
Major Tweedy's Neighborhood
www.majortweedy.com

No part of this book may be reprinted or reproduced or utilized in any form or by any electronic, mechanical, or other means, now known or hereafter invented, including photocopying and recording, or in any information storage or retrieval system, without permission in writing from the publisher.

Back cover art from a painting by Frank Dadd. The work is in the public domain in its country of origin and other countries and areas where the copyright term is the author's life plus seventy years or fewer. Image from Wikimedia Commons, Share-Alike License.

Quotation on back cover from *Ulysses*. Bloom in the Westland Row Post Office looking at army recruiting posters. U (Gabler) Lotus Eaters 5:66-68.

Acknowledgements

I thank Aisling O'Malley of The Institution of Engineering & Technology, London, who saved me a trip to the institution's office which allowed for more time at the British Library and The National Archives. I am especially grateful to Alan Renton of the Cable & Wireless Archives, Telegraph Museum Porthcurno, Cornwall, for his invaluable assistance.

Peter L. Fishback

Permissions

Images for plans of barracks other than Richmond Barracks, Dublin, are courtesy of the Military Archives, Defence Forces Ireland. Reproduction and distribution of such graphics without the authorization of the Officer-In-Charge, Military Archives, Republic of Ireland, is prohibited.

Notes on Sources

To the best of the author's knowledge, all quoted and reproduced material in this book is either from works in the public domain or is fair usage of copyright material.

British Parliamentary Papers and census reports cited were viewed in their digital versions. Digitized versions of Gibraltar census journals were viewed through the website of the Gibraltar National Archives. Poor quality digital scans of such material may have caused transcription errors in this book.

Sources not readily accessible by the reading public are archival state documents in the British Library and documents in The National Archives (UK), the James Joyce Collection at the University of Cornell, the Richard Ellmann Collection at the University of Tulsa, the Institution of Engineering & Technology, London, and the Cable & Wireless Archives, Telegraph Museum, Porthcurno, Cornwall. Where a referenced work gave Stanislaus Joyce's Triestine diary as a source, such claim was checked against the photocopy in the Richard Ellmann Collection.

Contents

Volume I

List of Maps	2
Abbreviations	4
Introduction	7
1. History of Irish Part-Time Soldiery	13
2. History of the British Army, Cromwell to 1853	53
3. The Crimean War	84
4. Late Victorian Military Campaigns and Army Reform	118
5. The Armies of the British East India Company	161
6. Army Life and Retirement: Officers	181
7. Army Life and Retirement: Other Ranks	210
8. Officers and Soldiers of the Auxiliary Forces	270

Appendices

A. The British Army in 1904	305
B. Battalion and Company Establishments	348
C. Prestige Ranking of Army Regiments	359
D. New Cavalry Officers' Mess at the Curragh Camp	368
E. Badges of Rank	371
Index	375

Maps

Unless noted otherwise, landmass boundary, river, and water body data are from the GSHHG 2.3.7 (2017) collection of map files developed and maintained jointly by NOAA and the University of Hawaii. Irish county boundary data from the Ordnance Survey Ireland and the Ordnance Survey for Northern Ireland.

Insurrection of 1798 35
Waterbody data from the Environmental Protection Agency, Republic of Ireland.

Jacobite Rising of 1745, Scotland 68
Digitized elevation map (edited), from C. Amante and B.W. Eakins, 2009,
ETOP1, National Geophysical Data Center, N.O.A.A.

Russo-Turkish (Crimean) War, 1854-1856 88
Digitized elevation map (edited) from C. Amante and B.W. Eakins,
2009, ETOP1, National Geophysical Data Center, NOAA.

Invasion of Crimea 97
Digitized elevation map (edited) from C. Amante and B.W. Eakins,
2009, ETOP1, National Geophysical Data Center, NOAA.

Sevastopol: Allied Positions on November 1, 1854 104

Indian Mutiny, 1857-1858 125

South Africa, 1795-1840 134

South Africa, September, 1899 140

1898 Maneuvers, Positions on September 5th 215

British Army Recruiting, Ireland 320

British Army Abroad, 190 333

British Army Home Command Districts, 1904 335
Historic county boundary data for Great Britain from the Ordnance Survey UK.

U.K. Coastal Fortifications, 1904 343

London Defence Positions 345
Relief map from Natural Earth, www.naturalearthdata.com.

You can download full-color versions of the maps, plus many of the other figures, from the book's companion website:

Major Tweedy's Neighborhood

www.majortweedy.com

Abbreviations

Army Ranks

FM	Field Marshal
GEN	General
LTG	Lieutenant-General
MG	Major-General
BG	Brigadier and Brigadier-General
COL	Colonel
LTC	Lieutenant-Colonel
MAJ	Major
CPT	Captain
LT, 1LT	Lieutenant, First-Lieutenant
2LT	Second-Lieutenant, Cornet, and Ensign
SM	Sergeant-Major and equivalent ranks
QMS	Quartermaster-Sergeant and equivalent ranks
CSG	Colour-Sergeant and equivalent ranks
SGT	Sergeant and equivalent ranks
CPL	Corporal and equivalent ranks
LCP	Lance Corporal and equivalent ranks
PVT	Private and equivalent ranks

Army Units

Div.	Division
Bde.	Brigade
Reg.	Regiment
Batt.	Battalion
Sqd.	Squadron
Coy.	Company
Btty.	Battery

Other Military

ASC	Army Service Corps	NCOIC	NCO in Charge
AWOL	Absent Without Leave	RAMC	Royal Army Medical Corps
Atty.	Artillery	RFA	Royal Field Artillery
Cav.	Cavalry	RGA	Royal Garrison Artillery
CINC	Commander-in-Chief	RHA	Royal Horse Artillery
CSM	Company Sergeant-Major	RSM	Regimental Sergeant-Major
Inf.	Infantry	S.M.E.	School of Military Engineering
NCO	Non-Commissioned Officer.	WO	Warrant officer

General

BSAC	British South Africa Company
CDA	Contagious Diseases Act
Co.	County (Ireland)
EIC	British East India Company
MP	Member of Parliament
U	*Ulysses* by James Joyce, Gabler Edition (New York, Vintage, 1986)
VOC	Dutch East India Company (*Vereenigde Oostindiche Compagnie*)

British Currency Notation, Examples

£3 11s., 7d.	3 Pounds, 11 Shillings, 7 Pence
3/11/ 7	3 Pounds, 11 Shillings, 7 Pence
8/-	8 Shillings, 0 Pence

Introduction

James Joyce spent a good deal of his youth, and all his university years, in a British Army garrison city: Dublin. Throughout that period, 4,500 to 5,500 soldiers were quartered in that city of 250,000 residents.[1] Barracks and former barracks were situated all over "dear, dirty Dublin" and probably one-in-eleven of the young men out in town during the evening and late afternoon and was in uniform.[2] The British Army was a major part of Dublin life and so it appears throughout *Ulysses* in characters, places, and references to wars and battles. Additionally, Joyce worked on *Ulysses* between 1912 and 1922.[3] During that period, two wars were fought in the Balkans in 1913, and a "Great War" raged throughout Europe from 1914 through 1918. These conflicts, particularly the Great War, certainly influenced Joyce and his writing. Finally, while Joyce was at university, the Boer War took place in South Africa. Joyce and his father keenly followed that conflict's events and like nearly all Irish nationalists, opposed Britain's policy of imperial conquest. As noted by Greg Winston, it is not surprising that in Joyce's writings the martial element is frequent and ubiquitous as before reaching age forty, the author had lived in four militarized societies and had frequent run-ins with police and military officials.[4]

Though opposed to militarism, Joyce had an interest in, and knowledge of, military and naval matters. His Trieste library included *What the Irish Regiments Have Done* by S. Parnell Kerr, and *The Ways of War* by Thomas Kettle, a long-time friend.[5] In 1907, Stanislaus Joyce noted that his brother, James, knew the fleet sizes of the European naval powers.[6] Also,

---

[1] In 1901, the last year of the Boer War, there were 5,339 enlisted personnel of the army resident in the Dublin Metropolis. *Census of Ireland, 1901*. In 1904, the Dublin garrison consisted of 4,483 army personnel, all ranks, excluding the permanent staff of the militia. *Army Medical Department Report for 1904*, 1906, [Cd. 2700].

[2] For the City of Dublin, 7.9% of males 18 to 30 were soldiers. Including the close-in southern suburbs, 6.6% of young males were soldiers. *Census of Ireland, 1901*; War Office, *General Annual Report on the British Army, 1904*, 1905, [Cd. 2269]. In that soldiers of the Edwardian Era had more free time and spending money than most young Dublin civilians, the 9.0% figure used in the text is reasonable.

[3] Luca Crispi, "Manuscript Timeline," *Genetic Joyce Studies* 4 (Spring 2004).

[4] Greg Winston, *Joyce and Militarism* (Gainesville: Univ. Press of Florida, 2012), 9-12.

[5] James Joyce Trieste Library, Harry Ransom Center, University of Texas at Austin:

What the Irish Regiments Have Done (London: Unwin, 1916). A description of war-time endeavors of such regiments in France, the Balkans, and the Dardanelles. Introduction by John Redmond, leader of the Irish Parliamentary Party.

Ways of War (New York: Scribner, 1917). A commentary on the First World War and memoir of the author's participation as a British Army officer.

[6] The Triestine diary of Stanislaus Joyce: *Book of Days*, Richard Ellmann Papers, University of Tulsa, 1988.012.1.142.

most of Joyce's pupils in Pola were naval officers, and as there was military conscription in the Austro-Hungarian Empire, many of his Trieste pupils were army reservists.

The opening scene of *Ulysses* is set in a demilitarized, fortified gun emplacement: The Martello Tower at Sandycove, Kingstown, Co. Dublin, a residence in 1904. Later, the reader learns the tower remains War Office property and is letted at £12 per annum.[7] The novel's third spoken line contains an obvious military reference: "Back to barracks!" shouts Buck Mulligan to Stephen.[8] Joyce calls the three tower residents "messmates" as if they were officers of the same army regiment.[9] Before the first episode concludes, we learn that a friend or acquaintance of both Buck Mulligan and Stephen Dedalus is to take a commission in the British Army.[10] All but one of the novel's remaining seventeen episodes, "Nestor," contain a reference to the British Army either through a battle, a person, place, or military unit. "Nestor," however, is saturated with non-British military references, allusions (hockey as a metaphor for battle), battles, militarism, and possibly a foreshadowing of the First World War. Robert Spoo claims this episode, written during 1917, is the part of *Ulysses* most influenced by Joyce's view of the war. Spoo argues that "the war is in fact so pervasively present in 'Nestor' that the episode bears comparison with the contemporaneous poems of Wilfred Owen and Siegfried Sassoon" [noted war poetry].[11]

This book's purpose is to illuminate the numerous British Army and general military references in what is arguably the most important novel of the twentieth century, and to provide a greater understanding of such references. The book's first volume is a detailed, reference work for the general readership on the history of the British Army, the social composition and the life of its personnel, and that military organization's structure on June 16, 1904. The second volume provides an in-depth look at the military allusions and references in *Ulysses*, with an emphasis on the novel's military characters, especially Major Brian Tweedy and his daughter, Marion (Molly) Bloom.

Summary of Chapters

This volume begins with a history of the part-time, amateur military forces in Ireland: Militia, Volunteers, Yeomanry, and Imperial Yeomanry. While that seems an odd subject to explore first, note that the history of auxiliary military forces in Ireland is intertwined with the history of Ireland through 1904, especially with regard to three major antagonisms: Irish v. English, Catholic v. Protestant, and Unionist v. Separatist. Amateur soldiers in Ireland are part of Stephen Dedalus' "nightmare of history" and such men were in the Irish

---

[7] U (Gabler) Telemachus 1:539-40.

[8] Ibid., 1:19.

[9] Ibid., 1:363-64.

[10] Ibid, 1:695-704.

[11] Robert Spoo, " 'Nestor' and the Nightmare: The Presence of the Great War in *Ulysses*" in *Joyce and the Subject of History*, edited by Mark Wollaeger, Victor Luftig, and Robert Spoo (Ann Arbor: Univ. of Michigan Press, 1996).

consciousness of 1904. In "Aeolus," Myles Crawford, editor of the *Freeman's Journal,* shouts almost incoherently about the "North Cork Militia" and the "Irish volunteers."[12] Throughout a good portion of "Sirens," Ben Dollard is singing the *Ballad of the Croppy Boy,* one of whose two main characters is a Yeoman captain.[13]

The next three chapters present a history of the British Army from its creation in 1661 as the bodyguard of King Charles II of England, through Bloomsday. They include histories of the military departments of the Kingdoms of Ireland and Scotland. Those chapters are followed by one that is a brief history of the British East India Company's armies. Note that the Royal Dublin Fusiliers, the only regiment Joyce associates with Brian Tweedy, was a direct descendant of two European regiments of the EIC. Volume I concludes with three chapters on British military life with emphasis on the Late-Victorian and Edwardian Eras.

Chapter Organization

To facilitate this volume's use as a reference work, chapters are organized into clearly delineated sections, sub-sections, and subordinate sub-sections.

Section headings are set in Arial type and are underlined.

Sub-section headings are set in Arial type and not underlined.

Subordinate sub-sections are indented on both sides and have centered headings.

Spelling and Place Names

In nearly all cases, spelling follows the US conventions. There are; however, a few exceptions. For example, mobilized reservists are recalled "to the colours" (not colors) and levels of multi-level buildings are "storeys" (not stories).

Where British and Irish writings are quoted, naturally, the non-US spellings are used (mobilisation, labour, defence, *etc.*). Additionally, archaic spellings are retained for all quoted sources. For Irish names and places, the official English spelling of the time is used (Connaught, not Connaght).

For names of places and streets in Ireland, those in effect prior to independence are used. For example: Sackville Street, not O'Connell Street and Queenstown not Cobh. There is one exception to this rule: City of Londonderry, but County Derry.

---

[12] U (Gabler) Aeolus 7:359, 739.

[13] U (Gabler) Sirens 11:991-1141. *The Croppy Boy* is an 1845 poem by Carrol Malone (pseudonym of William B. McBurney of Belfast), which was sung to a traditional Irish tune. Charles Welsh, ed., *The Golden Treasury of Irish Songs and Lyrics*, Vol. 2 (New York: Dodge, 1907), 32-33; *The Boston Pilot,* November 1, 1845.

British Money and Its Value Over Time

Pre-Decimal Currency Units
12 pence to a shilling; 20 shillings to a pound.

Annual Income Requirement, 1904
Maintenance of Social Class Living Standard, UK

Social Class	Minimum	Maximum
Upper	£ 1,000	
Middle	300	£ 999
Lower-Middle	100	299
Upper-Working	100	130
Lower-Working	50	75

Source: Susie L. Steinbach, *Understanding the Victorians* (New York: Routledge, 2012).

Monetary Comparison
UK and US, 1850-2018

| Year | Price of Gold, Oz. | | $ to £ | Consumer Prices | |
	Sterling	US Dollars		In UK	In US
1850	4	20	5.00	1.0	1.0
1900	4	20	5.00	1.1	1.0
1950	13	35	2.80	3.9	2.9
2000	180	270	1.50	80.0	20.7
2020	1,450	1,800	1.25	118.7	31.0

Note: During the U.S. Civil War, the dollar fell to 10.00 per pound but had recovered to the pre-war 5.00 per pound by 1870.

Enlisted Rank Equivalents, Edwardian and Early 21st Century

British Army, 1904 Infantry Rank	Pct. of EM	UK Army, 2014 Rank / Grade	Pct. of EM	US Army, 2020 Rank/ Grade	Pct. of EM
Sergeant-Major	1	WO 1	2	Sergeant-Major	1
Quartermaster-Sergeant	1	WO 2	5	Master-Sergeant	3
Colour-Sergeant	3	Staff-Sgt.	7	Sergeant 1st Class	10
Sergeant	6	Sergeant	11	Staff Sergeant	15
Corporal	6	Corporal	17	Sergeant	18
Private, LCP, Drummer	83	OR 1-3	58	E 1-4	53

EM = Enlisted Men/Personnel

British Army Formations

Hierarchy of Army Formations
British Army, c. 1900

	Rank of Commander	**Composition**	**Nomenclature**
Field Army or Expeditionary Force	GEN or LTG	Varied. Large armies consisted of corps plus army-echelon units.	Name
Army Corps	LTG	HQ staff, 2 to 4 divisions, and other units (combat and support).	Roman Numeral or Name
Division	LTG or MG	HQ staff, 2 or 3 brigades, artillery, engineers, reconnaissance, and support units.	Ordinal Number or Name
Brigade	MG or BG	HQ staff, 3 or 4 battalions, and other units (combat and support.	Ordinal Number with Type
Battalion and Equivalent	LTC	HQ staff and 2 to 8 companies or equivalents.	Various
Company and Equivalent	MAJ or CPT	Cadre and 2 to 4 sub-units termed half-company, troop, or section.	Single Letter or Ordinal Number

Field Army or Expeditionary Force

Depending on size, units under the commander's direct control could be army corps, divisions, or brigades. The 1867 Abyssinia Force (13,000 men) consisted of brigades; the 1882 Egyptian Expeditionary Force (35,000 men) divisions, and the Army in South Africa 1899-1902 (peak strength of 250,000), corps.

Army Corps

These formations had a total strength of 25,000 to 35,000, all ranks and were denominated with a Roman numeral (I Corps, II Corps, *etc.*), geographic area (Natal Force), or the commander's name.

Division

These formations were labelled with an ordinal number (1st Division, 2nd Division, *etc.*), a descriptive name such as Irish Division, or the commander's name. Cavalry divisions totaled, all ranks, about 6,000 men; infantry from 10,000 to 13,000.

Brigade

By the time of the 1899-1902 Boer War, brigades were identified by ordinal number and type, such as 1st Cavalry, 1st Infantry, *etc*. Cavalry brigade strength, all ranks, totaled about 2,500; infantry 3,500 to 4,000.

Battalion and Battalion-Sized Unit

These formations in the cavalry were termed "regiment" and in the artillery "brigade." The senior NCO position was titled Regimental Sergeant-Major. Infantry battalions were identified by number and regimental name such as 2nd Battalion, Royal Dublin Fusiliers. Cavalry regiments were identified by number and historic type such as 10th Hussars. Field and horse artillery brigades were denominated by type and roman numeral: II Horse Artillery, XXV Field Artillery. Garrison artillery brigades were usually labeled with a place name such as the Gibraltar Garrison Artillery.

Authorized strengths of these units varied by type, location (home or abroad), and combat readiness. Generally, these formations had a total strength, all ranks, from 400 (artillery at peacetime minimum) to 1,100 (infantry at war strength).

Company and Company-Sized Unit

These units in the field and horse artillery were termed "batteries" and in the cavalry "squadrons." The senior NCO position was titled Company, Battery, or Squadron Sergeant-Major.

Company authorized strength varied widely by corps. Support companies, (Ordnance Corps, Army Service Corps), could have as few as 50 enlisted personnel. Field artillery batteries could have as many as 160 soldiers (gunners, drivers, specialists, NCOs). The usual establishment for infantry companies was 100 soldiers, including NCOs and drummers.

Chapter 1
History of Irish Part-Time Soldiery: The Militia, Volunteers, and Yeomanry

On June 16, 1904, the British Army had about 28,000 troops in Ireland. But they were not the only "redcoats" in the Emerald Isle. There were also present approximately 20,000 part-time, Irish soldiers of the Militia and Imperial Yeomanry.[1] It is unlikely that any of Leopold Bloom's friends and acquaintances would have been in these auxiliary forces, though at the time four militia and two Imperial Yeomanry units were based in Dublin[2]. Being of the lower-middle class their social status was too low for the commissioned ranks (the aristocracy, gentry, and a few Unionist members of the middle class), and too high for the other ranks (predominantly laborers with skilled tradesmen as NCOs).[3]

Several of the characters in *Ulysses* express, in historic context, knowledge of Ireland's part-time soldiery. Myles Crawford, Leopold Bloom, Buck Mulligan, and Matt Lenehan all reference either the militia, yeomanry, or Irish Volunteers.[4]

The Traditional Protestant Militias

The militia in the Ireland of *Ulysses* had its roots in the part-time defense forces of English settlers "planted" in Ireland by the Crown and English land corporations 350 years before Bloomsday. In the 1550s, during the reign of Queen Mary I, English settlers in the newly created royal counties of King's and Queen's (Offaly and Laoise since 1922) were required by terms of their land grants to equip themselves with weapons and be prepared to serve without wages against all enemies of the realm. Fifty years later the royal grants that enabled

---

[1] *General Annual Report on the British Army for the Year Ending 30th September, 1904*, 1905, [Cd. 2268]. In 1904, 11.5% of the Regular Army's soldiers claimed Irish nationality, indicating 3,200 full-time "greencoats" were then in Ireland. That number is far fewer than the 20,000 Irishmen in the auxiliary forces. Of the total 23,200 Irish soldiers in Ireland on Bloomsday, about 83% were militiamen and 4% were Imperial Yeomen.

[2] Imperial Yeomanry: 2 squadrons. Militia: 2 battalions infantry, 1 artillery brigade, and 1 medical company. *Monthly Army List, July 1904*.

[3] The novel is populated chiefly by lower-middle class characters. Budgen, *James Joyce and the Making of Ulysses*, 67. Nearly all the senior militia officers were of the landed class, while the junior officers were either landed or attempting to secure regular army commissions. Depending on the unit, 20% to 35% of the officers were born in England. Butler, *The Irish Amateur Military Tradition*, 52-55; Testimony of Senior Officers, Militia in Ireland, *Minutes of Evidence taken before the Royal Commission on the Militia and Volunteers*, 1904, [Cd. 2062, Cd. 2063].

[4] Crawford: "North Cork Militia" U (Gabler) Aeolus 7:359-360, "Irish Volunteers" U (Gabler) Aeolus 7:739. Bloom, while listening to the Croppy Boy U (Gabler) Sirens. Mulligan: "the fencibles" as a reference to the militia U(Gabler) Oxen of the Sun 14:655. Lenehan: "imperial yeomanry" U (Gabler) Cyclops 12:1318.

History of Part-Time Irish Soldiery

the Ulster plantations required the new landowners to arm their Protestant tenants, many of whom were former soldiers.[5] These Ulster lands, formerly owned by Irish nobles, were forfeited to the Crown in the aftermath of the English provoked Nine Years War, Flight of the Earls, and O'Dougherty Rebellion.[6]

In 1641 a series of "risings" took place first in Ulster, then in other parts of Ireland. They were instigated by those Catholic nobles whose ancestral lands were forfeited to the Crown. The nobles encouraged demonstrations by their former tenants against the Irish Parliament and the Crown. The Catholic nobles lost control of what became armed remonstrances and several massacres of Protestant settlers resulted. Protestant survivors formed *ad-hoc* defense forces and reprisals by these militias and the English Army followed. These irregular militias subsequently fought for the Crown against the Catholic Confederation of Ireland, and then for Cromwell's Parliamentary army against Charles I and his son, Charles II.[7]

The Protestant militia remained intact, though ill-organized and untrained, through the Stuart restoration. After Prince William of Orange invaded England to dethrone James II, this militia was disarmed by the Catholic Lord Deputy of Ireland, Richard Talbot (1st Earl of Tyrconnell). To assist William, son-in-law of James II and commander of British forces in the Netherlands, Protestant civilians in Ulster arrayed themselves into new militia units. By the summer of 1690, this reconstituted militia totaled about 15,000 men.[8] The Stuart forces in Ireland surrendered on October 3, 1691, to the army of King William and his English wife, Queen Mary.

The experience of the Williamite War and persistent Jacobite intent to restore a Stuart to the throne, prompted the Irish Parliament to establish a regulated, permanent militia. Political machinations to create a statutory militia began in 1692. Nearly 25 years later, on June 20, 1716, an Irish Militia bill received royal assent.[9]

The Protestant Statutory Militias, 1716 - 1792

The First Statutory Militia

The Militia Act, 1716 created the first statutory militia in Ireland.[10] This auxiliary force was a Protestant-only organization whose stated purpose was to suppress Catholic rebellion and repel foreign invasion, and was aimed specifically at those in service of any Stuart pretender to the throne. All physically fit Protestant males, age 16 to 60, were subject to service in the militia for the initial two-year term of the statute. While all Protestants could

---

[5] Garnham, *The Militia in Eighteenth Century Ireland*, 4-6.

[6] Hayes-McCoy, "The Completion of the Tudor Conquest;" Clarke, "Pacification, Plantation, and the Catholic Question.".

[7] Garnham, *The Militia in Eighteenth Century Ireland*, 7-9.

[8] Garnham, "The Establishment of a Statutory Militia in Ireland."

[9] Garnham, *The Militia in Eighteenth Century Ireland*, 7-9.

[10] 2 Geo. 1 (Ireland), c. 9.

serve in the enlisted ranks, non-Anglicans (Dissenters) could only be commissioned as officers if they swore an Anglican oath of allegiance as required by the Sacramental Test Act, 1704.[11]

The militia was organized on a county basis with newly appointed royal governors as the peacetime commanders. Each governor nominated to the Crown men to serve as officers; selected which men would serve in the other ranks; and supervised the equipping and administration of militia units within his county. The governor organized the county force into regiments and smaller independent companies (infantry) and squadrons (cavalry). In cities and towns, municipal authorities performed the governor's militia duties as his agents. The militia was funded through county militia taxes, collected in the same manner as other county taxes, with Catholics required to pay at twice the rate set for Protestants. Weapons and ammunition were provided by the Irish Parliament.

If there were insufficient Protestant volunteers to fill a militia unit, then the county and city authorities could compel service by Protestant men. A man selected for compulsory service could provide a substitute to serve in his place. If a selectee did not provide a substitute, he could still avoid service by payment of a fine. In addition to direct drafts of Protestant men, the militia ranks would also be filled by nominees of Catholic householders resident in cities and towns. Such householder, if selected by the municipal authorities, had to arrange for a Protestant volunteer. If the Catholic householder was unable to provide a suitable inductee, he was fined at twice the rate of a Protestant who refused militia service.

Militiamen trained four times per year at one-day musters. Each militia unit was subject to array (call up) by the county governor, the British government, and the Lord Lieutenant of Ireland. If arrayed for general military service, militia units fell under the command of Crown military authorities. Though militia units were subject to control by the British Army, they could be employed only in Ireland. Overseas deployment, including to the United Kingdom, was not authorized by the statute. In March 1719, the Lord Lieutenant for Ireland arrayed the entire militia in response to a threatened Spanish invasion. Approximately 40,000 troops were mustered.[12]

In 1757, an amendment to the Militia Act authorized commissions for Dissenters without their taking the Anglican oath.[13] In Ireland at the time, Dissenters were nearly all Presbyterians and were concentrated in Ulster, particularly the eastern counties of Antrim and Down.

By the 1770s, the Irish Militia was a large, but poorly trained and equipped force in service to Britain and the Protestant Ascendancy in Ireland.[14] Despite its shortcomings, in 1775 the militia took on a new importance to the defense of Ireland, at least in the eyes of

---

[11] 2 Ann. (Ireland), c. 6, An Act to prevent the further growth of popery, 4 *Statutes at Large Passed in the Parliaments Held in Ireland*, 12-31.

[12] Garnham, *The Militia in Eighteenth Century Ireland*, 37. Garnham derived this figure from returns of units arrayed and reported to the Irish Parliament. 3 *Journals of the House of Commons of the Kingdom of Ireland*, appxs., clxiii-clxiv.

[13] 29 Geo. 2 (Ireland), c. 24.

[14] Garnham, *The Militia in Eighteenth Century Ireland*, 58.

Members of the Irish Parliament. That year the British government dispatched 4,000 troops from Ireland to North America where colonials had taken control of New England.[15] All that remained for the defense of Ireland, and the Anglo-Irish, Protestant Ascendancy, were 8,000 British troops and the untrained, poorly equipped, inexperienced, and likely demoralized Irish Militia.

In the years after the general array of 1756, the only training received by militiamen was whatever little could be provided during the four annual one-day musters.[16] An additional disability for the militia was that its weaponry, where there was any, had been provided by the state over 20 years earlier and was now in poor condition.[17] The militia's condition was probably portrayed accurately by the many public commentators of the time who complained of lack of useable arms, absence of discipline, dislike of officers by the militiamen, and low morale (due to unnecessary hardships endured when arrayed locally). "Overall, there seemed to be a general disgust with the militia, or at best a disinterest in it."[18] While the militia was of little military value, it was relatively large and probably adequate, regarding the Protestant Ascendancy, as a counter-balance to the Catholic majority.[19]

Alarmed by the reduction of British forces in Ireland, Members of Parliament and state officials acted to strengthen the militia. In early 1776, the Irish Parliament passed a bill to provide state funding of the militia and amend and extend the about to expire Militia Act, 1716. The bill was rejected by the British government.[20] There were several reasons for this veto: cost, perceived lack of necessity, and distrust of the Presbyterians in Ulster.

Westminster believed the peacetime cost of a new militia, £17,000 per year, was beyond the means of Ireland.[21] The proposed militia funding would strain the Irish state budget and would lead to spending cuts elsewhere; cuts that could be harmful to British interests (such as funding the British Army in Ireland). On March 31, 1775, the end of the fiscal year, outstanding state debt was £866,400.[22] The cumulative deficit was increasing as expenditures

---

[15] 17 *Journals of the House of Commons of the Kingdom of Ireland*, 97. The parliamentary return on movement of British forces in to, and out of, Ireland showed a net loss of 4,020 soldiers through deployment of one dragoon regiment (cavalry) and eight foot regiments (infantry).

[16] Smyth, " 'Our Cloud-Cap't Grenadiers'." Britain declared war against France on May 17, 1756. This was the start of the Seven Years War that eventually pitted Britain, Portugal, and several German states (including Prussia) against France, Spain, the Hapsburg Empire, Sweden and Russia.

[17] Higgins, "'Let us Play the Men'," 193.

[18] Smyth, " 'Our Cloud-Cap't Grenadiers'."

[19] Garnham, *The Militia in Eighteenth Century Ireland,* 46, 58. Garnham notes that in the mid-eighteenth century, the British army in Ireland never numbered more than 12,000 while the Irish Militia probably numbered 40,000.

[20] On March 22, 1776, the English Privy Council rejected the bill. *Journal of the Royal Privy Council*, 1776, at 482. UK National Archives, PC 2/119.

[21] Smyth, " 'Our Cloud-Cap't Grenadiers'."

[22] Clarendon, *Revenue and Finances of Ireland*. Clarendon tabulated financial data from the published accounts of the Irish government for the years 1729-1789.

for the half-year ended September 29, 1775, exceeded by £30,964 tax revenue of £407,397.[23] Additionally, the fiscal strain of a militia could lead to Irish demands for British subsidies. A noted anti-Union critic later expressed outrage that at a time when the Irish treasury ran a deficit and withheld payment of state pensions and salaries, the Irish Parliament sought to further burden the Kingdom with the additional expense of a state-funded militia.[24]

Some in the British government viewed an Irish Militia as unnecessary as there was no threat of foreign invasion, and they were confident the reduced British force, on its own, could suppress any Catholic rebellion. Army units on the Irish Military Establishment deployed to North America came primarily from the loyalist north. The relatively large regular force in the overwhelmingly Catholic south remained in place. Furthermore, they argued, in the event of a major rebellion, a militia in the south would be of little use to the army because of the paucity of Protestants in the region.[25]

Suspicion of Irish Presbyterians was another reason Westminster blocked regeneration of the Irish Militia.[26] Many government officials no doubt thought that to arm the Presbyterians of Ulster while many were vocal in their dissatisfaction with British rule, posed a risk to British dominion over Ireland. They were probably aware that the grievances of the Ulster Presbyterians were like those of the rebellious American colonials.

The Second Statutory Militia

The Militia Act, 1716 expired on April 4, 1776 and for a little over two years the Irish Militia existed as an informal, militarily irrelevant force with no statutory authority.[27] On July 1, 1778 royal assent was given to a new militia bill, one that substantially revised previous law.[28] All Protestants age 18 to 50 were subject to militia service. County officials were directed to register such men and note which were "under infirmity of body or mind, whereby such person is incapacitated to serve in the militia." The statute required that county governors establish company-sized units and commission three junior officers per unit. Governors were also directed to commission a county militia adjutant (chief administrative officer of each county's militia) and received authority to commission field grade officers to command groups of militia companies and fill staff positions.

Officers had to meet property requirements. Colonels needed a landed income of £300 per year, lieutenant colonels, majors, and captains £200 per year, and lieutenants £100 per year. City militia officer candidates could qualify on the annual tax value of their residences;

---

[23] 17 *Journals of the House of Commons of the Kingdom of Ireland*, 79, 160.

[24] MacNevin, *The History of the Volunteers*, 57, 67, 71. MacNevin, a Dublin physician, was a proponent of home-rule for Ireland with a parliament modeled on the American Congress. He was a leader of the Society of United Irishmen in Dublin.

[25] Smyth, " 'Our Cloud-Cap't Grenadiers'."

[26] Garnham, *The Militia in Eighteenth Century Ireland*, 328.

[27] The last extension by the Irish Parliament continued the Militia Act "… for ten years from the twenty fifth of March One Thousand Seven Hundred and Sixty-Six… 5 Geo. 3 (Ireland) c. 15, §30; Miller, "Non-professional Soldiers, c. 1600-1800."

[28] 17 & 18 Geo. 3 (Ireland), c. 13.

colonels £50, lieutenant colonels and majors £30, captains £25, and lieutenants £10. These were substantial sums. Agricultural laborers of the time, the largest sector of the Irish workforce, earned only £7 to £12 per year.[29] Tradesmen with full-time, year-round employment earned £29 to £39 per year.[30] Accordingly, only "gentlemen" of the upper and middle classes could serve as militia officers.

Militia rank-and-file positions were to be filled first by volunteers. If company authorized strength could not be attained with volunteers, then eligible Protestants would be compelled to serve, by "ballot" (selection by lot). Such selectees could avoid service by payment of a £1 fine. Subsequent vacancies were to be filled in the same manner as the companies were established.

Unlike the old statutory militia, this militia was funded wholly by the Irish Parliament. The counties simply managed the peacetime establishment through the governor and his militia adjutant. The enlistment term was limited to three years and men who completed a full term were not obligated to serve again. Authorized training under the new statute was much more extensive than under previous law. Training, dependent always on Parliamentary appropriations, was an annual muster held between May 1st and July 31st not to exceed 28 days, plus drills of up to 2 days per month. Total training was limited to 46 days per year. During training, privates were paid 8d. per day and sergeants 12d. per day. Such pay was commensurate with daily pay for agricultural laborers.[31] Fines and jail sentences were fixed for militiamen who did not report for scheduled training.

The size of each county militia was set at between 100 and 500 men, organized into companies of at least 20 men of all ranks. Three city militias were also established: Dublin with 100 to 1,000 men, Cork and Limerick each with 100 to 500 men. By this formula, the minimum total militia strength would be 3,500; the maximum total militia strength would be 18,000; far fewer than the 45,000 arrayed in 1756.

The peacetime militia establishment was limited in organization to companies; however, the county governor could assemble such companies into regiments when the militia was arrayed "for the purpose of protecting the peace and enforcing the execution of the laws" or to repel invasion. The county governor would fill higher echelon command and staff positions with previously commissioned senior officers. Once arrayed, a militia unit could serve only in its home county unless Ireland was invaded. In such case the militia could be employed anywhere in the island. The militia was specifically barred from service outside of

---

[29] Kennedy and Dowling, "Prices and Wages in Ireland, 1700-1850." Data was obtained from records such as construction project accountings and estate management ledgers of the gentry; Murray, *History of the Commercial and Financial Relations between England and Ireland*, 356-58. According to Murray, day laborers typically earned 10d. per day but worked only six to eight months per year. Excluding Sundays, this produced an annual income of £6.5 to £9.0. Full-time, live-in farmhands, then called cottiers, earned 8d. per day. Excluding Sundays, this produced an annual gross income of about £10.5, from which the farmer deducted the cost of housing.

[30] Kennedy and Dowling, "Prices and Wages in Ireland, 1700-1850."

[31] Agricultural laborers received 6.5d. to 12.0d. per day; skilled tradesmen received 22d. to 30d. per day. Kennedy and Dowling, "Prices and Wages in Ireland, 1700-1850."

Ireland. Authority to array the militia for defense of the realm was vested in the Lord Lieutenant; in effect, only the British government could mobilize the militia for war.

The Irish Parliament amended and extended The Militia Act of 1778 until the creation of a new non-sectarian Irish militia in 1793.[32]

The Volunteer Movement, 1778 - 1792

Rise of the Volunteers

In 1778, though Ireland had an organized militia, the Lord Lieutenant did not mobilize it to replace the regular army units sent to North America. The reason was mostly financial; the Irish government could not afford a general mobilization. The Kingdom's operating deficit for 1777 was £140,000, and by May 1778, the state's credit standing had deteriorated to where Dublin bankers rejected the government's request for a £20,000 loan.[33] The financial situation was so dire that the Lord Lieutenant asked the British government for a £50,000 loan secured by future tax revenues.[34] Note that the mobilized militia would become part of the Irish Military Establishment which was funded by Irish tax revenues. Additionally, the militia was an all-Protestant force at a time when the army was non-sectarian. Militia mobilization would likely cause unrest among the Catholic majority which for historic reasons, was apprehensive of Protestant, military formations.

In February 1778, when France and the United States of America entered the Treaty of Alliance, a feeling of vulnerability became widespread among Irish Protestants. These fears were justified in that the British Army in Ireland was below authorized strength. Many Protestants, especially those in Ulster, felt that Britain had abandoned them, and Ireland, to their own devices.[35] In May 1779, fearful of French invasion, "inhabitants of Belfast and Carrickfergus applied for assistance to Government, but as only sixty troopers could be sent to them, they formed themselves into three armed companies."[36] "The system of volunteering spread quickly. The companies which had been formed in Belfast in early 1779, soon saw their example followed throughout the kingdom."[37]

---

[32] The last continuation was granted through 29 Geo. 3 (Ireland), c. 40.

[33] Murray, *History of the Commercial and Financial Relations between England and Ireland,* 200-01; *Public Income and Expenditure, 1688-1869,* 1868-69, H.C. Accounts & Papers, Nos. 366, 366-I, hereafter cited *Public Income and Expenditure;* MacNevin, *The History of the Volunteers,* 57, 67.

Note that Ireland had its own currency which traded at £1.08 to 1.10 Irish = £1.00 British. Joseph Johnston, "Commercial Restriction and Monetary Deflation in 18th Century Ireland," *Hermathena* 28, no. 53 (May 1939): 79-87. All amounts from *Public Income and Expenditure* given in this chapter are in Irish Pounds translated at the 1.08 exchange rate.

[34] Murray, *History of the Commercial and Financial Relations between England and Ireland,* 194.

[35] Higgins, " 'Let us Play the Men'."

[36] Murray, *History of the Commercial and Financial Relations between England and Ireland,* 201.

[37] MacNevin, *The History of the Volunteers,* 77.

For the years 1778-1783, during which Britain was at war with both the United States and France, the British Army in Ireland was augmented not by the statutory militia, but by overwhelmingly Protestant, constitutional "Volunteer" units. By the end of 1779, there were 40,000 armed Volunteers in Ireland, men enthusiastically loyal; however, not answerable to the Crown and the Irish government, but to their unit sponsors and self-elected officers.[38]

The Volunteers and their Political Aspirations

Initially, Volunteers were from the property-owning, Protestant middle class. The poor were precluded from Volunteer participation due to the cost of uniforms and weapons, and lack of free time for military training. Leadership came from the gentry or the urban commercial class. Most Volunteer units in Ulster were Presbyterian and were led by the non-landed Presbyterian elite: Belfast manufacturers and linen entrepreneurs.[39] The Volunteers were strongest in Ulster. Units raised there accounted for 38% of the 88,827 Volunteers claimed by delegates to the Dungannon Convention of Volunteers, February 15, 1782.[40] At the time, the British Army in Ireland totaled approximately 12,000 troops.[41]

As a large, wartime body of men outside the apparatus of state control, the Volunteers became an important political force.[42] Volunteers were highly visible to the population through continual public celebration, display, and festivity. Additionally, military reviews, complicated exercises, and mock battles became a familiar sight throughout the country.[43]

Volunteers, through unit assemblies and conventions of delegates from multiple units, expressed dissatisfaction with the state of the nation and espoused political views. Nearly all volunteers desired greater Irish independence from "decayed" Britain, the removal of British-imposed economic constraints, and governmental reform, especially with respect to Parliament.[44] Volunteers put forth a wide variety of state solutions to Ireland's economic problems.[45] "Liberty Volunteers," led by Dublin radicals, also sought electoral reform,

---

[38] Kelly, "A Secret Return of the Volunteers of Ireland;" Murray, *History of the Commercial and Financial Relations between England and Ireland*, 202.

[39] Miller, "Non-professional Soldiers, c. 1600-1800."

[40] "Abstract of return of Volunteers presented by the Convention to the House of Commons of Ireland, April 16, 1782," in MacNevin, *The History of the Volunteers*, 220-22. Internal inconsistencies in the return indicate the total arrived at was an overstatement of actual strength. Kelly, "A Secret Return of the Volunteers of Ireland."

[41] *Return of the Military Establishment, 25 July 1782*, 20 *Journals of the House of Commons of the Kingdom of Ireland*, 409.

[42] Small, *Political Thought in Ireland 1776-1798*, 84-89.

[43] Miller, "Non-professional Soldiers, c. 1600-1800."

[44] Dungannon Conventions, February 1782, and September 1783; Ballinasloe Conventions, March and July 1782; Dublin Conventions, November and December 1783. MacNevin, *The History of the Volunteers*, 96, 118, 143, 153.

[45] Small, *Political Thought in Ireland 1776-1798*, 98-112.

including extension of the franchise.[46] The "Patriot" opposition in the Irish Parliament, led by Henry Gratton, Barry Yelverton, and Lord Charlemont, secured Volunteer support for their campaigns for free trade and legislative independence.[47] Their efforts were partly successful. In 1780, Great Britain lifted its import ban on Irish woolen goods and glass, and allowed into its colonies imports from Ireland on the same terms as imports from Britain. In furtherance of legislative independence, the Irish Parliament's bill to amend most of Poyning's Law (the 1495 Irish law that gave Britain control of Irish legislation) received royal assent in 1782. Finally, in April 1783, the British Parliament renounced its legislative authority over Ireland and its right to overturn judicial decisions of the Irish House of Lords.[48] The alliance between Patriot politicians and the northern Presbyterian Volunteers also produced a change in the Sacramental Test Act.[49] In 1780, the Irish Parliament amended the act to allow Dissenters to hold municipal offices.[50]

British Misuse of Irish Tax Revenues

Also of concern to the Volunteers, was long-standing, *de facto* British control over Irish state finances. With such control, the British government effectively channeled Irish tax revenues to the Crown, Britain, individual Englishmen, and local supporters of British interests. Egregious British misuse of Irish tax revenue was demonstrated by the heavy

---

[46] Ibid., 106-07; MacNevin, *The History of the Volunteers*, 190-98; Morley, *Irish Opinion and the American Revolution*, 298-99.

The Irish House of Commons had 150 constituencies and each returned two Members. Voters were as follows:

   32 counties: Owners of land with an annual, net rental value of at least £2.

    7 manor boroughs: Same as for counties.

   53 corporation boroughs: Self-electing voting councils.

   11 "potwalloper" boroughs: Owners of a house with an annual, net rental value of at least £5, or land with an annual, net rental value of at least £2.

    1 Trinity College: The teaching staff, senior administrators, and post-graduate degree holders.

   46 freeman boroughs as follows: In 40, voters were appointed by a statutory borough patron whose right to appoint voters was saleable. In the other 6, "freedom" to vote was either hereditary, through membership in a guild, or attained by municipal corporation appointment. A freeman did not have to reside in the borough for which he voted.

   Edward and Annie G. Porritt, *The Unreformed House of Commons,* Vol. 2 (Cambridge: Cambridge Univ. Press, 1909), 290-374.

[47] Kelly, "A Secret Return of the Volunteers of Ireland."

[48] 20 Geo. 3, cc. 10, 18; 21 & 22 Geo. 3 (Ireland), c. 47; 23 Geo. 3, c. 28.

[49] Miller, "Non-professional Soldiers, c. 1600-1800."

[50] 19 & 20 Geo. 3 (Ireland), c.6.

expenditure for royal pensions, maintenance of an over-sized Irish Military Establishment, and the numerous highly-paid civil service and military staff positions, many of which were sinecures.

Civil Pensions

Throughout the eighteenth century Westminster viewed Ireland's tax revenues as the Crown's personal "hereditary income." This claim was founded on the feudal argument that significant territory in Ireland had been Crown lands and therefore the income of such territory was the monarch's and not the state's.[51] The concept of hereditary income was accepted by successive Irish governments.[52]

George III made use of this fiscal device to bestow numerous gifts upon family members and court favorites which Irish governments funded as pensions on the Civil Establishment. For the fiscal year ended March 25, 1775, "pension" expense of £94,000 represented 30% of total civil expenditure.[53] Pensions included £10,000 to George III's sister Augusta and £12,000 to her husband Prince Ferdinand of Brunswick-Wolfenbuttel. (Upon his father's death Ferdinand became Duke of Brunswick and later commanded the armies of the First Coalition against the French Republic. Ferdinand authored the notorious Brunswick Manifesto of 1792, *infra*.) Other pensions were £8,000 to the Estate of the Countess of Yarmouth (Amalie von Wallmoden, mistress of George II), £6,000 to the Duke of Gloucester (a brother of George III), and £6,000 to the Duke of Cumberland (another brother of George III). Ten titled persons accounted for £55,500 of the £144,275 in pension obligations for the two years ended March 25, 1775.[54] A peculiar case was the £2,000 annual pension to Christian Schroeder.[55] According to Murray, the name is an alias and payments to "Schroeder" persisted for about 20 years.[56] Royal pension payments continued throughout the war years.

---

[51] The hereditary income was the "ancient patrimony of the Crown." This included perpetual rents on monastery lands seized by Henry VIII as well as lands forfeited subsequently as penalty for rebellion. The Crown's claim to customs and excise revenues are even older; they date to 1372. Clarendon, *Revenue and Finances of Ireland*, 7-12.

[52] Murray, *History of the Commercial and Financial Relations between England and Ireland*, 165-166. Murray argues that this funding was unconstitutional in that nearly all statutory taxes were earmarked for specific state purposes. She claims the monarch's personal income from the Kingdom of Ireland was only £15,000 per year. Her position is based on material in the booklet by Alexander McAuley, *Inquiry into the Legality of Pensions on the Irish Establishment* (London: J. Wilkie, 1763).

[53] *Public Income and Expenditure, Part 1*, 313, translated to Irish Pounds.

[54] 17 *Journals of the House of Commons of the Kingdom of Ireland*, 103.

[55] Ibid.

[56] Murray, *History of the Commercial and Financial Relations between England and Ireland*, 170.

The Patriots were unable to rein in these personal, royal grants. After fiscal year 1775-76, civil pension payments increased progressively and peaked in fiscal year 1787-88 at £149,108.[57] In 1793, the British government revoked the sovereign's hereditary Irish income through an act of the Irish Parliament.[58] It did, however, allow Irish civil pension payments for the life of George III, so long as they did not exceed annually, £1,200 per person for other than royal family members, and in aggregate, £145,000. By the next fiscal year, civil pension payments were down to £44,508; 7% of the total civil expenditure of £665,554.[59]

The Military Department of Ireland [60]

Irish subjects, compared to British subjects, bore a disproportionately large share of the British Army's cost. Throughout the eighteenth century, the British government maintained a peacetime home force of 12,000 troops in Ireland and 14,000 troops in Great Britain. The British Army in Ireland (the "Irish Military Establishment") was funded with Irish taxes. Ireland was also responsible for construction and maintenance of barracks, forts, and military warehouses. Ireland, with 25% to 32% of the combined population of the two kingdoms, paid for 45% of the British Army's home garrison.[61] In peacetime, Ireland's annual military expenditure was typically 2.5 times its annual civil expenditure due to the large number of soldiers stationed in the island.[62] For example, in 1767, of the British Army's 112 regiments, 36 were in Ireland, 42 in Great Britain and 34 were abroad.[63]

Not only did the Irish pay for the British Army's upkeep in Ireland, they paid for maintenance of "Irish" regiments deployed abroad. During the American Revolutionary War, Irish revenue funded half the cost of 4,000 Irish Military Establishment troops employed in North America. During the Seven

---

[57] 26 *Journals of the House of Commons of the Kingdom of Ireland*, appx., cixvii.

[58] 33 Geo. 3 (Ireland), c. 34.

[59] 31 *Journals of the House of Commons of the Kingdom of Ireland*, appx., xxi; *Public Income and Expenditure, Part 1*, 341, translated to Irish Pounds.

[60] Murray, *History of the Commercial and Financial Relations between England and Ireland*, 161-65.

[61] Population estimates from Michael Anderson, "Population change in north-western Europe 1750-1800," in *British Population History*, edited by Michael Anderson (Cambridge: Cambridge Univ. Press, 1996).

[62] For the five years ended March 25, 1774, military expenditure totaled £2,755,459 while civil expenditure totaled £1,114, 902. *Public Income and Expenditure, Part 1*, 307-11, translated to Irish Pounds.

[63] *Army List, 1767* (London: J. Millan). Cavalry Regiment Disposition: Britain 18, Ireland 12. Infantry Regiment Disposition: Britain 24, Ireland 24, North America 15, Mediterranean 12, Caribbean 6, Africa 1. Infantry regiments in Britain included 6 regiments of over-age veterans, fit only for internal security and guard duty.

Years War (1756-1763), Irish revenue paid for 4,000 troops added to the Irish Establishment. Additionally, the Irish treasury funded in part Britain's wartime military and naval operations abroad.[64]

State Sinecures

The executive branch of the Kingdom of Ireland had two components, the Military Department and the Civil Department. Both were subordinate to the British-appointed Lord Lieutenant, who was the Chief Governor of Ireland, nominal Commander-in-Chief of British forces in Ireland, and Viceroy. The Lord Lieutenant was assisted by permanent secretaries and their staffs. This Irish executive establishment was headquartered in Dublin Castle, located south of the River Liffey on Dame Street.

The executive apparatus was composed of numerous highly paid offices, civil and military, that were often little more than sinecures. Additionally, many high office-holders never resided in Ireland.[65] A quick perusal of the *Journals of Parliament* reveals expenditures for positions such as Commissioner of the Imprest Account (of which there were five, each with an annual salary of £500), Chaplain of Dublin Castle (£200 per year), and Constable of the Castle (who's lodging allowance was £60 per year).[66] Dublin Castle routinely awarded state offices to nominees of MPs in return for their votes on pro-British bills.[67] Highly-paid positions on the Lord Lieutenant's personal staff included two Gentlemen of the Bed Chamber, four Gentlemen at Large, Master of the Horse, Master of the King's Riding House, a Master of Revels, and nine military aides-de-camp.

Demise of the Volunteers

As the likelihood of French invasion declined, Westminster's support for Irish governmental reform waned. By 1782, Patriot-sponsored bills no longer received backing from members of the pro-British, majority bloc in the Irish Parliament. With no prospect for further political and economic reform, and the end of the war with France and the United States (Treaty of Paris, 1783), the Volunteer movement had passed its peak.[68] By Autumn 1784, Volunteers totaled at most 18,000 men.[69] Enthusiasm for Volunteering continued to

---

[64] Murray, *History of the Commercial and Financial Relations between England and Ireland*, 161-63.

[65] Ibid., 153-86.

[66] 17 *Journals of the House of Commons of the Kingdom of Ireland*, 151-58.

[67] Murray, *History of the Commercial and Financial Relations between England and Ireland*, 153-86.

[68] Miller, "Non-professional Soldiers, c. 1600-1800;" Kelly, "A Secret Return of the Volunteers of Ireland.".

[69] Kelly, "A Secret Return of the Volunteers of Ireland." The author presents a secret report on the Volunteers, the product of an investigation ordered by the Lord Lieutenant in 1784. NLI, Bolton Papers, MS 15891/3 and NMI, MS 22A-1938.

wane and by 1786, Dublin Castle regarded it as a harmless activity.[70] In early 1793, there were at most 63 Volunteer units in Ireland, nearly all in counties Antrim, Derry, and Dublin.[71]

The state actively suppressed Volunteerism after enactment of the Gunpowder and Convention Acts, both of 1793.[72] The Gunpowder Act forbade private persons to possess more than four pounds of gunpowder, possess artillery, and to sell military weapons, without a license from the state. The Convention Act made unlawful all extra-governmental delegate assemblies to express "grievances in church and state." By the end of 1793, the armed Volunteer units had vanished from Ireland.

Though the Volunteers were gone, their former political influence and threat to the Anglo-Irish establishment remained in the nationalist consciousness. In 1808, John Wilson Croker, a prominent nationalist Irish MP, wrote:

> "The Volunteers, a great body of all religions, heated by popular discussions in military assemblies - confiding in their arms and numbers - bold in their impunity, and infected with licentious politics, had wishes which they dared not speak, and would gladly have taken what it were treason to demand."[73]

The Nonsectarian Statutory Militia, 1793-1800

Prelude to War with Revolutionary France [74]

After the French Revolution transformed France into a constitutional monarchy, the Austrian Hapsburg Emperor, Francis II (brother of the French Queen Marie Antoinette), took steps to restore the Bourbon Dynasty to its former power. Francis II garnered first Prussia, then other German states and the Savoyard Kingdom of Sardinia, into a Coalition supportive of French royal prerogative. By early 1792 a substantial Coalition army was in place along the Rhine. In response, on April 20, 1792, France, claiming self-defense, declared war on the Hapsburg Empire and the Kingdom of Prussia. On April 28, 1792, the French "Army of the North" invaded Austrian-controlled Belgium (Lower Netherlands). The attack failed and the French withdrew to their northern defense line which ran from the fortified port of Dunkirk, to the fortified city of Lille, and then to its terminus, the fortified town of Maubuege. The Austrian army in Belgium, 56,000 troops, did not counterattack until June 11th when it advanced on Lille.

---

[70] Ibid., 275.

[71] Ó Snodáigh, *The Irish Volunteers: A List of the Units*. Through medals, manuscripts, newspaper reports, pamphlets, and journals, the author identified about 1,700 Volunteer units that were active at some time during the eighteenth century. For the year 1793, there was existential evidence for only 63 units.

[72] 33 Geo. 3 (Ireland), cc. 2, 29.

[73] Croker, *Sketch of the state of Ireland*.

[74] Sources for this sub-section are listed at the end of the chapter bibliography.

On June 21, 1791, Louis XVI, head of the French state, secretly fled Paris with his family for Coblenz, where French counter-revolutionaries had assembled. (Coblenz was in the Electorate of Trier, an ecclesiastical German principality that was nominally part of the Holy Roman Empire.) He was captured the next day by French security forces who placed him under house arrest in the Tuileries Palace from which he had fled.

Two months later, France, through its Ambassador François-Bernard de Chauvelin, asked Britain to dissuade its allies from joining the royalist Coalition. De Chauvelin assured Foreign Minister William Grenville that France had no territorial ambitions. Grenville responded that Britain would remain neutral and neither assist, nor impede, the Hapsburg Emperor in the execution of his foreign policy.

On July 25, 1792 Charles William Ferdinand (Duke of Brunswick and brother-in-law of George III), commander of Coalition forces in Germany, issued a "war manifesto" in which he demanded, *inter alia,* that the people of Paris

> "… submit at once and without delay to the king, to place that prince in full and complete liberty, and to assure to him … the inviolability and respect which the law of nature and of nations demands of subjects toward sovereigns… If the chateau of the Tuileries is entered by force or attacked, if the least violence be offered to their Majesties, and if their safety and their liberty are not immediately assured, [we] will inflict an ever memorable vengeance by delivering over the city of Paris to military execution and complete destruction, and the rebels guilty of the said outrages to the punishment that they merit …"

Shortly thereafter, on August 10, 1792, a French mob attacked the royal residence, the Tuileries Palace. The attackers killed two-thirds of the 900 royal Swiss Guards and captured the remaining third (who were subsequently executed). The royal family; however, escaped to the protection of the National Assembly. The following day, the National Assembly deposed King Louis, declared a republic, imprisoned the royal family, and arranged for the election of a "National Convention" to be the government of the new republic. Britain in response, recalled its ambassador and suspended its recognition of de Chauvelin as French ambassador. The Foreign Office reasoned that since France was now a republic, de Chauvelin, as an emissary of the King of France, lacked proper diplomatic credentials.

On August 19, 1792 Coalition forces invaded northeastern France. The Coalition army's northern wing (Austrian) besieged Lille, while its southern wing (mostly German) quickly captured the French fortress towns of Longwy and Verdun in its advance towards Paris. On September 20th, the French stopped the Germans at the town of Valmy and a campaign of attrition set in. Ferdinand's German force, initially 80,000 troops, was now down to only 40,000 effectives, due mostly to dysentery. Also, the army was struggling to subsist at the end of a 150-kilometer supply line to its base in Trier. In late October, Ferdinand withdrew his depleted and exhausted force back into Germany to rebuild during the winter. The French then counter-attacked the Coalition army's northern wing and lifted the siege of Lille.

In early November, French troops advanced into Belgium, quickly taking Mons and Brussels. The British government responded with a ban on grain exports to France. By the end of November, French forces occupied nearly all of Belgium, including the dormant port city of Antwerp. This led to fears by the British that France would attack the Dutch Republic (northern Netherlands) to gain control of the Scheldt Estuary and open the port of Antwerp.[75] (Antwerp had been closed to shipping since 1648, under terms of the Treaty of Westphalia.) On December 27th de Chauvelin asked the British government to remain neutral in the war between France and the Coalition. He also stated that France had no further territorial ambitions in the Netherlands. Unbeknownst to de Chauvelin, the British government was engaged in negotiations with the Dutch for joint military operations in defense of the northern Netherlands.

In December, the British government, apparently to make clear its strong disapproval of the French occupation of Belgium, banned French merchants from Britain. On January 7, 1793, by letter to the Foreign Office, de Chauvelin objected to the ban and noted that it violated the bilateral Treaty of Navigation and Commerce of 1786. Four days later, de Chauvelin notified the Foreign Office that the French government declared the 1786 treaty null and void. On December 27th, Grenville wrote to de Chauvelin and expressed Britain's displeasure with France, noting "the unhappy events of the 10th of August" and the duplicity of the French government in its invasion of Belgium only five months after de Chauvelin assured him that France had no territorial ambitions. Grenville stated further that the British government viewed the National Convention's "Edict of Fraternity with subject peoples" proclaimed November 19, 1792, as a call for revolution throughout Europe. (Grenville may have confused the National Convention's Edict of Fraternity with its more bellicose Proclamation of December 15th.) Grenville also reminded de Chauvelin that the port of Antwerp was closed pursuant to a European treaty to which France was a signatory.

By letter to the Foreign Office dated January 13, 1793, de Chauvelin informed Britain that France viewed the Treaty of Westphalia's provisions that closed the port of Antwerp an infringement of the Belgian peoples' rights. He reminded Grenville that the treaty left Antwerp in Austrian hands and inferred that the shipping ban benefited Britain and the Dutch Republic (whose territory included the port of Amsterdam) at Belgium's expense. In effect, de Chauvelin posited that French military action to open Antwerp's port would be within international law.

On January 21, 1793, the French National Convention executed King Louis XVI, convicted of treason by that body seven days earlier. Two days after the execution, Britain expelled the French ambassador. On February 1st, France declared war on Great Britain.

Now that Britain was at war, the government prepared the bulk of its army for deployment to the continent, primarily to support the Dutch Republic. With a soon to be weakened home army, fears of French invasion arose in Great Britain and Ireland. Threat of invasion prompted the Irish Parliament to both strengthen the militia and address Catholic

---

[75] Though the United Netherlands confederation was styled a republic, the House of Orange was the executive power (*stadtholder*) in most of the constituent provinces.

grievances. Irish officials worried that under current conditions, where many of the "Penal Laws" that deprived Catholics of numerous civil rights were still in effect, republican French invaders would receive armed support from Irish Catholics.[76]

The Militia Act of 1793

On February 4, 1793 Robert Hobart, MP Armagh Borough, introduced a bill to lessen the civil disabilities of Irish Catholics. The next day, Arthur Hill, MP County Down, introduced a bill to reorganize and strengthen the Irish Militia. The militia bill allowed Irish Catholics to serve as militiamen, but not as officers. With both bills favored by Westminster, they passed quickly through the Irish Parliament and received royal assent on April 9, 1793.[77] The 1793 Catholic relief act for Ireland resembles the relief act for England and Scotland of two years earlier.[78] The Irish statute permitted Catholics to vote for members of parliament, enter the legal profession, hold municipal office, receive university degrees from Trinity College, and if property requirements were met, to possess weapons. Catholics with a landed annual income of £100, or personal property with a net value of £1,000, received an unconditional right to possess weapons. Those with a landed annual income of £10, or personal property with a net value of £100, could possess weapons only upon taking a special oath of allegiance. Catholics viewed these grants as "a panic-struck capitulation – a sacrifice of ancient monopoly, given up reluctantly to the command of a superior, and in obedience to the advancing dangers of the times."[79]

The Militia Act of 1793 established a force of about 17,000, all ranks, organized into 38 regiments of specified strengths. The regiments were North County Mayo, South County Mayo, North County Cork, South County Cork, Dublin City, Cork City, Limerick City, Drogheda, and 30 additional county-wide units. Each county governor had to establish territorial companies of from 50 to 100 militiamen (sergeants, corporals, drummers, and privates). Men resident in a company's territory would be subject to serve with such company. The companies were to be grouped permanently into regiments, styled "battalions" for groupings of fewer than eight companies. For regiments of six or more companies the chief executive officer was no longer the county governor but a new official with the rank of Lieutenant Colonel Commandant (appointed by the Lord Lieutenant). For smaller units, the county governor remained the chief executive officer.

---

[76] "But, before the close of 1792, a new scene was opened. The French armies defeated their enemies at every point. The Netherlands were conquered, and a torrent of republicanism, driven on by military power, threatened every State in Europe. The cannon of the battle of Gemappe were heard at St. James…It was the deep interest of the British Government to detach the wealth and intelligence of the Catholics of Ireland from the republican party." O'Connell, *An Historical Memoir of Ireland*, 25-26.

[77] 33 Geo. 3 (Ireland), cc. 21, 22.

[78] 31 Geo. 3, c. 32.

[79] Wyse, *Historical Sketch of the Late Catholic Association*, Vol. 1, 129.

The chief executive officer commissioned militia officers, subject to veto by the Lord Lieutenant.[80] Officers, other than peers with a county residence, had to meet annual landed income requirements: lieutenant colonel £1,200, major £300, captain £200, lieutenant £50, Ensign (Second Lieutenant) £20. The requirement for a lieutenant colonel commandant was £2,000. The number of company-grade officers was limited to three per company, plus officers of the regimental staff. The number of field-grade officers was set at two per regiment. Regimental commanders appointed sergeants and corporals in the ratio of 3 sergeants and 3 corporals to every 20 privates. Two drummers were authorized for each company. The regimental commander appointed the regiment's sergeant-major and drum-major from the pool of sergeants and drummers, respectively. Each regiment was to train annually for one 28-day period as scheduled by the county governor. As under previous law, militiamen who failed to attend annual training were subject to fine and imprisonment. During training, the daily pay of militiamen was sergeants 1s., corporals and drummers 8d., and privates 6d. Pay was subject to deductions (stoppages) for negligent damage to uniforms and personal equipment.

Unlike under previous law, the militia was now open to all men, not just Protestants. Service was limited to those aged 18 to 45 and "fit to carry arms." As with previous law, company quotas were to be met first by volunteers, then by men selected for compulsory service. Men were compelled to serve by lot drawn from names of eligible men not exempt from service ("the ballot"). The practice of substitutes for balloted men continued; however, if a balloted man did not provide a substitute the fine to avoid service was now £10 (formerly £1). After four years, the enlistment term, men who "bought out" their obligation were again liable for service. Exempt from compulsory service were peers, university students, clergy, seamen, teachers, constables, peace officers, lawyers, civilian employees of the Military Department, and tax-exempt poor men with more than three legitimate children.

County governors and magistrates could call out the militia to aid the civil authority in law enforcement. For such service, militiamen were to be paid at the training rates. The Lord Lieutenant could call out the militia to repel invasion or suppress insurrection ("embodiment"). If so embodied, each militiaman was entitled to a bounty of one guinea (one pound plus one shilling, a total of 252d.); however, company commanders could defer payment until the militia was disembodied, but in no case longer than three years. During such "war service" militiamen were to be paid the same as soldiers of the British Army.

The Militia Riots of 1793

Widespread opposition to the compulsory service provisions of the Militia Act triggered disturbances in half the counties of Ireland. The public understood that because of the war with France, militia service would be a full-time endeavor. Many Irish, especially Catholics, viewed the Militia Act as a back-door attempt by Dublin Castle to compel service in the British Army. Also, there was great concern as to how families would support themselves if

---

[80] The power to commission officers was taken from the county governors to reduce political patronage. Nelson, *The Irish Militia, 1793-1802,* 44.

the breadwinner was in full-time military service. The contemporary press described the disturbances of the Spring and Summer 1793 as "militia riots."[81]

Violent opposition to the militia occurred primarily in the heavily Catholic south. The Lord Lieutenant described the anti-militia activities in County Kerry, where regular army troops killed twelve rioters, as "an insurrection" while the press applied the same label to violent militia resistance in County Limerick. There were also disturbances in County Tipperary but not of the same magnitude as in Limerick. It was in the historic province of Connaught where the most serious resistance to the militia took place. There, unlike elsewhere, several gentry houses were plundered, and there were clashes between organized bands of armed civilians and army units. One battle purportedly left 36 rioters dead. In all, there were 116 reported militia disturbances from May through June.[82] Grievances over the Militia Act became associated with general dissatisfaction amongst the rural population. One report for County Kilkenny stated that the rioters' resistance to the Militia Act was joined with calls for lower rents and abolition of tithes to support the established church. From County Carlow and Queen's County government and press reports stated that rioters vowed to not pay rents, Church of Ireland tithes, and taxes.

Several measures by the Irish government lessened public opposition to the militia. Forced militia service through the ballot was quietly, and unofficially, dropped. It is unlikely that any Irishmen were forced to serve in that there were enough volunteers and substitutes to fill the militia quotas. Typically, after volunteers were taken, counties repeatedly balloted until the ranks were filled with paid substitutes.[83] Through advertisements in the press and other means, the Irish government made clear to the public that the embodied militia would not serve outside of Ireland. Additionally, new legislation provided family separation allowances for balloted militiamen.[84] These measures, together with Dublin Castle's vigorous military response to rebellious actions, calmed the country into a sullen and resentful peace.

The New Irish Militia at Inception

By October 1793, the militia was near authorized strength of about 17,000. Only two regiments were incomplete, those of Cavan and Monaghan. All but three regiments had lieutenant-colonel commandants as their chief executive officer. Except for the South Mayo militia, the lieutenant-colonel commandant was either a Member of Parliament, an Alderman of Dublin, or a Peer. Each regimental staff had six officers: a field commander with rank of lieutenant colonel, a second-in-command with rank of major, an adjutant, a chaplain, a quartermaster, and a surgeon. Adjutants, by statute, were either regular army officers seconded to the militia, or former regular army officers with at least three years of service. The newly formed militia was embodied in its entirety for full-time service for the duration

---

[81] Source: Bartlett, "An End to Moral Economy."

[82] Nelson, *The Irish Militia, 1793-1802,* 58-59.

[83] Ibid., 63-70.

[84] 33 Geo. 3 (Ireland), c. 28.

of the war with France.⁸⁵ By the end of 1793, the regular army in Ireland plus the Irish Militia totaled about 25,000 soldiers.⁸⁶

The Yeomanry

In 1795, Ireland was wracked with internal disturbances caused by agrarian grievances. In Ulster, these disturbances had a pronounced sectarian tone. Armed organizations, most notably the Protestant Peep O'Day Boys (precursor to The Loyal Order of Orange) and the Catholic Defenders, took to the field. In County Armagh, armed Protestants forced thousands of Catholics from their homes. Open battles occurred between armed bands of Catholics and Protestants.⁸⁷ With Dublin Castle apparently unable to maintain law-and-order, some Protestants formed themselves into paramilitary police units. The rural disturbances, the resurgence of self-appointed militias, and the underground presence of seditious organizations, caused alarm within Dublin Castle and amongst the landed elite.⁸⁸ As the militia and army were occupied with anti-invasion preparations, there might not be sufficient means available to maintain order.

Conditions in Ulster worsened and by the Summer of 1796, the Catholic areas were in a state of "smothered rebellion."⁸⁹ On October 17, 1796 Thomas Pelham, British Secretary of State for Ireland, introduced (through the English Privy Council) a bill in the Irish Parliament to establish a volunteer force "for the protection of property and preservation of the peace." It passed quickly through the legislature and royal assent was given three weeks after the bill was introduced.⁹⁰ The Irish Volunteer bill was modeled on a British statute of 1794 that established a new armed force of "men as shall voluntarily enrol themselves for the defence of their counties, town or coasts, or for the general defence of the Kingdom during the present war." That statute included provision for a mounted force called "Gentlemen and Yeomanry Cavalry." In Britain, yeomen provided their own horses and served without pay. Their officers, aristocrats and gentry, were commissioned by county officials. Parliament provided the yeomanry and Volunteers with arms and the counties furnished uniforms and accoutrements. In 1795, county authorities made use of the yeomanry to suppress politically inspired "riots and tumults" that occurred throughout the English countryside.⁹¹

At the beginning of 1797, the new Irish volunteer force, known commonly as "the Yeomanry," stood at 30,000 of which 18,000 were cavalry. This force, organized in volunteer

---

⁸⁵ The statutory term of militia service was four years. Effective April 24, 1797 recruits enlisted for the greater of four years and the duration of the war plus two months. 37 Geo. 3 (Ireland), c.19.

⁸⁶ Nelson, *The Irish Militia, 1793-1802,* 248. Quoting official returns, Nelson shows 8,514 Regular Army troops in Ireland.

⁸⁷ McDowell, "The Age of the United Irishmen.".

⁸⁸ Bartlett, "An End to Moral Economy;" Madden, *The United Irishmen,* 223-24.

⁸⁹ Blackstock, *An Ascendancy Army,* 234-43.

⁹⁰ 37 Geo. 3 (Ireland), c. 2.

⁹¹ Mileham, *The Yeomanry Regiments,* 1-16; Beckett, *The Amateur Military Tradition,* 72-79.

companies of 50 to 80 men, had 199 units in Leinster, 118 in Ulster, 115 in Munster, and 45 in Connaught. In Leinster, the City of Dublin provided 46 companies. Like the Volunteers of 1782, Protestants predominated in the yeomanry. At first, Catholics made up a significant minority of the yeomanry, but by Spring 1798, Catholic membership was negligible. This was due to coercion by the United Irishmen, the discriminatory and brutal manner in which the yeomanry enforced the sedition laws against Catholics, and expulsion of Catholics by commanders who questioned their loyalty.[92] In Ulster, Anglicans at first were the majority of yeomen, but as new units formed the force there became predominantly Presbyterian. By Spring 1978, a significant number of yeomen throughout Ireland were members of the Loyal Order of Orange. In the yeomanry of the northeastern counties of Antrim, Down, and Armagh, Orangemen were in the majority.[93] The yeomanry, in under two years, had become a Protestant force with an anti-Catholic bias. In effect, it was the unpaid, part-time army of the Protestant Ascendancy.

In January 1798, British forces in Ireland consisted of 22,728 embodied militia, 10,751 fencibles, 1,906 Regulars, plus the paramilitary yeomanry.[94] The Irish Militia, which accounted for 64% of the Crown's 35,385 troops in Ireland, would accordingly play an important role in the insurrection of 1798. Unlike the Irish Yeomanry, the Irish Militia was a mostly Catholic force. Protestants; however, were over-represented in the militia ranks and in Ulster accounted for most of the troops. The following table illustrates the composition of some Irish Militia formations by religious adherence.[95]

Table 1.
Protestant Representation in County Militias
(January 1798)

County	Population, Protestant %	Militiamen, Protestant %	Militia Protestant % to Pop. Protestant %
Fermanagh	33	85	2.57
Monaghan	20	75	3.75
King's	12	14	1.17
Cork	8	20	2.50
Kilkenny	4	12	3.00
Kerry	2	16	8.00

Source: Nelson, *The Irish Militia, 1793-1802*, 124-25.

---

[92] Blackstock, *An Ascendancy Army,* 271-78, 286-87.

[93] Blackstock, "A Dangerous Species of Ally."

[94] Nelson, *The Irish Militia, 1793-1802,* 180-82, from returns of the Irish Military Establishment. "Fencibles" ("defensible") were locally-recruited, home defence regiments formed for the duration of a war. Officers and senior NCOs were regulars seconded from their regiments.

[95] The Irish government did not record the religious profession of militiamen. The table's data is from contemporary accounts. Nelson, *The Irish Militia, 1793-1802,* 124-25.

The United Irishmen and the '98

The Society of United Irishmen sprung from the ashes of the Volunteers. The Society arose in 1791 when a 28-year-old Belfast barrister, Theobald Wolfe Tone, organized its first chapter. The United Irishmen's stated goal was the establishment of national government through equal representation of all the people in a radically reformed Irish Parliament.[96] From its outset, the United Irishmen sought full equality of rights among all the Irish. Tone wrote that he wanted "to unite the whole people of Ireland ... and to substitute the common name of Irishman in place of the denominations of Protestant, Catholic, and Dissenter."[97] By year's end, a second chapter formed in Dublin under the leadership of Simon Butler and James Napper Tandy.[98] The movement was bi-polar with one pole in Belfast, Protestant and predominantly Presbyterian, and the other pole in Dublin, about equally Protestant and Catholic. While Society members held diverse political views, the need for legislative independence with electoral reform was the common, unifying belief.

By early 1794, Dublin Castle perceived the United Irishmen as a serious threat to the government and planned action to suppress the movement. On May 4, 1794 Dublin police raided a meeting of the local United Irishmen chapter and seized its papers. Subsequently, there were raids elsewhere and Dublin Castle began to infiltrate informers into the Society.[99] In response, the movement went underground; however, in Ulster it maintained an open presence. By May 1795, the United Irishmen had reconstituted itself as a revolutionary, republican leaning organization whose members had to swear oaths of allegiance and secrecy.[100] Because of the oath of allegiance, the United Irishmen became an outlawed organization under the Insurrection Act of March 24, 1796. The act made it a capital offense to administer an oath of any society formed for seditious purposes. Those who took such oaths were liable to transportation for life.[101]

In the mid-1790s, the United Irishmen senior leadership included Wolfe Tone, Lord Edward Fitzgerald, MP for County Kildare, Arthur O'Connor, MP for Phillipstown Borough, Thomas Adis Emmet, a Dublin barrister, William MacNevin, a Dublin physician, and Oliver Bond, a Dublin woolens merchant; all were Protestant.[102] At the beginning of 1798, Fitzgerald claimed the Society had 280,000 sworn members, with Ulster accounting for 40% of the membership. A memorandum prepared by Fitzgerald and delivered to Dublin Castle by Thomas Reynolds, a police informant, showed the Society's armed men as follows:

---

[96] Madden, *The United Irishmen*, 222-23.

[97] O'Faolain, *The Autobiography of Wolfe Tone*, 36.

[98] Madden, *The United Irishmen*, 223-24.

[99] Information provided by informants led to the arrest of several Belfast leaders on September 16, 1796.

[100] Madden, *The United Irishmen*, 263-69; Curtin, "The transformation of the Society of United Irishmen."

[101] 36 Geo. 3 (Ireland), c.20.

[102] Madden, *The United Irishmen*, 264-65; Maxwell, *History of the Irish Rebellion in 1798*, 12-21.

Ulster 110,990, Munster 100,634, and the counties of Dublin 5,177, Kildare 10,863, Wicklow 12,895, Queens 11,689, Kings 3,600, Carlow 9,414, Kilkenny 624, and Meath 1,400.[103]

The United Irishmen, having first obtained French assurance of armed support (initial landings of 3,000 troops with an additional 8,000 troops in a floating reserve), set in motion a plan to forcibly take control of the Irish state. On the night of May 23, 1798, the Dublin chapter would seize the offices of government, while in the city's environs rebels would overwhelm the isolated detachments of the British Army. The next day, in support of these coup-like actions, the United Irishmen would mobilize in the adjoining counties and prepare defensive positions to protect Dublin from an expected British relief force. Armed uprisings planned for Ulster and Leinster would place the newly established revolutionary government in control of a *de facto* state. Finally, French troops would land, link up with the insurgents, and the combined force would defeat the British Army. The Society's plan was doomed at inception. The United Irishmen was riddled with Dublin Castle informants and the government had detailed knowledge of the planned revolt. On May 19th, Fitzgerald, the United Irishmen's military leader and architect of the planned revolt, was arrested. Other arrests followed, and the Dublin seizure never came off. Additionally, the revolt in the Dublin environs was put down quickly by British troops and yeomanry. The capital remained firmly in government control.

With exceptions in Ulster and the Midlands, the revolt of the United Irishmen failed at the outset. In Ulster, they seized most of the countryside but Belfast and other towns, such as Downpatrick, remained in government hands. British forces attacked rebel-held areas and quickly re-established government control of Ulster except for County Down. There a 5,000 strong force held out until June 13th when it was defeated at Ballynahinch. In the Midlands, the United Irishmen army of 3,000 troops captured towns in County Kildare and County Meath. On June 19th, at Clonard, County Meath, the United Irishmen suffered a major military defeat and by the end of the month, the Midlands force had dispersed. By Summer 1798, the organized insurgency directed by the United Irishmen leadership, was effectively over. Though many Catholics participated in the revolt, the Society's leadership was predominantly Protestant. For example, of the twenty senior leaders imprisoned after the revolt in Kilmainham Gaol, ten professed to the Church of Ireland, six Presbyterian, and only four Catholic.[104]

For early twentieth century Catholic, Irish nationalists, the year 1798 is memorable for the Wexford Rising.

> "Wexford rose, not in obedience to any call from the United Irish organization, but purely and solely from the instinct of self-preservation. … It was the wild rush to arms of a tortured peasantry, unprepared, unorganized, unarmed."[105]

---

[103] Madden, *The United Irishmen,* 283-84.

[104] *Ibid.,* appx. x, 585.

[105] P. W. Joyce, A. M. Sullivan, and P. D. Nunan, eds., *Atlas and Cyclopedia of Ireland* (New York: Murphy & McCarthy, 1900), Part II, 219.

History of Part-Time Irish Soldiery

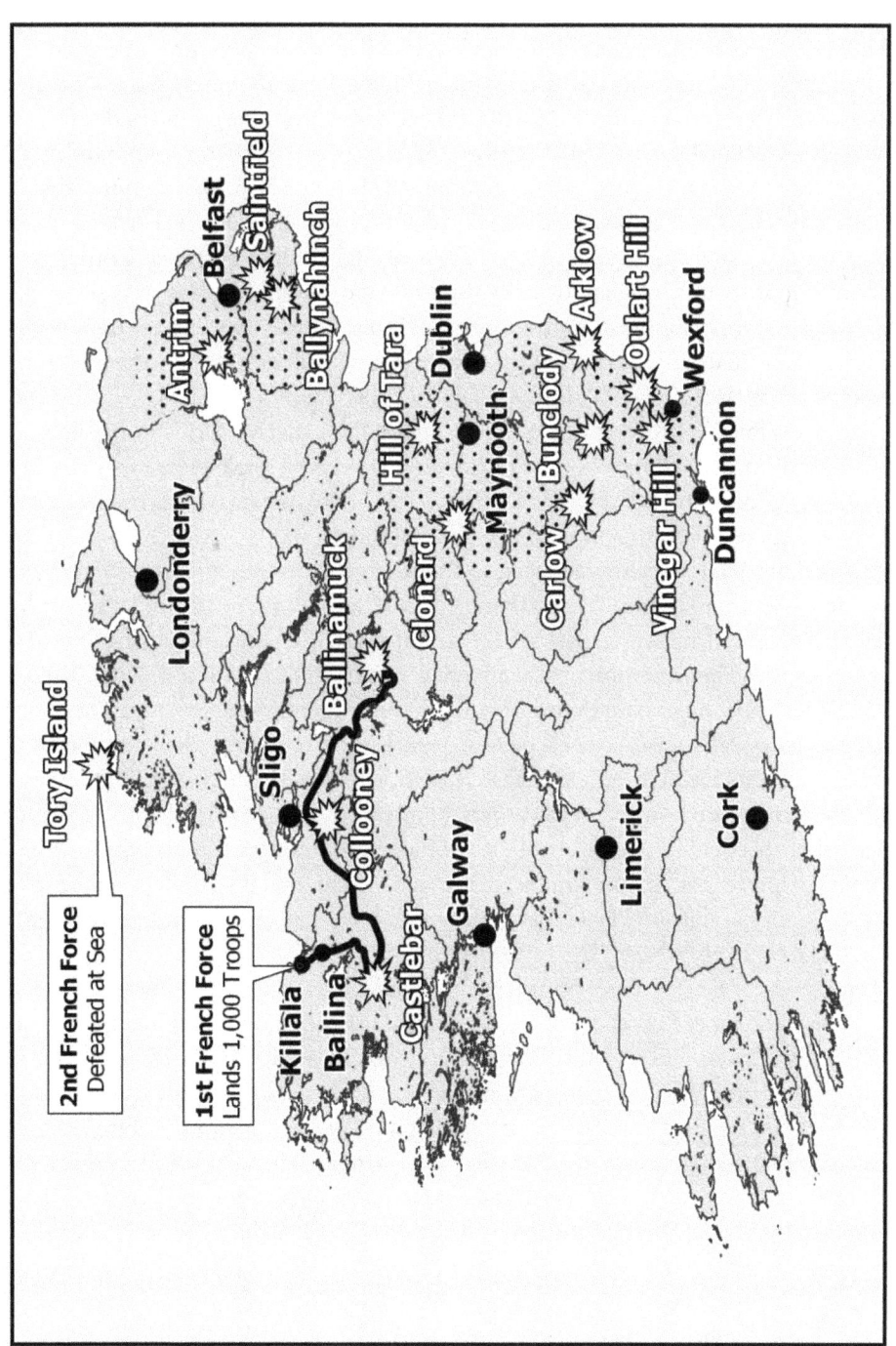

In May 1798, the North Cork Militia was the only British force in County Wexford.[106] Most of the regiment was scattered about the county in isolated companies and sections. At the rising's inception, the insurgents successfully attacked the component units of the North Cork. The British command, fearful of losing the Town of Wexford along with the countryside, dispatched elements of the Donegal Militia from its barracks at Duncannon Fort to the town. The remnants of the North Cork withdrew from the countryside and joined the Donegal troops in the Town of Wexford. The Wexford garrison commander, LTC Maxwell of the Donegal Militia, soon determined his position to be untenable and accordingly, the remnants of the North Cork, together with the Donegal troops, made a fighting retreat to Duncannon Fort. With the only local British forces bottled up in Duncannon, the insurgents took control of County Wexford. Once its base was secured, the Wexford "peasant army" moved north towards Dublin along two axes. The advancing western wing met with defeat at Bunclody (County Carlow) and the eastern wing was stopped at Arklow (County Wicklow). With their advance on Dublin blocked, the insurgents withdrew into County Wexford and prepared to meet the expected British onslaught. The culminating battle of the British campaign to retake Wexford took place on June 21st at Vinegar Hill near Enniscorthy. After the British victory there, the peasant army dwindled to isolated bands kept intact by the hope of a French landing.

By mid-July, the insurgency was reduced to small, largely inactive, and isolated, groups of rebels awaiting French assistance. That assistance came when a French force of about 1,000 troops landed on the northwest coast of Ireland at Killala, County Mayo, from where it marched southward. The government assembled troops from throughout the island and met the French force, now augmented with about 3,000 insurgents, at Castlebar. The Franco-Irish army, commanded by the French general Jean Joseph Humbert, defeated the British, then moved northeast into Ulster and defeated a British force at Collooney. On September 5th Humbert's army turned southeast towards the highland headwaters of the River Shannon. Dogged by British counterattacks, short of supplies, outnumbered and without hope of reinforcement, locally or from France, Humbert surrendered on September 8th at Ballinamuk, County Longford. "His Irish auxiliaries got no quarter, nor did those left behind in Ballina, Killala, and elsewhere."[107]

On September 16, 1798 the French government, unaware of Humbert's surrender, dispatched another amphibious force to Ireland. This army of 3,000 troops, in ships commanded by Commodore Jean-Baptiste-Francois Bompart, was to link up with Humbert's troops in Ulster. Accompanying Bompart was Wolfe Tone, who had fled previously to France. After several minor engagements with the Royal Navy, Bompart's squadron arrived off Lough Swilly on October 10th. Bompart was unable to locate a suitable landing site for the expeditionary force and withdrew from the coast. The next day his squadron was soundly defeated by the Royal Navy near Tory Island, off the northern coast

---

[106] After the general embodiment, Dublin Castle decided that no militia regiment would be stationed in its home county or any county adjacent thereto.

[107] O'Connell, *The Irish Wars,* 138.

of County Donegal. Of Bombart's squadron of ten ships, six were captured and one sunk. Among the French sailors and soldiers taken prisoner by the British was Wolfe Tone.[108]

Throughout 1798, Catholic militiamen, with very few exceptions, remained loyal to the Crown.[109] They fought the United Irishmen rebels throughout Ireland, the Catholic insurgents in County Wexford, and the French invasion force in the northwest. The heavily Catholic Kerry, Kilkenny, and Longford militia regiments were a major component of the British force that opposed Humbert's Franco-Irish army.

The Irish Militia's greatest defeat in 1798 took place on May 27th at Oulart Hill, about 25 kilometers north of Wexford town. There, a force of 110 to 120 North Cork Militia, about a quarter of the regiment's strength, attacked an insurgent force which numbered anywhere from 3,000 to 5,000. The North Cork force was led by the regimental commander, a Lieutenant-Colonel Foote, who was accompanied by several staff officers. Foote foolhardily ordered his troops to attack uphill against a force that outnumbered his by 25 to 1 or more. The North Cork were quickly enveloped and massacred. Foote was one of the handful of militia survivors.

The yeomanry fought with the militia and army against the insurgents of 1798. Among Catholics, the yeomanry gained a reputation for depredations and commission of atrocities.[110] Two years after Union between Ireland and Great Britain, Parliament affirmed the status of the yeomanry in Ireland and the Volunteers in England and Scotland.[111]

---

[108] Maxwell, *History of the Irish Rebellion in 1798,* 304-15.

[109] Approximately 60 militiamen were court-martialed for treason. They were primarily Protestant United Irishmen in Ulster regiments, and Catholics in the Longford and Westmeath regiments. Karsten, "Irish Soldiers in the British Army," 31-64. Nelson claims 67 militiamen joined the insurgency, 5 from Kilkenny, 16 from Meath, and 50 from Longford. Nelson, *The Irish Militia, 1793-1802,* 222, 232.

[110] Francis Plowden, a Catholic, English lawyer, characterized the yeomanry as Orange-dominated and quick to commit atrocities against Catholics. His history of Ireland recites numerous incidents in support of his contention. Plowden, *The History of Ireland from its Union with Great Britain,* 70-75, 92-100, 112-13. Shortly after the book's publication, the Crown convicted Plowden, then living in Dublin, of criminal libel. He was fined the then enormous sum of £5,000 and in response fled to Paris. He remained there until his death in 1829. *Dictionary of Irish Biography,* s.v. "Plowden, Francis."

Thirty-two years later, Richard Robert Madden, in his sympathetic account of the revolutionaries, echoed Plowden's sentiments. Madden, *The United Irishmen,* 303, 306, 310-11, 318, 322-24, 346, 351. For example, he writes of "… the savagery of the Carlow slaughter and conflagrations, chiefly by the Yeomanry, after the defeat and flight or concealment of the rebels…", Ibid., 343. Madden, born in Dublin and baptized an Anglican, was a physician who during his travels abroad became a noted abolitionist. Politically, he held republican views and strongly opposed the Protestant Ascendancy. *Dictionary of Irish Biography,* s.v. "Madden, Richard Robert."

For modern-day accounts of yeomanry misconduct see Blackstock, "A Forgotten Army" and Patterson, "White Terror: Counter-Revolutionary Violence in South Leinster, 1798-1801."

[111] Yeomanry (Ireland) Act, 1802, 42 Geo. 3, c. 68; Volunteer Act, 1802, 42 Geo. 3, c. 66.

The Militia, 1800-1898

Under the Act of Union, 1800 the" Irish Militia" effectively became the "Militia in Ireland." At first, the Militia in Ireland operated in accordance with the old statutes of the Kingdom of Ireland; however, in practice those statutes had little bearing on the militia. The entire Militia in Ireland was embodied and effectively part of the British Army until 1816 (a year after the Napoleonic Wars concluded).

The first statutory militia change after Union was the Militia (Ireland) Act, 1802 which officially ended compulsory service by ballot.[112] Further changes were made by four bills all enacted in 1809, during the Napoleonic Wars.[113] The first act of 1809 authorized militiamen in Irish units to enlist in the regular British Army, provided that no more than 40% of the strength of any one regiment or battalion so enlisted. The second increased by 20% the authorized strength of each unit of the Militia in Ireland. The third granted family allowances to all embodied militiamen, not just those compelled to serve by ballot. Payments were 1s. per week for each legitimate child under age 10, 2s. per week for the wife of a balloted militiaman, and 1s. per week for the wife of a volunteer militiaman. The maximum family allowance was 4s. per week (£10.4 per year). The fourth act of 1809 consolidated and restated the law for the Militia in Ireland and redesignated lieutenant-colonel commandants as colonels.

After demobilization in 1816, the government placed the Militia in Ireland into suspended animation. While county governors still commissioned new regimental colonels (who in turn commissioned new subordinate officers), and small depots were maintained, rank and file recruitment and training ceased.[114] As enlistments expired the muster rolls shrank, but the officer corps remained near authorized strength. Many of the officers by 1854; however, were practically elderly having been commissioned prior to 1815. For example, in the Donegal Militia, all but one of its lieutenants and all its ensigns had been commissioned before 1815. The numbers for the Dublin City Militia were similar to those for Donegal; four of eight lieutenants and three of six ensigns had been commissioned prior to 1815.[115] The permanent staffs, which maintained and guarded militia stores (uniforms, accoutrements, and weapons) dwindled through ever-decreasing Parliamentary appropriations. The stores were maintained at state expense in case the government decided to revive the militia.

In this period of dormancy, 1817 through 1853, the 38 Irish militia regiments had only cadre staffs of full-time adjutants, NCOs, and drummers.[116] By 1853, adjutants, with the rank

---

[112] 43 Geo. 3, c. 2.

[113] 49 Geo. 3, cc. 5, 56, 86, 120.

[114] Bowman and Butler, "Ireland" in *Citizen Soldiers and the British Empire*.

[115] *Thom's Irish Almanac and Official Directory, 1852.*

[116] *Returns of the Militia Staff in Great Britain; of the Militia Staff in Ireland, of the Establishments of Officers and Men of each Militia Corps in Great Britain; and of the Sum paid by the Public for each Corps of Militia in the United Kingdom*, 1828, H.C. Accounts & Papers, No. 183.

of captain, were paid 8s. per day, sergeant-majors 1s. 10d., sergeants 1s. 6d., and drummers 1s.[117] Beginning in 1823, Westminster progressively reduced the number of permanent militia positions.[118] An 1834 army inspection of the permanent staff in Ireland (783 of all ranks), revealed that nearly all the NCOs were over 40 years of age (many in their 50s) and possessed little or no military skills. A preponderance of English-named staffers in the report indicates that Protestants were most of the cadre force. Though the staff was paid at a daily rate, there were few military duties, so many staffers also held civilian employment.[119] By 1853, the total permanent staff of the Militia in Ireland was down to 26 adjutants, 24 sergeant-majors, and 120 sergeants, which equates to 4 or 5 men per regiment.[120]

The Militia in Ireland was reconstituted during the Crimean War and reorganized under the Militia (Ireland) Act, 1854.[121] The law's provisions included five-year enlistment terms, authorization of enlistment bonuses to be set by the War Office, an increase of authorized strength to 30,000 privates, and specific authority for embodiment during time of war, not just threatened invasion. The act also gave the Lord Lieutenant greater control over the militia than under previous statutes. For example, Dublin Castle became the sole source of both officer commissions and NCO appointments. Annual training was set at 21 days, but the Lord Lieutenant received authority to reduce it to not fewer than 3 days, or to increase it to not more than 56 days. As was the case since 1802, there was no provision for compulsory service.

Recruits into the new militia understood, as in 1794, that they were not enlisting into a part-time force. With Britain at war with Russia, the militia would be embodied for the duration of the conflict and be a major component of the home army.

During the Crimean War, the government embodied nearly the entire militia to free regular army units at home for active service abroad.[122] From May 1854, through September 1856, embodied militia units were stationed throughout the United Kingdom.[123] Of the 37

---

[117] *Militia Estimates for the Year Ending 31 March 1854*, 1852/53, H.C. Accounts and Papers, No. 777.

[118] *Abstract of the Sums Voted and the Amount Actually Expended for the Militia of the United Kingdom*, 1834, H.C. Accounts & Papers, No. 231; *An Abstract of the Sums Voted and the Amount Actually Expended for the Militia of the United Kingdom, in each Year [from 1834 through 1842]*, 1843, H.C. Accounts & Papers, No. 600.

[119] In 1834, 293 of the permanent staff of 735 sergeants and drummers apparently had full-time civilian employment. None of the 38 officers, the regimental adjutants, had outside employment. *Reports of the Officers appointed on the Recommendation of the Committee on the Militia Estimates of the last Session of Parliament, on the State of the Staff of the Disembodied Militia of the United Kingdom*, 1835, H.C. Accounts & Papers, No. 201.

[120] *Militia Estimates for the Year Ending 31 March 1854*, 1852/53, H.C. Accounts & Papers, No. 777.

[121] 17 & 18 Vict., c. 107.

[122] Of the 164 militia regiments in the United Kingdom, 146 were embodied by July 1855. *Return of Regiments of Militia Embodied in each Month to Present Time*, 1854/5, H.L. Other Papers, No. 266.

[123] *Hart's New Annual Army List, 1856*.

embodied Irish regiments, 13 were stationed in England.[124] As the war progressed, and the demand for troops in the theatre of war continued, Parliament authorized militiamen to volunteer for service in the colonies with their units. These militia regiments would free regular army garrison units for active service in Crimea.[125] Irish regiments accounted for 13 of the 49 militia regiments that volunteered for colonial postings.[126] Only 10 of the 49 volunteering units were deployed abroad.[127] After the conclusion of the Crimean War, the militia was demobilized between May and September 1856.[128]

In 1857, the Indian "Mutiny" prompted the government to again embody the militia to free regular army units at home for active service abroad. This mobilization was not as extensive as the one for the Crimean War.[129] Of the 15 embodied Irish regiments, 11 were stationed in England and 1 in Scotland.[130] Again, the government asked militiamen to volunteer for foreign service with their units. In response, nine of the embodied Irish militia regiments agreed to serve abroad. Of those nine, four had also volunteered for overseas service during the Crimean War (North Cork, Antrim, Armagh, and Roscommon). By the end of 1861, all the embodied Irish regiments had been demobilized.[131]

In the latter part of the nineteenth century, Dublin Castle had serious doubts as to the loyalty of Catholic militiamen. Accordingly, the government suspended recruitment and all training from 1866 through 1870 (threat of Fenian infiltration of units), and again from 1881 through 1882 (fears that the militia would undertake "rural agitation" - the "Land Wars").[132]

In 1871, county authorities lost their power to appoint militia officers. County "Governors" now styled "Lieutenants," could only nominate officer candidates to the War

---

[124] Ibid.

[125] 18 & 19 Vict., c. 1.

[126] *Return of the Militia Regiments of the United Kingdom which Volunteered for Foreign Service during the Crimean War and during the Indian Mutiny*, 1868/69, H.C. Accounts & Papers, No. 318.

[127] *Hart's New Annual Army List 1856*, shows two militia units stationed in Gibraltar, two in Malta, and six in the Ionian Islands (Corfu, Zante, Cephalonia). All Irish militia units were in the United Kingdom.

[128] *Return of Total Expense Incurred on Account of the Embodied Militia Showing Period During Which Embodied*, 1860/61, H.L. Other Papers, No. 157.

[129] In 1857, the War Office embodied 47 infantry regiments while during the subsequent two years it embodied a further 10 regiments, all artillery. *Return of the Total Expenses incurred on account of the Embodied Militia in each Year since 1854 showing the Strength of the Force during the part of the Year it has been Embodied*, 1860, H.C. Accounts & Papers, No. 380. As of February 1859, 33 militia regiments remained embodied (18 English, 3 Scottish, 12 Irish). *Returns of the Number of Volunteers given by each Regiment of Embodied Militia to the Regular Army during the Year 1858*, 1859 Sess. 1, H.C. Accounts & Papers, No. 158.

[130] *Hart's New Annual Army List, 1858*.

[131] *Hart's New Annual Army List, 1862*.

[132] Bowman and Butler, "Ireland" in *Citizen Soldiers and the British Empire*.

Office. The War Office had final say as to who would receive a Queen's militia commission.[133]

In 1882, Parliament put into a single, comprehensive statute the governing authority for the militias of England & Wales, Scotland, and Ireland.[134] At about the same time, by War Office directive, militia infantry regiments were restyled "battalions" and linked with regular battalions into territorial regiments that shared a common basic training depot establishment.

The Yeomanry, 1800-1834

From 1800 through the early 1830s, Irish county authorities used the yeomanry as a part-time, paramilitary police force. Irish nationalists, especially Catholics, came to view this force as an instrument of British oppression. In June 1831, an incident in Newtownbarry, County Wexford, brought the Irish Yeomanry to the attention of the British public. The incident, termed by many "the Newtownbarry Massacre" became notorious and triggered a Parliamentary investigation. In Newtownbarry, irate townsmen tried to release one or more heifers taken by county officials as a statutory tithe for the Church of Ireland. The magistrates called for the yeomanry. In the ensuing tumult the yeomen opened fire on the crowd and caused multiple civilian casualties. The conservative *Dublin Evening Mail* reported the incident as a mob attack on the authorities that necessitated a firm response. It reported one yeoman and 18 "insurgents" killed, and a great number of yeomen gravely wounded. The radical *London Globe*; however, reported that "One of the most sanguinary and brutal outrages that ever gave pain to the eye, or sadness to the heart, took place yesterday at Newtownbarry." The account claimed 13 dead, including 1 yeoman (most likely killed accidentally by his own party), and 23 gravely wounded. The Globe portrayed the incident as an attack by "barbarous assailants" who fired indiscriminately in response to stones thrown at them by some boys.[135]

After the Newtownbarry incident, British government support for the part-time force, by then half-hearted, waned further. In 1834, Westminster disbanded the Irish Yeomanry. "The force was widely seen to be sectarian in nature and was certainly largely Protestant in composition. In any case, its function as a part-time constabulary force had been rendered redundant both by the concentration of the yeomanry in Ulster, far removed from some of the worst 'disturbed districts' and the formation of the Irish Constabulary in 1822."[136]

The Militia and the War in South Africa, 1899-1902

Unlike 45 years earlier when the Crimean War began, the United Kingdom at the start of the war in South Africa (the 2nd Boer War), had a pool of recently discharged soldiers who

---

[133] Regulation of the Forces Act, 1871, 34 & 35 Vict. c. 86.

[134] The Militia Act, 1882, 45 & 46 Vict., c. 49.

[135] *The Spectator*, June 25, 1831, 16.

[136] Bowman and Butler, "Ireland" in *Citizen Soldiers and the British Empire*, 41.

were obligated to reinforce the army when so ordered by the War Office. This manpower pool was the Army Reserve. With the Army Enlistment Act of 1870, Parliament authorized "short-service" enlistment.[137] Recruits served three to nine years with the Regular Army then spent the balance of their twelve-year engagement in the Army Reserve. Reservists received "reserve pay" simply for their reserve status, plus additional pay for the occasional training muster. Pay was 1s. per day during the first year in the Reserve, then 4d. per day thereafter.[138] At the start of the Boer War, the Army Reserve totaled 79,000 men: 61,000 were sent to South Africa, 5,000 remained at home, and 13,000 the army found unfit for service (16.5%).[139]

There was also another source of reinforcements and replacements for the army: the Militia Reserve. This reserve pool consisted of militiamen who after two annual trainings volunteered as reservists for the Regular Army. Such reservists remained on the rolls of their militia units and were subject to both militia embodiment and reserve mobilization. At the start of the Boer War, the Militia Reserve totaled approximately 30,000 men, 27.5% of the entire militia.[140]

The Army and Militia Reserves, called to the colours in October, did not provide enough men to meet war requirements and the government had to look elsewhere for the needed troops.[141] Among the measures taken to obtain more soldiers was mobilization of the militia. The War Office issued the first embodiment orders on November 4, 1899, which called up 38 infantry battalions (of which 7 were Irish). Further embodiments were soon ordered and by the end of January 1900, 40,000 of the available 110,000 militiamen were embodied.[142] As with the Crimean War and the Indian Mutiny, the main purpose of the embodiment was to free regular army units at home for active service at the front. Also, as during earlier wars, militiamen could volunteer to serve abroad with their units; however, for the Boer War militia units were not barred from the theatre of war. On April 1, 1900, 19,000 of the by then 44,000 embodied militiamen were in South Africa.[143] By year's end, every militia unit was embodied

---

[137] 33 & 34 Vict., c. 67.

[138] *Royal Warrant for the Pay, Appointment, Promotion, and Non-Effective Pay of the Army, 1899*, Arts. 1289-1306A.

[139] *Report of His Majesty's Commissioners Appointed to Inquire into the Military Preparations and Other Matters Connected with the War in South Africa,*1903, [Cd.1789], at 40. Hereafter cited as *South African War Report*.

[140] Henry Jenkyns, "Constitution of the Military Forces of the Crown;" Testimony of LTG T. Kelly-Kenny, *Minutes of Evidence Vol. 1, South African War Report*, 1903, [Cd. 1790], qq. 4624-4625, 4629.

[141] The reserve pool would be exhausted by the end of September 1900. *South African War Report*, 1903, [Cd. 1789], at 40.

[142] *Appendix to the Minutes of Evidence, South African War Report*, 1903, [Cd. 1792], nos. 10, 14; *General Annual Report on the British Army for the Year Ending 30th September, 1904*, 1905, [Cd. 2268].

[143] Total Strength on April 1, 1900, *Appendix to the Minutes of Evidence, South African War Report,* 1903, [Cd. 1792], no. 10.

and 30 militia infantry battalions were in, or in transit to, South Africa.[144] Though the United Kingdom was under no threat of invasion, all of the Militia Royal Garrison Artillery was embodied and dispatched to coastal forts. Three Irish RGA units manned the fortifications of the Royal Navy dispersal anchorages of Lough Swilly, Berehaven, and Cork Harbor. The nine other RGA units manned coastal fortifications in England. Note that of the first 16 Irish infantry battalions embodied 12 were stationed in England.[145]

During the Boer War, a total of 68 militia infantry battalions served abroad, all but 8 in South Africa. Among the units that saw overseas service were seven Irish battalions, one in Malta and six in South Africa.[146] The War Office asked 72 militia infantry battalions to volunteer for overseas service. The army deemed a battalion to have volunteered if enough men agreed to serve abroad with the unit.[147] Only those individuals who so volunteered would depart with the headquarters staff and receive the £5 foreign service bonus. Those who did not volunteer were stationed at the battalion's training depot. All 54 English battalions that were asked to serve abroad had enough volunteers to qualify. Of 8 Scottish units asked, 1 failed to qualify (3/Cameron Highlanders, at its depot, Inverness), and of 10 Irish units asked, 3 failed to qualify. The three Irish units that did not produce enough volunteers for foreign service were as follows:

4th Battalion, Royal Irish Regiment, Clonmel, Co. Tipperary, at Aldershot, England.

4th Battalion, Connaught Rangers, at its Galway City depot.

6th Battalion, Royal Irish Rifles, Dundalk, Co. Louth, at Sheffield, England.

According to the then Inspector-General of Recruiting, two days after the Royal Irish Regiment battalion failed to qualify for foreign service, its commander notified him that he now had the needed number of volunteers. The battalion; however, never went abroad as the War Office, in those two days, had filled the quota with another unit.[148]

---

[144] *Harts Annual Army List 1904*; Return of Militia Battalions Abroad, *Appendix to the Minutes of Evidence, South African War Report*, 1903, [Cd. 1792], no. 16.

[145] Army Order 112, *Appendix to the Minutes of Evidence, South African War Report*, 1903, [Cd. 1792], no. 14.

[146] Return of Militia Battalions Abroad, *Appendix to the Minutes of Evidence, South African War Report*, 1903, [Cd. 1792], no. 16. The Irish contingent in South Africa included two of the three militia battalions of the Royal Dublin Fusiliers, Brian Tweedy's regiment that is mentioned several times in *Ulysses*.

[147] Army Orders, 1899, No. 93 issued under 61 & 62 Vict. c. 9. Typically, a unit would not be sent overseas unless it could deploy 400 men. Testimony of MG G. Barton, *Minutes of Evidence Vol. 2, South African War Report*, 1903, [Cd. 1791], q. 16343.

[148] Testimony of MG H.C. Borrett, *Minutes of Evidence Vol. 1, South African War Report*, 1903, [Cd. 1790], qq. 5306-08, 5323-32.

Prior to the amalgamations of 1881, 6/Royal Irish Rifles was styled the "Louth Militia" and its failure to volunteer for service abroad became notorious as "The Louth Mutiny."[149] On December 1, 1899 the 716 militiamen and 25 officers of the battalion were embodied at the unit's depot in Dondalk, County Louth. Of the militiamen on the roll, 17 were unfit for service and another 22 were absent without leave. Five days later the War Office moved the battalion to Sheffield, England.

During the next two months, the unit's 237 militiamen who were also Militia Reservists were sent to Regular Army units. In that same time period, 25 recruits finished their training and joined the battalion. On February 9, 1900, the War Office asked the now 465 enlisted men of the Louth Militia to volunteer for foreign service. Only 290 did so, a number insufficient to qualify the unit to serve abroad. Later, 95 of the 175 men who refused foreign service recanted, but that changed nothing as the War Office had stricken the battalion from the list for overseas service.[150] The battalion's failure to qualify for overseas service became a subject of newspaper articles and debate in Parliament. That was because some Louth Militiamen claimed the battalion's officers obtained consents to foreign service through deceit or coercion.

Among Catholic nationalists, there was widespread belief that militia officers in Ireland, of which about 85% were Protestant and either Anglo-Irish or English, coerced or deceived uneducated, subservient, rural Catholic militiamen into acceptance of foreign service.[151] Throughout February 1900, several times Irish MPs voiced concerns in Commons regarding Irish militia volunteers for foreign service."[152] For example, on February 15, 1900, in Commons during Questions, the following exchanges occurred among the Under-Secretary of State for War and several Irish MPs:[153]

> Patrick O'Brien, Kilkenny:
> I beg to ask the Under Secretary of State-for War whether, when Irish Militia regiments are brought to England and the men are asked to volunteer for service at the front, he will provide Members of this House who may wish to visit Militiamen belonging to their constituencies who are in such regiments, with the opportunity of ascertaining in what way they were asked to volunteer for the front, and whether they were free agents or not, and with the necessary authority to interview them on these points.

---

[149] Hall, "The Louth Militia Mutiny.".

[150] Even if the War Office had reconsidered the battalion's decision to volunteer, it is unlikely the Louth Militia would have been sent abroad with only 385 militiamen.

[151] Butler, *The Irish Amateur Military Tradition*, 62.

[152] 78 *Parl. Deb.* (4th ser.) (1900) February 2, 8; 79 *Parl. Deb.* (4th ser.) (1900) February 15, 19, 20, 22, 27.

[153] 79 Parl. Deb. (4th ser.) (1900) 62-66.

COL Edward James Saunderson, Armagh, North:
Arising out of that question, may I ask my hon. friend whether he is aware that the hon. Member for Kilkenny on a recent occasion is reported to have advised Irish soldiers to shoot their English comrades and join the Boers?

J.G. Swift MacNeill, Donegal, South:
Don't stand that.

Speaker:
Order, order! That does not arise out of the question.

George Wyndham, Under-Secretary of State for War:
In reply to the hon. Member who asked the original question, I have to say that this would be quite contrary to military discipline.

Eugene Crean, Queen's Co.:
Is the hon. gentleman aware that several of the soldiers have written to their friends saying that they have been coerced into volunteering?

Wyndham:
No, Sir; I have no information to that effect, and I believe the case has been misrepresented.

• • •

O'Brien:
I beg to ask the Under Secretary of State for War whether nearly 50 per cent, of the 6th Battalion Royal Irish Rifles (Louth Militia), stationed at Sheffield Barracks, have signified their objection to go to the front in South Africa; and if so, whether the men are within their rights in refusing for foreign service; and whether he will see that no undue pressure is put upon them to compel or induce them to go.

Wyndham:
I have no information to the effect stated in the question. But, as I have before informed the hon. Member, commanding officers have received the most explicit orders not to place any pressure upon their men to accept service in South Africa, and the Secretary of State has no reason to believe that such pressure is being exercised.

O'Brien:
May I ask the hon. Gentleman whether, considering he asked me to postpone the question, he instituted the inquiries which he promised to make?

Wyndham:
I cannot accept the statement that I promised to make any inquiry. I have on more than one occasion given the answer I have given this afternoon.

O'Brien:
I am in the recollection of the House—

Speaker:
Order, order!

Swift MacNeill:
Move the adjournment.

Speaker:
The question has been fully answered.

O'Brien:
I will take the earliest opportunity to raise the question of the kidnapping of Irish Militiamen to send them abroad. I beg to ask the Under Secretary of State for War whether he can explain the method adopted by the colonel of the 3rd Royal Munster Fusiliers (Militia), now stationed at Dover, en route to South Africa on the 21st inst., to ascertain whether any of the men were willing to volunteer for the front; whether he will inquire if the men were asked en masse while on parade; whether he is aware that very few of them understood the colonel's words; also that some men who were not on parade at the time afterwards explained to their officers that they were not willing to volunteer or be bound by what happened on parade on the occasion, and were then told that the majority had volunteered, and that they were bound to go with them; whether he will order, before this regiment is sent to the front on the 21st inst., that each man is asked separately if he wishes to volunteer, and allowed reasonable time to give his reply, and that ho is protected against any attempt to influence him in his decision; and whether he will give the necessary authority to Irish Members of this House to interview any of their constituents who are amongst the 3rd Royal Munster Fusiliers.

Wyndham:
There is no information on the subject in the War Office; but strict orders are given that no pressure is to be put upon the men; and the Secretary of State is not disposed to interfere with the commanding officer, who, he does not doubt, did his duty.

• • •

Richard M'Ghee, Louth, South:
I wish to ask the Under Secretary of State for War a question of which I have given him private notice — namely, whether his attention has been directed to a letter dated 12th February, in the Irish press, from Dr. Logue, the Cardinal Primate of Ireland, with reference to the case of a Militiaman named Duffy, under orders, as he believe, for military service in South Africa, without having been given any option of accepting or declining active service; and whether, having regard to the statement of the Cardinal on the subject of the grievances of this man, that "he knows as little of what he is about as a bullock being led to the shambles" —

Speaker:
Order, order!

At the outbreak of hostilities, the militia had 109,551 men, including Militia Reservists.[154] During the course of the war a further 85,834 men joined. Of the total 195,385 militiamen that served during the Boer War, 22.5% went to South Africa.[155] Excluding Militia Reservists, 28,474 ordinary militiamen from throughout the United Kingdom were in the war zone.[156] This indicates that about 3,300 Irish militiamen volunteered after the declaration of war against the Boer republics.[157]

By September 1900, the British Army had defeated the Boer field forces and captured the principal towns of the Boer republics including Bloemfontein, the capital of the Orange Free State, and Pretoria, the capital of the Transvaal Republic. In October 1900, the Boer's initiated guerilla warfare that continued until their capitulation in May 1902.

The British government began to stand down the militia in late 1900. Of the twelve Irish garrison artillery regiments, the War Office disembodied five in October 1900, and the remaining seven the following month. Of the 28 Irish infantry battalions, 17 were disembodied in the last quarter of 1900 leaving 7 in England and 4 in South Africa.[158] The War Office disembodied a further seven Irish battalions in 1901 and four in 1902. The last Irish battalion to return to part-time status was 5/Royal Irish Rifles (South Down Light Infantry), which was disembodied in July 1902.[159]

The Imperial Yeomanry

Shortly after the start of hostilities, the War Office concluded that the army lacked sufficient mounted infantry to meet military needs in South Africa. On December 24, 1899, the War Office called for volunteers for active service in the soon to be formed Imperial Yeomanry units. The Imperial Yeomanry was open to troopers of the existing "home" yeomanry (the part-time cavalry force in Great Britain), members of the volunteer force, and any civilian provided "… that he is a good rider and a marksman according to yeomanry standard."[160] As mounted infantry, not cavalry, the new force's basic element was denominated "company" and not "squadron." The first 10,242 Imperial Yeomen were

---

[154] *General Annual Report on the British Army for the Year Ending 30th September, 1904*, 1905, [Cd. 2268], Part XI.

[155] From October 1, 1899, through December 31, 1901, 85,384 men joined the Militia. 43,875 Militiamen, including Militia Reservists, went to South Africa. Grant, *History of the War in South Africa*, Vol. 4, 677-78.

[156] Deployments to South Africa, *Appendix to the Minutes of Evidence, South African War Report*, 1903, [Cd. 1792], no. 5.

[157] Based on 7 of the 60 volunteered militia battalions in South Africa were Irish (11.7% of total).

[158] *Hart's Annual Army List, 1901*.

[159] *Hart's Annual Army List, 1904*

[160] Formation of Imperial Yeomanry, *Appendix to the Minutes of Evidence, South African War Report*, 1903, [Cd. 1792], no. 14. The yeomanry, like the volunteer force and militia, by statute, could not be compelled to serve abroad. There were no home yeomanry or volunteer units in Ireland.

dispatched to South Africa in 1900 after two to three months training with their units. About 30% of the men were of the home yeomanry or Volunteers. Many in the first draft could not ride well and few were proficient with rifles. A further 16,597 Imperial Yeomen were sent to South Africa in 1901. This second draft consisted almost entirely of civilians of whom only a quarter had ever ridden a horse and nearly none could demonstrate any marksmanship ability. The men of the second draft were trained hastily in South Africa then assigned to secondary combat roles: guard duty and protection of lines of communication.[161]

The initial contingent of Imperial Yeomanry in South Africa included five Irish companies, each of 121 men. The ranks were filled primarily from the Anglo-Irish elite. The 45th Company, commanded by Thomas Pakenham (Earl of Longford) recruited Dublin professionals and members of hunt clubs located throughout Ireland. It was dubbed by the press "The Irish Hunt." The 47th Company, raised in London by the Earl of Donoughmore, consisted of rich, Anglo-Irish men-about-town, who each paid £130 for the cost of his horse, equipment, and passage to South Africa.[162] The 46th and 54th Companies were raised in Belfast, the 60th Company in other parts of Ulster. All but the 60th were regimented into the 13th Imperial Yeomanry Battalion, commanded by a regular army officer, LTC Basil Spraage. On May 27, 1900 that 500-man battalion was nearly surrounded at Lindley in the Boer Republic of the Orange Free State. At the time, the battalion could have made good an escape, but Spraage ordered his men to dig in and await relief. On May 30th, after heavy fighting, Spraage surrendered the battalion to the Boers after it took 80 casualties. Two days later the relieving British troops arrived to find only the left behind bodies of Spraage's men killed-in-action.[163] As the war progressed six additional Irish companies formed. One of these, the 74th Company, incurred casualties of 25% of its initial strength.[164]

The Irish Imperial Yeomanry units returned home in 1901. In June of that year, the conservative Anglo-Irish journal *Irish Society*, to honor the returning Irish soldiers, published the following poem, "Ode of Welcome" by J.R.S.

>The Gallant Irish yeoman
>Home from the war has come
>Each victory gained o'er foeman
>Why should our bards be dumb.
>
>How shall we sing their praises
>Our glory in their deeds
>Renowned their worth amazes
>Empire their prowess needs.

---

[161] *South African War Report*, 1903, [Cd. 1792], at 71-72; War Recruiting, *Appendix to the Minutes of Evidence, South African War Report*, 1903, [Cd. 1792], no. 13.

[162] Doherty, *North Irish Horse*, 4-8; Formation of Imperial Yeomanry, *Appendix to the Minutes of Evidence, South African War Report*, 1903, [Cd. 1792], no. 14.

[163] Maurice, *History of the War in South Africa* Vol. 3, 115-25.

[164] *Minutes of Evidence Vol.1, South African War Report*, 1903, [Cd. 1790], appx. D.

So to Old Ireland's hearts and homes
We welcome now our own brave boys
In cot and Hall; neath lordly domes
Love's heroes share once more our joys.

Love is the Lord of all just now
Be he the husband, lover, son,
Each dauntless soul recalls the vow
By which not fame, but love was won.

United now in fond embrace
Salute with joy each well-loved face
Yeoman: in women's hearts you hold the place.

The poem is an acrostic where the first letter of each line spells out THE WHORES WILL BE BUSY. The purported author was Oliver St. John Gogarty, Joyce's former friend who appears in *Ulysses* as Malachi Roland St. John (Buck) Mulligan.[165]

After the war, the government reorganized the old "home" yeomanry and christened it "Imperial Yeomanry."[166] The new force had two Irish regiments: The North of Ireland and the South of Ireland Imperial Yeomanry.[167] The "North Irish Horse" was headquartered in Belfast, the "South Irish Horse" in Limerick. Detached squadrons were in Londonderry, Enniskillen, Dundalk, Cork, and Dublin.[168] The regiments were manned with many veterans of the eleven Irish Imperial Yeomanry companies that fought in South Africa. These units retained the elite social status of the first five wartime Irish companies of Imperial Yeomanry. On Bloomsday, the North Irish Horse was com-manded by the Earl of Shaftesbury, the South Irish Horse by the Marquess of Waterford. Of the 33 Irish Imperial Yeomanry officers in 1904, 10 were titled and among them were 4 peers.[169]

The Imperial Yeomanry in Ireland, like elsewhere in the United Kingdom, proved immensely popular. Unlike the militia, the Imperial Yeomanry on Bloomsday was only a few men short of its authorized enlisted strength, which for Ireland was 888.

---

[165] Ulick O'Connor, *Oliver St John Gogarty* (London: Cape, 1964), 22-23. During the war, Gogarty contributed articles to *Sinn Fein* that opposed British Army recruitment of Irishmen. Gogarty became a prominent otolaryngologist, a noted man of letters, and served in the Irish senate until its dissolution in 1936.

[166] Militia and Yeomanry Act, 1901, 1 Edw. 7, c. 14.

[167] *London Gazette*, January 7, 1902.

[168] *Monthly Army List, December 1904*; Vaugh, The South Irish Horse; Baillie-Stewart, The North Irish Horse.

[169] *Monthly Army List, December 1904*.

Chapter Bibliography

Anderson, Michael. "Population change in north-western Europe 1750-1800." In *British Population History*, edited by Michael Anderson. Cambridge: Cambridge Univ. Press, 1996.

Baillie-Stewart, Gareth. The North Irish Horse Regimental Association. www.northirishhorse.com.

Bartlett, Thomas. "An End to Moral Economy: The Irish Militia Disturbances of 1793." *Past and Present* 99, no. 1 (May 1983): 41-64.

Beckett, Ian F.W. *The Amateur Military Tradition*. Manchester: Univ. of Manch. Press, 1991.

Blackstock, Allan F. "A Forgotten Army." *History Ireland* 4, no. 4 (Winter 1996): 28-33.

—— "A Dangerous Species of Ally: Orangeism and the Irish Yeomanry." *Irish Historical Studies* 30, no. 119 (May 1997): 393-405.

—— *An Ascendancy Army*. Dublin: Four Courts, 1998.

Bowman, Timothy and William Butler. "Ireland." In *Citizen Soldiers and the British Empire, 1837-1902* edited by Ian Beckett. London: Pickering & Chatto, 2012.

Budgen, Frank. *James Joyce and the Making of Ulysses*. London: Indiana Univ. Press, 1960.

Butler, William. *The Irish Amateur Military Tradition in the British Army, 1854-1992*. Manchester: Manch. Univ. Press, 2016.

Clarendon, R.V. *Revenue and Finances of Ireland*. London: Lowndes & Debrett, 1791.

Clarke, Aidan with R. Dudley Edwards. "Pacification, Plantation, and the Catholic Question." In *A New History of Modern Ireland,* Vol. 3, edited by F. X. Martin, F. J. Byrne, W. E. Vaughan, A. Cosgrove, J. R. Hill, and T. W. Moody. Oxford: Oxford Univ. Press, 1976.

Croker, John Wilson. *Sketch of the State of Ireland, Past and Present*. Dublin: M.N. Mahon, 1822.

Curtin, Nancy J. "The transformation of the Society of United Irishmen into a mass-based revolutionary organization, 1794-6." *Irish Historical Studies* 24, no. 96 (November 1985): 463-92.

Doherty, Richard. *North Irish Horse*. Staplehurst, UK: Spellmount, 2002.

Garnham, Neal. "The Establishment of a Statutory Militia in Ireland, 1692-1716: Legislative Processes and Protestant Mentalities." *Historical Research* 84, no. 224 (May 2011): 266-87.

—— *The Militia in Eighteenth Century Ireland*. Woodbridge, UK: Boydell, 2012.

Grant, Maurice Harold. *History of the War in South Africa*. Vol. 4, London: Hurst & Blackett, 1910.

Hall, Donal. "The Louth Militia Mutiny of 1900." *Journal of the County Louth Archaeological and Historical Society* 24, no. 2 (1998): 281-95.

Hayes-McCoy, G. A. "The Completion of the Tudor Conquest and the Advance of the Counter-Reformation." In *A New History of Ireland*, Vol. 3, edited by T. W. Moody, F. X. Martin, and F. J. Byrne. Oxford: Oxford Univ. Press, 2009.

Higgins, Padhraig. "'Let us Play the Men': Masculinity and the Citizen-Soldier in Late Eighteenth-Century Ireland." In *Soldiering in Britain and Ireland, 1750-1850*, edited by Catriona Kennedy and Matthew McCormack. New York: Palgrave Macmillan, 2013.

Jenkyns, Henry. "Constitution of the Military Forces of the Crown." In *Manual of Military Law*. London: HMSO, 1907.

Karsten, Peter. "Irish Soldiers in the British Army, 1792-1922: Suborned or Subordinate?" *Journal of Social History* 17, no. 1 (October 1983): 31-64.

Jeffrey, Keith, "The Irish Soldier in the Boer War." In *The Boer War: Direction, Experience and Image* edited by John Gooch. New York: Routledge 2013.

Kelly, James. "A Secret Return of the Volunteers of Ireland in 1784." *Irish Historical Studies* 26, no. 103 (May 1989): 268-92.

Kennedy, Liam and Martin W. Dowling. "Prices and Wages in Ireland, 1700-1850." *Irish Economic and Social History* 24, no. 1 (1997): 62-104.

McDowell, R.B. "The Age of the United Irishmen: Reform and Reaction, 1789-94." In *A New History of Ireland*, Vol. 4, edited by T.W. Moody and W.E. Vaughan. Oxford, Oxford Univ. Press, 2009.

MacNevin, Thomas. *The History of the Volunteers of 1782*. Dublin: James Duffy, 1845.

Madden, Richard R. *The United Irishmen, Their Lives and Times*. Dublin: James Duffy, 1858.

Maurice, Frederick. *History of the War in South Africa*. Vol. 3, London: Hurst & Blackett, 1908.

Maxwell, W. H. *History of the Irish Rebellion in 1798*. London: H. G. Bohn, 1848.

Mileham, Patrick. *The Yeomanry Regiments*. London: Spellmount, 2003.

Miller, David. "Non-professional Soldiers, c. 1600-1800." In *A Military History of Ireland* edited by Thomas Bartlett and Keith Jeffery. Cambridge: Cambridge Univ. Press, 1996.

Morley, Vincent. *Irish Opinion and the American Revolution*. Cambridge: Cambridge Univ. Press, 2002.

Murray, Alice Effie. *History of the Commercial and Financial Relations between England and Ireland*. London: P.S. King, 1903.

Nelson, Ivan F. *The Irish Militia, 1793-1802*. Dublin: Four Courts, 2007.

O'Faolain, Sean, ed. *The Autobiography of Wolfe Tone*. London: Thomas Nelson, 1937.

O'Connell, Daniel. *An Historical Memoir of Ireland and the Irish*. Dublin: James Duffy, 1869.

O'Connell, J.J. *The Irish Wars*. Dublin: Martin Lester, 1920.

Ó Snodàigh, Padraig. *The Irish Volunteers: A List of the Units 1715-1793*. Blackrock, Ireland: Irish Academic Press, 1995.

Patterson, James G. "White Terror: Counter-Revolutionary Violence in South Leinster, 1798-1801." *Eighteenth Century Ireland* 15 (2000): 38-53.

Plowden, Francis. *The History of Ireland from its Union with Great Britain.* Dublin: John Boyce, 1811.

Small, Stephen. *Political Thought in Ireland 1776-1798: Republicanism, Patriotism, and Radicalism.* Oxford: Oxford Univ. Press, 2002.

Smyth, Peter. "'Our Cloud-Cap't Grenadiers': the Volunteers as a Military Force." In *Essays from the* Irish Sword, Vol. 1, edited by Thomas Bartlett and Harman Murtagh. Dublin: Irish Academic Press, 2006.

Vaugh, Doug and Hugh Vaugh. The South Irish Horse. www.southirishhorse.com.

Wyse, Thomas. *Historical Sketch of the Late Catholic Association of Ireland.* London: Henry Colburn, 1829.

Sources for Prelude to War with Revolutionary France:

Blanning, T.C.W. *The French Revolutionary Wars, 1787-1802.* London: Arnold, 1996, 58-95.

Kelly, Christopher. *History of the French Revolution.* London: Thomas Kelly, 1820, 62-87.

Correspondence between M. Chauvelin and Lord Grenville, May 12, 1792 - January 24, 1793, H.L. Papers, Correspondence, 1792-12-13 through 1793-6-21, 569-658

Declaration of Pillnitz, August 27, 1791, Robinson, James H., ed. *Readings in European History.* Vol. 2, Boston: Ginn, 1906, 432-33.

Brunswick War Manifesto, July 25, 1792, Ibid., 443-45

Edict of Fraternity, November 19, 1792. Anderson, Frank Maloy, ed. *The Constitutions and Other Select Documents Illustrative of the History of France.* Minneapolis: H. W. Wilson, 1904, 130.

Proclamation of December 15, 1792, Ibid., 130-32.

Chapter 2
History of the British Army, Cromwell to 1853

The British Army of the *Ulysses* characters Privates Carr and Compton, was in many respects different from the mid-Victorian army in which Molly Bloom's father served. Soldiers' living conditions were much improved, recruits were better educated, pay was higher, and flogging had been abolished. In the more fundamental aspects; however, it was the same red-coated force. Unlike nearly all other European armies, the British Army remained a volunteer force. As in the past, it was led by gentlemen officers and its rank and file were of the lower-strata of the working class. Equally unchanged was the common soldier's status in the eyes of the civilian populace, which was just barely above that of the mentally defective, the chronically drunk, and the habitually criminal. Though to the residents of Dublin the Army seemed an ancient institution (few Irish of the time called the force the "British" Army, it was just "the Army"), by European standards it was fairly young. In fact, in 1904, the British Army, through its antecedent English Army, was only 243 years old. Compared to the United States Army, established 1775, and the Canadian Army, established 1867, the British Army was a long-standing force.[1] But Dublin Castle predates the Army by 430 years, and Dublin City by about 700 years.

Prior to 1652, there was no permanent English army. The peacetime armed forces of England were the sovereign's guards (Yeomen of the Guard, Sergeants-at-Arms, Gentlemen Pensioners), a few voluntary associations (such as The Honourable Artillery Company of London), castle and fortress caretakers, the Office of Ordnance (for the acquisition of artillery and maintenance of military stores), and the armed retainers of the wealthier nobles.[2]

In 1652, after cessation of hostilities in the English Civil War, Oliver Cromwell and his parliamentary supporters kept in place their victorious military force, the red-coated New Model Army. This was a radical departure from custom in that since the earliest times English armies disbanded on the conclusion of hostilities (after Parliament cut the purse strings). Though the New Model was a standing army, it differed in some important respects from the British Army of 1904. Of greatest difference is that Cromwell's army from 1651 through 1660, was primarily a political force. For many Englishmen, the Interregnum was not a period of peace, but a continuation of the Civil War and the Commonwealth government had to rely upon the army for its very existence.[3] The first standing army in England in the nature of the modern British Army was that of Charles II. His royal army, formed in 1661, was the true ancestor of the British Army as it was a non-political body. Its concerns were limited to the execution of the civil authority's wishes regarding national defense and the preservation of internal law and order.[4] Though at the time there was no United Kingdom, or even a Great Britain, Charles' army was the first "British" Army.

---

[1] From the time of Confederation until 1940, the Canadian standing army was styled "Permanent Active Militia."

[2] Jenkyns, "History of the Military Forces of the Crown.".

[3] Childs, *The Army of Charles II*, 1.

[4] Ibid.

The English Army

The British Army has its roots in both the republican New Model Army and the royalist military units of Charles II formed abroad during his nine-year exile. In May 1659, Richard Cromwell, successor to his father Oliver, as Lord Protector (head of state and government), resigned after the military leadership forced him to dissolve Parliament. Generals Charles Fleetwood and John Lambert, members of the Commonwealth's Council of State, then took control of the government. In response, George Monck, commander of the New Model Army in Scotland, and like Lambert and Fleetwood, a former lieutenant of Oliver Cromwell, marched the bulk of his troops towards London. Lambert led an army north to stop Monck's force. Lambert's troops deserted en masse, and he returned to London to await his fate. Monck and his army entered the capital unopposed, arrested Lambert, and restored Parliament. In early 1660, Parliament and the New Model Army leadership invited the exiled king, Charles II, to take up the throne. Charles, along with his mounted bodyguard, arrived in England in May 1660. Parliament recognized Charles as the lawful head of state and in accordance with custom voted to disband all the armed forces then present in England.[5] At the time Parliament also passed a bill that authorized an annual appropriation of £1,200,000 to Charles for life, to support the royal establishment and fund administration of the realm. This appropriation necessitated that £381,000 in other taxes be earmarked for the King as Parliament's estimate of Crown revenue was £819,000 (primarily "customs, farms and rents, and the composition for the court of the wards").[6] For the year ended March 1859, the state expenses of England, other than those for the army, were £454,000 for the navy and £329,000 for the civil establishment.[7] That would leave Charles £417,000 for the maintenance of forts, castles, and numerous royal residences, pay of his retainers, customary gifts to foreign and domestic supporters, and maintenance of the lavish life of a major, European monarch.

The Disbandment Act called for dissolution of all field forces in England and Scotland. It did not address the New Model units that kept Ireland under Westminster's control. The statute allowed the monarch to maintain some forces at his own expense. These were three infantry regiments and one cavalry squadron in Scotland plus the fort and castle garrisons in England, Wales, and the Channel Islands that existed in 1637. The units to be disbanded last were the horse and foot guards of Monck, newly created Duke of Albemarle (an infantry regiment and a cavalry squadron), and the Duke of York's Troop of Horse (one of two bodyguard units that accompanied Charles to England).

Charles had no intention to denude himself of troops. He planned for a small army of two infantry and two cavalry regiments and would circumvent the Disbandment Act by

---

[5] An Act for the speedy disbanding of the Army and Garrisons of this Kingdome, 12 Car. 2, c. 15.

[6] 4 Parliamentary History (Cobbett) cc. 117-120.

[7] Reports from the Committee for the Inspection of the Accounts, 7 *Journal of the House of Commons,* 628-30.

designating the force a personal bodyguard.[8] This guard would be paid for out of the annual, governance appropriation of £1.2 million. At the beginning of December 1660, after nearly all regiments were disbanded, the English Army at home and in Dunkirk (captured from Spain in 1657 by the New Model Army with French assistance), consisted of the following units:[9]

<u>In England</u>

Duke of Albemarle's Coldstream Regiment of Foot
Russell's Royal Regiment of Foot (formed by Charles in November, 1660)
Duke of Albemarle's Life Guards Troop of Horse
King's Guards Troop of Horse
Duke of York's Guards Troop of Horse

Castle and Fort Garrisons

<u>In Dunkirk</u>

Royal Forces Raised Abroad During the Exile of Charles II
    Lord Wentworth's Regiment of Foot

New Model Army:
    Farrell's Regiment of Foot
    Harley's Regiment of Foot
    Rutherford's Regiment of Foot
    Falkland's Regiment of Foot
    Taafe's Regiment of Foot
    3 Troops of Horse

On 26 January 1661 Charles II, by decree, established a "bodyguard" of two infantry and two cavalry regiments.[10] They were the 1st Foot Guards (Russell's Regiment of Foot), 2nd Foot Guards (Monck's Coldstream Guards), the Life Guards, (3 named squadrons: King's Own Troop, Duke of York's Troop, Lord General's Troop which was formerly Monck's Life Guards), and the Royal Regiment of Horse Guards, a newly raised formation.

In February 1661, as required by law, Charles disbanded Monck's units but then immediately enlisted the former soldiers into his new guards regiments.[11] As Russell's Regiment and the Horse Guards were formed as "bodyguards" after the Disbandment Act

---

[8] Hallam, *The Constitutional History of England*, 314-15; Packe, *The Royal Regiment of Horse Guards*, 2; Childs, *The Army of Charles II*, 6-17.

[9] Childs, *The Army of Charles II*, 1-20, End Note, appx. A.

[10] Ibid.,16-17 which references the State Papers (Domestic) of Charles II, SP 29/29, Nos. 45-47.

[11] Grose, *Military Antiquities*, Vol. 1, 60-61. Grose cites an article in the *Mercurius Politicus*, February 20, 1661, that describes the ceremonial disbandment and relevying of Monck's two units.

took effect, they remained standing.[12] Neither was the King's Troop of Horse disbanded as Parliament viewed its 150 to 200 soldiers as legitimate bodyguards.[13] Charles did not disband The Duke of York's Troop of Horse as he had sent it to Dunkirk in late December 1660. He brought it back to England early in 1661 as part of the bodyguard cavalry regiment designated "Life Guards."[14]

To allay fears Parliament may have had about the purpose of the new force, Charles appointed George Monck its commander (titled "Lord General") and William Clarke, MP as Secretary-at-War. He also appointed Members of Parliament to the force's administrative offices: Paymaster-General (accounts and disbursement of funds), Judge Advocate General (legal), Master of the Ordnance (artillery and munitions acquisition, fortification maintenance, stores), and Commissary-General of the Musters (audit of regimental rolls). The modern British Army had been born.[15]

The Irish Military Establishment

The Crown's standing army in Ireland began in 1535 when Henry VIII left in Dublin a regiment of 700 soldiers from the army he dispatched in 1534 to suppress the Kildare Rebellion.[16] In 1537, Henry disbanded this regiment but engaged 340 of its men as "retinues" for the Lord Deputy and the Treasurer of Ireland. This guard was in effect an English standing army in Ireland but was not in the direct lineage of the modern British Army. By the end of the sixteenth century, the commander of this force was known as "Marshal of the King's Army" and the positions of "Master of the Ordnance" (artillery, munitions, arms) and "Surveyor General of Victuals" (food and other supplies) had come into being.[17] As part of the executive in Ireland, the retinue was funded from Irish revenues. During the late 1500s, the permanent force fluctuated between 1,000 and 1,500 troops, with about three-fourths being infantry.[18]

---

[12] Hamilton, *The Origin and History of the First or Grenadier Guards*, Vol. 1, 44-45; Packe, *The Royal Regiment of Horse Guards*, 3.

[13] Horse Guards, *Historical Record of the Life Guards*, 9. Only the Duke of York's Troop was specified for dissolution in the Disbandment Bill.

[14] Childs, *The Army of Charles II*, 264, n. 18; Horse Guards, *Historical Record of the Life Guards*, 1-19.

[15] Childs, *The Army of Charles II*, 260-61; Roper, *The Records of the War Office and Related Departments*.

[16] In June 1534, "Silken" Thomas Fitzgerald, Lord Deputy of Ireland, renounced allegiance to the Tudor state after Henry imprisoned in the Tower of London, his father, the Earl of Kildare, and his uncles. Fitzgerald, after a failed attempt to capture Dublin Castle, was unable to sustain his rebellion and surrendered to Henry's army in October 1535. Fitzgerald was subsequently convicted of treason and executed.

[17] Ferguson, "The Army in Ireland from the Restoration to Act of Union," 2; Ellis, "The Tudors and the origins of the modern Irish state."

[18] Ferguson, "The Army in Ireland from the Restoration to Act of Union," 5; Brady, "The captains' games."

The Irish Military Establishment functioned as a guard for the administration in Dublin Castle and a paramilitary police force in the provinces. In 1611, the paramilitary force consisted of 8 cavalry detachments of about 25 men each, and 27 infantry companies of 50 men each. The Irish Establishment in the early seventeenth century numbered from 1,500 to 2,000 troops.[19]

Dublin Castle lacked the financial resources to administer the state, maintain British regiments in Ireland, and pay for the upkeep of forts and barracks administered by the Military Department.[20] Soldiers' pay often went months into arrears. Voluntary contributions from loyalist landholders were needed to house the troops, and subsidies from the English government were often sought and obtained.[21] At the outbreak of the Civil War in 1641, the forces of Charles I in Ireland totaled 943 cavalry and 2,297 infantry.[22] These troops were defeated by the soldiers of Oliver Cromwell and the old "Irish Military Establishment" came to an end.

During the Interregnum, the English Parliament kept separate accounts for each of the three nations of the Commonwealth in the vain hope that Scotland and Ireland would somehow be self-supporting. For the seven and one-half years ended November 1656, the Commonwealth spent £3.5 million to administer Ireland and 45% of those funds came from England.[23] Maintenance of the army accounted for 85% of state expenditure.[24] For 1659, the annual cost to maintain the 12,000 New Model Army troops in Ireland was £312,000.[25] The cost of civil administration for that year was £35,000.[26] As Irish annual revenue was only £208,000, a large English subsidy was required to maintain this reconstituted British force in Ireland. The subsidy appropriated for that year was £104,000, far short of the £139,000 anticipated deficit. Though Parliament could easily vote an appropriation, to find the revenue to meet the spending obligation was another matter. Taxes, though high for an England at peace, were insufficient to both fund the armed forces and maintain the civil apparatus.

At the restoration of the monarchy, the New Model Army in Ireland was down to approximately 10,000 troops. By the end of 1661, Charles further reduced this force to about

---

[19] Ferguson, "The Army in Ireland from the Restoration to Act of Union," 5.

[20] At the time, the Crown received all Irish tax revenue and directed all Irish state expenditure. The Irish parliament had no financial powers.

[21] Ferguson, "The Army in Ireland from the Restoration to Act of Union," 6-7.

[22] Guy, "The Irish Military Establishment, 1660-1776."

[23] An Abstract of all the monies received and paid for the public service in Ireland. Dunlop, *Ireland Under the Commonwealth*, 638-39.

[24] Ibid., 639-43.

[25] Reports from the Committee for the Inspection of the Accounts, 7 *Journal of the House of Commons*, 628-30.

[26] Instructions for James Standish, Esq. from the Lord Deputy. Dunlop, *Ireland Under the Commonwealth*, 676-77.

2,500 cavalry and 6,000 infantry, organized into 30 cavalry squadrons and 66 infantry companies. These units were not fully funded. By the end of 1659, total arrears for the army in Ireland were equivalent to one year's pay.[27]

Army units in Ireland were not regimented but dispersed throughout the island as an internal security force. As part of the reorganization and reduction, Monck, Lord General of the English Army and by then also Lord Lieutenant of Ireland, replaced republican officers with royalists.[28] The new Irish Military Establishment was to be funded in full by Irish revenues.

In 1662, unsure of the loyalty of the former New Model Army soldiers of the Irish Establishment, Charles recruited in England a regiment of Irish Foot Guards. This new regiment, some 1,200 strong, arrived in Dublin in 1663.[29] The following year Charles added a royal cavalry squadron to the Irish Establishment, the Irish Troop of Horse Guards.[30] As guards, both units were funded from the annual royal appropriation.

Due to insufficient tax revenue and corruption, the Irish government was unable to properly maintain the army units in Ireland. In 1664, the Irish Military Establishment began to whither. As former New Model officers resigned their commissions they were replaced with inexperienced and untrained gentlemen officers who served only part-time. Most such officers effectively embezzled whatever funds their units received from Dublin Castle.[31]

In 1672, the units on the Irish Establishment were grouped on paper into six cavalry and six infantry regiments. As there were insufficient funds to provide for proper regimental staffs and infrastructure, the army remained no more than a collection of scattered small units; the new formations existing in name only.[32] By 1676, nearly all soldiers were on extended furlough because there was no money to pay them. Additionally, there were few serviceable weapons and little in the way of munitions. The only effective units in Ireland were the Crown-financed regiment of Foot Guards and the squadron of Horse Guards, in all about 1,300 to 1,400 troops. In essence, the Guards were the army in Ireland.[33]

In 1684, James II, successor to Charles II, decided to re-establish the army in Ireland. In 1685, James installed a new commander in Ireland, Richard Talbot, whom he had previously

---

[27] Division of the annual expense of 312,000, by the cumulative arrears of 299,000 for the army in Ireland.

[28] Ferguson, "The Army in Ireland from the Restoration to Act of Union," 9.

[29] Guy, "The Irish Military Establishment, 1660-1776."

[30] Childs, *The Army of Charles II*, 204.

[31] Ibid., 207-09.

[32] Ibid., 204.

[33] Ibid. 205-07. The Troop of Horse Guards was disbanded by James II in 1685. The Foot Guards was stricken from the rolls of the Irish Establishment after James II was defeated in Ireland by King William (the Dutch Prince William of Orange and son-in-law of James). Ferguson, "The Army in Ireland from the Restoration to Act of Union."

created Baron of Talbotstown, and charged him with the reinvigoration of the forces. Upon Talbot's appointment, Charles elevated him to the Earldom of Tyrconnell. Talbot opened the army to Catholics and replaced many of the often corrupt Protestant officers with Catholics loyal to James II. He also took measures to ensure that army funds reached the troops, presumed loyal to James. Of further benefit to the army in Ireland, annual tax revenue from 1684 through 1687 averaged 10% higher than for 1683.[34] By the summer of 1686, the Irish Establishment totaled 8,364 men, all ranks, of which 5,460 were Catholic.[35]

Three years after the forces of James II were defeated in Ireland by those of King William (and his English co-sovereign Mary) the annual cost of the British army in Ireland was £167,000, nearly seven times that of Ireland's civil administration. For the year ended June 30, 1689 Ireland ran a deficit of £38,000, which represented 25% of its revenue of £153,000.[36]

In 1699, Parliament disbanded the regiments raised to support William in the "Glorious Revolution" against James II, and later deployed abroad in The Nine Years' War. The Disbanding Act of 1699; however, encompassed more than just an army reduction-in-force.[37] Article IV set the size of the Irish Military Establishment at 12,000 and limited its recruits to "His Majesties natural born subjects" (to prohibit service by foreign mercenaries). Article VI mandated that the troops in Ireland not disbanded be maintained "at the sole charge of the said Kingdom of Ireland." Throughout the eighteenth century during times of peace, the Irish Establishment totaled about 12,000 men, the minimum imposed by the English statute of 1699. This was approximately 45% of the British Army garrisoned at home. In wartime, Irish revenue paid for both the British Army in Ireland and Irish-raised regiments deployed abroad.[38]

Prior to the Union, the British Army in Ireland consisted of a permanent establishment known as the Military Department of the Kingdom of Ireland, a staff, and formations rotated on and off the Irish Military Establishment. The Military Department built and maintained barracks and forts, provisioned the army, and maintained stores of munitions, arms, and equipment. Though British regiments could remain in Ireland for years, between 1714 and 1745 Westminster prohibited local recruitment. In 1745, the War Office opened enlistment to Irish Protestants and in 1771 to Irish Catholics.[39] The 1801 Union between Ireland and Great Britain ended the legal fiction of an Irish Army. The staff of the Irish Military Establishment became the staff of the British Army's Ireland District and the military infrastructure in Ireland was acquired by the British Board of Ordnance.

---

[34] Miller, "The Earl of Tyrconnell and James II's Irish Policy."

[35] Ibid., 818.

[36] *Public Income and Expenditure, 1688-1869, Part 1*, 1868-69, H.C. Accounts & Papers, No. 366, at 228-29.

[37] 10 Will. 3, c. 1.

[38] Murray, *History of the Commercial and Financial Relations between England and Ireland*, 160-65.

[39] Ferguson, "The Army in Ireland from the Restoration to Act of Union," 68-74.

The Scottish Military Establishment

Unlike Ireland, Scotland has a history as a united (at least nominally), independent nation; however, as a nation it never possessed a standing army. Historically, the only full-time soldiers in Scotland in time of peace were the armed retainers of the nobles and Highland clan chiefs. When the need arose, Parliament would raise an army through an "Act of Levy." Such an act would assign a quota of cavalry and infantry to each shyre (county) and burgh (town). The local authorities, in turn, would apportion the quota among the resident heritors (landowners) and liferenters (lessees with life tenancies). These men of property had the obligation to procure the assigned number of horses and enlistees. The heritors and liferenters were subject to heavy fines for failure to meet their obligation (£268 per horseman with horse, £100 per foot soldier).[40] Many men of property, to avoid the fines for unfilled quotas, paid the mobilization officer to record falsely that they had attained their recruitment numbers. That this practice was widespread is indicated by provisions in the Act of Levy of June 25, 1650, which imposed severe penalties for such bribery.[41] The language that precedes the recitation of penalties is "… upon consideration of the great abuses in former levies committed by officers in taking of money from heritors and others aforesaid who are lyable in putting forth of horse-men and foot …" The penalty for such abuse by officers was immediate cashierment, and "branding with one mark of infamy and disgrace," and a fine of quadruple the amount of the bribe received, with the convicted officer imprisoned until such fine was paid.

The levied recruits were organized into foot (infantry) companies and horse troops (cavalry squadrons), grouped into regiments as set forth in the Act of Levy. Daily pay for foot soldiers was 6d; for horse soldiers who provided a horse, 16d.[42] Parliament appointed the regimental commanders and the troops were outfitted and armed by the shyres and burghs. Ongoing army funding was through "Acts of Maintenance" which directed the localities to assess additional property taxes in specified amounts.[43]

While Scotland possessed no standing army, it did have an ancient, statutory "militia" that was simply all able-bodied males between the ages of 16 and 60, known as "fensible persons."[44] Fensible persons could be compelled by the state to arm themselves and stand ready to suppress insurrection and oppose invasion. For example, on December 18, 1645,

---

[40] For example, after the execution of Charles I, the Scottish Parliament, on February 5, 1649, recognized his son Charles II as King of Scotland. Such recognition meant war with England and accordingly, an army of 13,600 infantry and 5,440 cavalry was summoned through an act of levy. Scot. Parl. Acts, Sess. January 1649, No. 128.

[41] Scot. Parl. Acts, Sess. May 1650, No. 150.

[42] Scot. Parl. Acts, Sess. January 1649, No. 128.

[43] For the levy of February 1649 Parliament passed in March a bill of maintenance to support the new army for three months. The act set forth specific assessments for each shyre and burgh. Scot. Parl. Acts, 2nd Sess. January 1649, No. 158.

[44] Term derived from "defensible" or "defencible" depending on whether the English-speaker resides in the United States of America.

Parliament directed the shyres "to make new Rolls of the whole fensible within each one of their several Shires, and to cause the Fourth man of the whole number of their fensible persons, be Trained, Armed, and formed in Companies, and made ready upon all occasions for the Publick Service."[45]

The statutory mechanism for Parliament to call into service the armed fensibles was passage of "An Act Putting the Kingdom in a Posture of Defense." Such *lévee en masse* by the Scottish government was often an act of desperation of little military value. On July 3, 1650, after three levies of armies, and as Cromwell's troops advanced on Edinburgh, the Scottish Parliament passed an "Act for putting the Kingdome in a Posture of Defence, to joyn with the present standing Forces, and new Levy." The statute proclaimed that "All fensible persons betwixt 60 and 16 are bound to rise in Arms to defend the King and Kingdome from Invasione" and summoned 2,542 horsemen and 25,862 men on foot.[46] Two months later Cromwell's force, the New Model Army, entered Edinburgh.

As with Ireland, the Commonwealth government hoped that the cost of the New Model Army in Scotland could be funded locally. Scotland; however, was poorer than Ireland and English subsidies were needed for the maintenance of the Scottish garrison. In 1649 the annual revenue of Scotland totaled £144,000 while the annual cost of the army stationed there was £271,000. For that year, the English appropriation in aid of Scotland was £148,000.[47]

In 1659, prior to Monck's march on London, the Scottish Military Establishment consisted of five cavalry regiments, four separate cavalry squadrons, eleven infantry regiments, one separate infantry company, and castle garrisons. Arrears in the soldiers' remuneration were equivalent to four-months' pay.[48] At the end of 1660 this force was down to one cavalry regiment, four infantry regiments, and castle garrison troops; in all 5,500 men.[49] Both Charles II and Parliament saw no need for field forces in Scotland. Accordingly, the New Model regiments in Scotland were disbanded with most of their soldiers sent to Tangiers and Portugal.[50]

By 1663, the Scottish Military Establishment consisted of 1,200 troops organized into one small infantry regiment, one small cavalry squadron, and garrisons for Edinburgh, Dumbarton, and Stirling Castles. This was the largest army that Scotland could afford.[51] Scotland's main military function for the Crown was to serve as a source of mercenary

---

[45] Scot. Parl. Acts, Sess. November 1645, No. 5.

[46] Scot. Parl. Acts, Sess. March 1650, No. 188.

[47] Reports from the Committee for the Inspection of the Accounts, 7 *Journal of the House of Commons*, 628-30.

[48] Ibid.

[49] Hallam, *The Constitutional History of England*, Vol. 2, 312. Monck had marched on London in 1859 with most of the army then in Scotland.

[50] In 1661 England received from Portugal the North African town of Tangiers. The grant was a wedding present for Charles II on his marriage to the Portuguese Princess, Catherine of Braganza.

[51] Childs, *The Army of Charles II*, 196-97.

soldiers for English foreign commitments. The Scottish Military Establishment was designed for ceremony and to keep order in the Lowlands. Maintenance of law and order in the Highlands was a lost cause and Scotland's military units never ventured into the area.[52]

In 1678, the Scottish Parliament formed an organized militia of 5,000 infantry and 500 cavalry.[53] By 1693, more than 30 years after the Restoration, the full-time strength of the Scottish Establishment had increased only slightly; in that year it totaled 1,558 troops. The standing army in Scotland was as follows:[54]

Scots Guards Troop of Horse	118
2nd Dragoons (2 squadrons)	120
Earl of Argyll's Regiment	780
9 garrison companies	540
Regular Troops, Total:	1,558

By 1702, the Scottish Military Establishment had increased to 2,934 men. Field units included a squadron of Life Guards (122 cavalry), a squadron of Horse Guards (64 cavalry), and a regiment of Foot Guards (806 infantry). These "guards" units, unlike in Ireland, were paid for by Scotland. There were also two small cavalry "regiments" of 200 men each, and three small infantry regiments of about 400 men each. Additional soldiers garrisoned Edinburgh, Stirling, Dumbarton, and Blackness Castles. The annual cost to Scotland of the Military Establishment was £65,740.[55]

In 1708, Union between Scotland and England ended the legal fiction of the Scottish Army and the staff of the Scottish Military Department became the staff of the British Army's Scotland District. The War Department moved the cavalry guards units to London where they merged into the Life Guards and Horse Guards. After Union, the strength of the former Scottish units increased to British Army standards.[56]

The Transitional British Army with Funding Sources and Cost, 1661

In early 1661, the English army consisted of the King's guards in England, former New Model Army units in Scotland and Ireland, and regiments abroad in Tangiers and Dunkirk. The army was deployed, organized, and funded as follows:[57]

---

[52] Ibid., 197-98.

[53] Ibid., 198.

[54] Walton, *History of the British Standing Army*, 501.

[55] Johnston, "The Scots Army in the Reign of Anne."

[56] Ibid.

[57] Childs, *The Army of Charles II*, 11-13, 17, 163-64, 196-97, 203, 233, 235-36, 256; Walton, *History of the British Standing Army*, appx. 87 from "An Abstract of His Majesty's Guards which were raised the 26th of January, 1660 and continued to the 1st of January, 1663."

England
1st Foot Guards
2nd Foot Guards
Life Guards (cavalry)
Horse Guards
28 fort and castle garrisons

Funded by the Crown, £189,000 out of £1,200,000 annual, royal appropriation.

Scotland
1 cavalry regiment, former New Model Army
4 infantry regiments, former New Model Army
3 castle garrisons
(In 1661 the cavalry regiment disbanded and one infantry regiment reorganized into garrison companies. In 1662 two infantry regiments went to Portugal.)

Funded with Scottish revenues.

Ireland
30 cavalry squadrons, former New Model Army
66 infantry companies, former New Model Army

Funded with Irish revenues.

Tangiers
1 infantry regiment, new formation
2 infantry regiments, former New Model
1 cavalry squadron, new formation

Funded by Parliament, £75,000 per year.

Dunkirk
3 infantry regiments, former New Model Army
3 cavalry squadrons, former New Model Army
1 infantry regiment, former Royal Army
1 infantry regiment, new formation

Funded by Parliament, £60,000 per year.

Since 1585, when Elizabeth I sent 5,000 men under Robert Dudley (the Earl of Leicester), to the United Provinces, a British corps had served the Dutch. These English and Scots mercenaries served in their own units as part of the Dutch Army. The force became known as the Anglo-Dutch Brigade. Its officers were commissioned by the Crown and its regiments were subject to disbandment or recall to England by the English government. As a mercenary formation, it was funded wholly by the Dutch and in 1661 totaled 3,000 troops. In 1665, Charles brought the brigade to England. He disbanded the brigade's regiments and placed most of the now ex-mercenaries into the newly formed Dutch Guards Regiment and the Admiral Regiment (formed a year earlier).[58]

---

[58] Childs, *The Army of Charles II*, 171-72.

The Army of Charles II: Parliament and the English Public

Politically and socially, the army worked for the sovereign, which the aristocracy, gentry, and middle class viewed as working against their interests. To Parliament, the army was at the core of the struggle between it and the Crown for state control. The great landholders, both titled and gentry, viewed the army as a threat to "their" militia. The ordinary people loathed the army as the police force of "the establishment." The conduct of the troops at home was appalling and they caused more trouble and disorder (by way of riots and assaults) than they were supposed to quell. Furthermore, the army often housed itself in private lodgings free of charge and recruited illegally through press gangs. Overall, the standing army of patrician officers and plebian soldiers was distrusted and disliked by all quarters. Additionally, its royal connection with Catholicism caused resentment and fear among the Protestant majority. In the countryside, the standing army was little-noticed by the populace, but in London and the provincial towns where the soldiers were quartered, the residents were always aware of the King's armed presence. Parliament, especially Commons, voiced constitutional and political objections to the King's "bodyguard" army.[59]

Though a majority of Parliament was against a standing army, without an armed force of their own they feared to take measures to reduce and rein in the Crown's troops. Charles II; however, did not press his political advantage as he well remembered that his father literally lost his head because of political over-reaching. Accordingly, an uneasy balance of power resulted. Parliament's safeguard against the army at home was that at its present size, Charles could just barely support it from the annual, royal appropriation.[60]

In June 1678, Charles asked Commons to increase his annual appropriation from £1.2 million to £1.5 million. Aware that the lifetime, annual general appropriation allowed the King to ignore the Disbandment Act of 1660 and maintain 5,000 soldiers in England and another 1,000 in nearby Ireland, the Members ignored the royal request.[61]

On April 1, 1679, to express its displeasure with the army, Commons passed a resolution (not a bill), that read as follows:

> "*Resolved*, &c. That the Continuing of any Standing Forces in this Nation, other than the Militia, is illegal; and a great Grievance and Vexation to the People."[62]

That was the last time that a Parliament of Charles II addressed the issue of a standing army. Throughout the King's reign the army never became a political force and Parliamentary fears subsided. The troops remained subordinate (though in an unruly

---

[59] Ibid., 213-17.

[60] Ibid., 218-21.

[61] Ibid., 227.

[62] 9 *Journal of the House of Commons*, 581.

manner) to the civilian government and such subordination became a precedent that the British Army would follow for centuries.[63]

In 1689, as part of the "Glorious Revolution" that terminated the Stuart Dynasty, the first Parliament under William and Mary passed the Mutiny Act.[64] This statute, renewed annually for nearly 200 years, effectively legalized the standing army and placed it under control of Parliament. That same Parliament, in its second session, enacted what came to be known as "The Bill of Rights."[65] Amongst its many provisions the act specified

> "That the raising or keeping a standing Army within the Kingdome in time
> of Peace unlesse it be with Consent of Parlyament is against Law."

These English statutes of the late seventeenth century clearly indicate that in the collective mind of Parliament of the time, the creation of the British Army by Charles II was unlawful.

War at Home and Abroad, 1692-1815

From King William's consolidation of power to the final fall of Napoleon at Waterloo, the British Army fought frequently both at home and abroad. In Ireland and Scotland, on four occasions, the government utilized the army to suppress rebellion. In England, the army was frequently called upon by the local, civil authority to suppress "riots." Abroad, the army fought on four continents in national wars and colonial campaigns. While the troops were engaged in battle and police actions, the army's administrative apparatus grew in a haphazard manner.

Several times in the eighteenth century the British Army was called upon to both maintain Westminster's hold on Ireland and thwart attempts to restore the Stuart Dynasty. In Ireland, the armed action in 1798 for independence by the United Irishmen, together with the collateral rising in Wexford, was a major challenge to British control. It was a bloody conflict that remained in the collective memory of Irish nationalists to Bloomsday and after.[66] In Scotland, there were several attempts to seize the crown of Great Britain for the exiled Stuarts.[67] These Jacobite rebellions were Lord Mar's Revolt of 1715, the Rising of 1719, and The '45 (1745). All these armed attempts to restore the Stuart Dynasty originated in the Scottish Highlands. Of the three Jacobite rebellions, The '45 was the bloodiest and most notorious.

---

[63] Childs, *The Army of Charles II*, 227-32.

[64] 1 W. & M., c. 5.

[65] 1 W. & M. Sess. 2, c.2.

[66] See, Chapter 2 of this work, History of Irish Part-Time Soldiery, *infra*.

[67] The Kingdom of Scotland merged with that of England & Wales through the 1706 and 1707 Acts of Union, 6 Ann., c. 11, 7 Ann. (Scotland), c. 7.

History of the British Army, Cromwell to 1853

Rebellions and Risings at Home: The '45 [68]

On July 25, 1745, while Britain was at war with France, Charles Edward Stuart, the 24-year old son of James Francis Stuart, pretender to the British throne, departed France and after evading the Royal Navy, landed in the Highlands of Scotland. His objective was to unseat the Hanoverian monarchy and install his father as British king. With the bulk of the British Army in Flanders fighting the War of the Austrian Succession, a Stuart seizure of Britain was plausible. There were no more than 3,000 British regulars in Scotland and at most 6,000 in England. Charles hoped to raise a substantial army among Stuart supporters in the Highlands then march south towards London. He was confident that while en route, Scottish nobles, clan chiefs, and the common people would rally to his cause. Charles expected that when Westminster moved its assembled forces northward to meet the Jacobite army, France would land its Scottish and Irish mercenaries somewhere south of the British. The Jacobites would then attack the British field force head-on while the French would attack the British rear. Charles believed he would crush the British Army and consequently Parliament would recognize his father as king.

In August, while in western Scotland, Charles asked eight clan leaders to support his cause. Only three committed to the Pretender. A fourth, Ranald MacDonald of Clan Ranald, did not join Charles formally, but allowed his son to raise volunteers for the Jacobite army. Charles appointed as his secretary of state Sir John Murray of Broughton, a noted Stuart supporter who was educated at the Universities of Edinburgh and Leyden. Murray's primary duty was to raise funds for the Jacobites. Charles' chief of staff was Captain John William O'Sullivan, an Irish mercenary in French service. Charles appointed O'Sullivan Quartermaster & Adjutant General even though the officer had no staff experience. The germ of Charles' army, about 1,200 soldiers, mustered in Glenfinnan on August 19th then marched towards the Scottish Lowlands.

The government's armed presence in Scotland was five battalions of foot, two regiments of dragoons, and the castle garrisons. There were no Royal Artillery gunners anywhere in the country. Castles were held by "Companies of Invalids" made up of retired soldiers and those discharged for medical reasons. While adequate for paramilitary police duties, those garrison troops could neither fight in the field nor properly man castle artillery. The army's field units were at about half-strength as regiments in Flanders had priority for receiving recruits. In July 1745, this force of about 3,000 soldiers, including castle garrisons, was commanded by Lieutenant-General John Cope. He decided to engage the Jacobite army while it was still in the Highlands and in mid-August assembled three of his infantry battalions at Stirling. Cope's force of about 1,200 men entered the Highlands along General Wade's Military Road. On August 26th, while about 35 kilometers southeast of Fort Augustus, Cope learned that the Jacobites were 10 kilometers to his west on the heights of Corrieyairack Pass. Rather than attack uphill along the military road, and fearful that the Jacobite army would infiltrate his rear, Cope marched his brigade north to Fort George in

---

[68] Sources for this sub-section are listed at the end of the chapter bibliography.

Inverness, held by a newly forming British regiment of Highlanders. The route to the Lowlands was now open to Charles's army.

Charles and his lieutenants went to Perth while the Jacobite army marched down the military road. At Perth, Charles obtained additional support for the Stuart cause. While the Pretender's son sought allies at Perth, Cope and his brigade marched to Aberdeen to await transport by ship. Cope left at Inverness the Highland 63rd Regiment of Foot, commanded by John Campbell (Earl of Loudon) who was also Adjutant-General, Scotland.

On September 12th, the Jacobite army marched southwest to Stirling. Charles appointed William Drummond (Viscount Strathallan) to command the Jacobite forces that remained in the Highlands. Strathallan was to contain the British troops isolated in strong points, recruit more volunteers, and provide the London-bound Jacobite force with arms, munitions, and supplies. From Stirling the roads led to both Glasgow and Edinburgh. As Glasgow was overwhelmingly pro-government, Charles marched on Edinburgh. The Jacobite army, now of 2,300 men, moved south between government-held Stirling Castle and the Firth of Forth, then swung eastward towards Edinburgh. When Cope received word that Edinburgh was threatened, his force embarked for the city. Unfavorable winds prevented entry into Edinburgh's port, and Cope's brigade sailed into Dunbar harbor, 45 kilometers east of the old capital.

Available to defend Edinburgh were the two regiments of dragoons sent earlier by Cope, the Edinburgh Castle garrison, about 300 Town and City Guards (paramilitary police), and 800 untrained volunteers. On September 15th, the 500 dragoons, augmented with 200 auxiliaries, marched west to block the Jacobite advance. The Lord Provost of Edinburgh hoped this force could hold out until Cope's brigade arrived. The next day, a Jacobite reconnaissance detachment fired on the encamped government troops who then panicked. The dragoons, commanded by Brigadier Thomas Fowke, retreated to Dunbar where Cope had landed that day. The auxiliaries simply went home. When news of the government defeat reached Edinburgh the newly organized volunteer companies melted away and a few hundred anti-Jacobite civilians took refuge in the Castle. On September 17th, the Lord Provost surrendered the city. At Hollyrood Palace, the former state residence of Scottish kings, Charles proclaimed his father King James VIII of Scotland and King James III of England.

Though the Highlanders now occupied Edinburgh, the Castle remained in government hands. The castle commandant was 85-year old Lieutenant-General Joshua Guest. His force consisted of about 450 civilians and 150 over-age or infirm castle guards. Though there were numerous guns at the castle, there were no Royal Artillery gunners. Guest could not hope to reclaim the city, but neither could Charles take the Castle as the Highlanders lacked the artillery to force the Castle's only accessible wall. Accordingly, an uneasy truce developed between the opposing forces, interrupted by sporadic, inaccurate artillery fire from the Castle. Guest finally agreed not to bombard the city and in return, Charles agreed not to besiege the Castle. In reality it was a unilateral decision by Guest to spare the city and not harm its many pro-government inhabitants. The Castle had provisions to hold out for months and the Jacobite army lacked the time and resources for a prolonged siege.

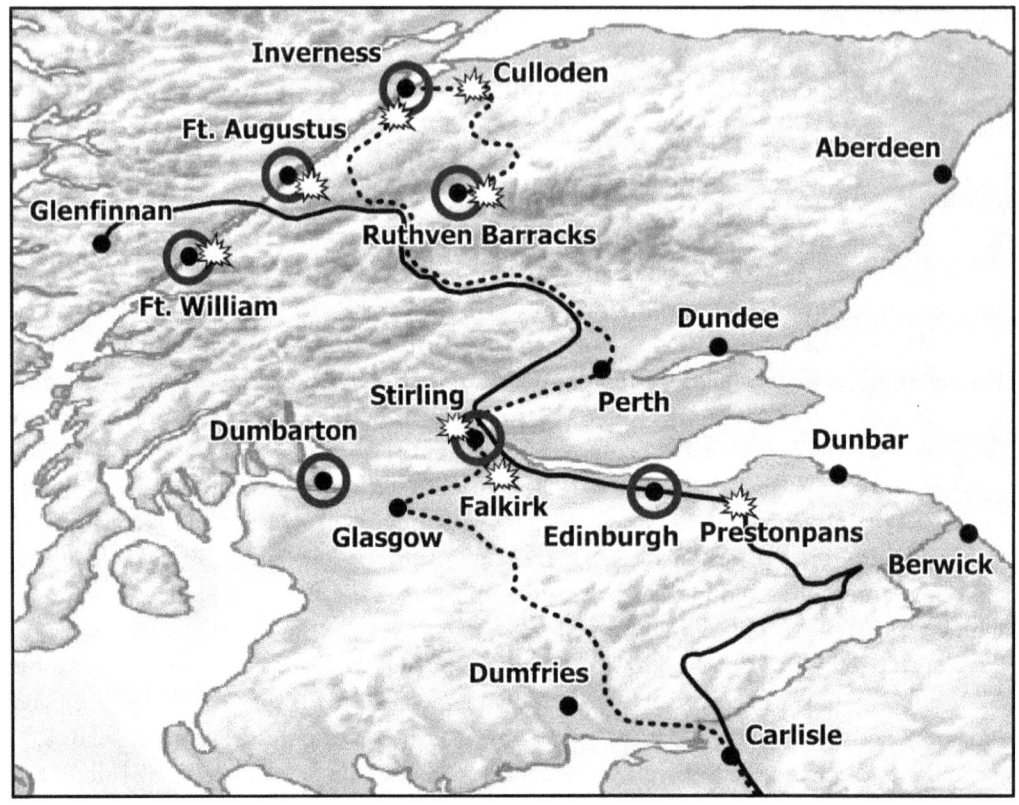

The Jacobite Rising of 1745, Scotland

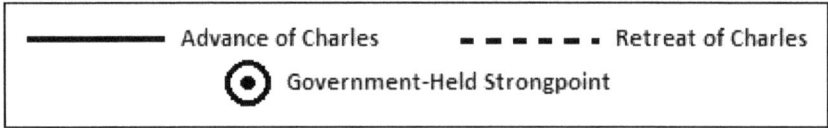

Sources: Scottish History Society, *A Map to Illustrate the Movements of Prince Charles Edward Stuart and His Armies*, 1897; Riding, *Jacobites: A New History of the '45 Rebellion*; Duffy, "The Jacobite Wars;" Blaikie, *Itinerary of Prince Charles*.

In July 1745, the government's field forces in Great Britain consisted of Cope's units in the north, a brigade of guards in London (four infantry battalions and a cavalry regiment), and nine line units (five infantry battalions and four cavalry regiments) scattered throughout the south. All units were much below authorized strength. The bulk of the British Army was in Flanders where it opposed the French. The British force there was commanded by the 24-year old William Augustus of Hanover (Duke of Cumberland), third son of King George

II. He was also commander-in-chief of the "Pragmatic Alliance" forces in Flanders.[69] When Cumberland learned that a Jacobite army had formed in the Highlands, he dispatched two British regiments to England. He also persuaded the Dutch Republic, then allied with Great Britain, to send nine infantry regiments to England (eight Dutch and one Swiss mercenary). In all, about 6,000 battle-hardened soldiers would soon leave the Continent for England.

On September 16th. the Jacobite army left its Edinburgh encampments to engage Cope's brigade as it approached the city. Charles' field commanders were George Murray (son of the Duke of Atholl) and James Drummond (Duke of Perth). Lord George was an experienced soldier who had held a British Army commission and took part in the two previous Jacobite risings. Perth had no military experience. The two forces fought the next day at Prestonpans where the Highlanders decimated the British. Cope's force of about 2,500 men, including volunteers, lost 300 killed and 1,500 captured. Jacobite casualties totaled about 25 killed and 75 wounded. The British general and his remaining regulars retreated to Berwick-upon-Tweed to recruit replacements and await reinforcement. The way was clear for Charles to enter England.

Except for seven British-held strongpoints (Forts William, Augustus, George, Ruthven Barracks, and the castles at Dumbarton, Stirling, and Edinburgh), the Highlands and the eastern Lowlands were now in Jacobite hands. Charles' triumphs at Edinburgh and Prestonpans spurred several clan chiefs and nobles to the Stuart cause. The Jacobite military successes also induced many individuals to volunteer for Charles' army. Charles' proclamation of October 9th that dissolved the union of Scotland and England further aided recruitment. At the end of October, the Jacobite army totaled about 9,500 troops. Most Scottish nobles and clan chiefs; however, did not support the Pretender and his son. They feared the autocratic tendencies of the Stuarts, were apprehensive of the Stuart's Catholicism, and against continuation of the dual monarchy. Most importantly, they believed that without substantial French intervention, or a general uprising against the Hanoverian regime, Charles would fail to conquer Great Britain.

In early November, the Jacobites marched south from Edinburgh and made a feint towards Cope's former brigade at Berwick-upon-Tweed. That force was now commanded by Lieutenant-General Roger Handasyde, as Westminster had summoned Cope to London for trial by court-martial. When Charles reached Kelso, about 35 kilometers southwest of Berwick-upon-Tweed, he turned west and headed for his true immediate objective, the English garrison town of Carlisle. Charles, with about 6,000 soldiers and 13 artillery guns, crossed into England on November 8th. A week later Handasyde entered Edinburgh unopposed.

Carlisle was defended by the "Invalids" of Carlisle Castle, and about 1,000 recently recruited volunteers. Like at Edinburgh, there were no Royal Artillery gunners at Carlisle. Though the Edinburgh Invalids were able to man a few guns, those at Carlisle were unable to operate any. The volunteers expected to fight as auxiliaries to a force of regulars. When no such regulars arrived, the volunteer officers refused to engage the Highlanders. On

---

[69] The Pragmatic Alliance consisted of the Dutch Republic, Hanover, Great Britain, the Austrian Empire, and several other states. In 1745 the four named nations had troops in Flanders fighting the French.

November 17th, the Jacobites entered Carlisle and the castle commandant surrendered. The artillery, personal weapons, and munitions stored at the castle were of great value to the Jacobite army; however, Charles obtained very few recruits among the English. Facing no opposition, the Jacobite army proceeded south through Cumberland, Westmoreland, and Lancashire. On November 28th, Charles entered Manchester. While the army awaited the town's surrender, Lord George gave up his general's epaulets to take personal command of the regiment he had raised. Charles appointed David Wemyss (Lord Elcho) his replacement. Perth continued as one of Charles' two lieutenant-generals. After Manchester surrendered, several senior Jacobite officers questioned whether their army (5,500 infantry, 500 cavalry, and 20 guns) should continue south as there had yet been a public uprising or French invasion.

As the grandson of James II moved through England, he received little armed support from the populace.[70] Apparently, neither did King George II. The small farmers and shopkeepers in the Jacobite army's path were not eager to fight the Pretender's forces. While thousands joined the government's volunteer regiments, these amateur soldiers refused to engage the enemy unless accompanied by substantial numbers of regulars. English civilians remained spectators to a royalist joust.

By mid-November, Westminster no longer viewed Charles as a military threat. The regular army at home totaled 21 infantry battalions and 9 cavalry regiments, with another 13 infantry battalions and 3 cavalry regiments en route from Flanders. Also under orders for England were 2 infantry battalions in Ireland. In addition to the British Army units, the government had at its disposal 13 Dutch infantry regiments at Newcastle-upon-Tyne. The British forces included about 24,000 regulars while the Jacobite irregulars totaled 5,000 to 6,000 at Manchester and 2,000 in the Highlands. Not all British regulars were seasoned soldiers. Prior to the Jacobite victories at Edinburgh, British regiments at home were at half-strength, while those in Flanders were also below establishment. After Charles' military success in Scotland, Parliament authorized special enlistments for the duration of the emergency. Generous enlistment bonuses, coupled with a limited term of service, stimulated recruitment and brought unit manning levels much closer to authorized strength. Accordingly, about 9,000 British Army soldiers were recent recruits with little or no training. The foreign mercenaries; however, were all combat veterans. The government deployed its British, Dutch, and Swiss regulars as follows:

> <u>Highland Strongpoints</u> - 1,000 isolated in Forts William, Augustus, and George plus Ruthven Barracks. Loudon was *de facto* commander of these troops.
>
> <u>Berwick-upon-Tyne and Edinburgh</u> - 2,000 under Handasyde.
>
> <u>Newcastle-upon-Tyne</u> - 10,000 under Field Marshal George Wade who was also Commander-in-Chief of the British Army. Half these regulars were troops sent by the Dutch Republic and the Marshal considered them unreliable:

---

[70] About 300 English recruits were obtained in Lancashire (the Manchester "regiment") and fewer than 100 in Cumberland and Westmoreland.

Should France recognize the Pretender as rightful King, the Treaty of Tournai would bar use of Dutch troops against the Jacobites.[71]

<u>Lichfield</u> - 5,000 under Cumberland. After the Battle of Prestonpans the govern-ment recalled Cumberland from Flanders and gave him command of the troops assembling in the Midlands. On October 19th, the Duke arrived in London and a month later was with the British field force encamped between Birmingham and Derby.

<u>London and Southeast England</u> - 6,000 as a strategic reserve.

Westminster's complacency did not last long for between November 24th and 28th French ships landed in northeastern Scotland and delivered troops, artillery, munitions, and gold to Strathallan. The soldiers were 600 Scots and 200 Irish of the French Army. While Louis XV did not openly acknowledge the Pretender as King James III, this material support was taken by all parties as French endorsement of the Jacobite cause. The British government now faced the strong possibility of a French invasion and could no longer utilize the 5,000 Dutch troops stationed in England. In response to the landing, Parliament appropriated funds to engage 6,000 German mercenaries from Hesse-Cassel. These troops would replace Marshal Wade's Dutch contingent.

On December 4th, the Jacobite army entered Derby, 200 kilometers north of London. Charles wanted to march on the capital but nearly all his senior officers opposed such venture. They believed that without a local rising of civilians, or a French invasion, their army was doomed if it remained in England. The officers were convinced that Cumberland would block the London advance and Wade would attack their army's rear. Sandwiched between two enemy forces the outnumbered Jacobites would be annihilated. Two days after Charles arrived in Derby, he ordered his army back to Scotland. Cumberland pursued Charles to Carlisle while Wade's force remained at Newcastle where it was snowed-in.

During the Jacobite retreat, Admiral Edward Vernon, commander of the Channel fleet, informed the government that the French were assembling an invasion force. He was certain it would land somewhere in southeast England and then march on London. On December 15th, Westminster ordered Cumberland and most of his troops south to deal with this new threat. Shortly thereafter the government modified this order and allowed Cumberland to pursue the Jacobites but with only his four cavalry regiments, two cavalry regiments from Marshal Wade, and about 1,000 foot soldiers. The bulk of Cumberland's army moved south. Westminster assigned the defense of London to General John Dalrymple (Earl of Stair), recently recalled from retirement.

---

[71] At the Battle of Fontenoy, May 11, 1745, the Pragmatic Alliance suffered a resounding defeat in which the French captured several thousand Dutch soldiers and were poised to invade Holland. The Dutch Republic, by treaty, withdrew from the Alliance. France released the Dutch captives and pledged not to enter Holland, and the Dutch agreed not to use its soldiers against France or her allies.

The Jacobite army arrived at Carlisle on December 19th. The next day, Charles continued north to Glasgow and left behind somewhat over 400 troops commanded by John Hamilton and Francis Townley. Their suicidal mission was to stand and delay Cumberland to allow the Jacobite army a leisurely march north. The vanguard of Cumberland's 4,000 troops reached Carlisle on December 21st. Nine days later, the Jacobites surrendered. Hamilton, Townley, and the other prisoners awaited their punishment for treason: Hanging, imprisonment, or transportation to the colonies. With Carlisle again in government hands, Cumberland left his troops to take command of the British force assembling in southeast England to oppose the expected French landing. During Cumberland's advance to Carlisle, Marshal Wade, citing advanced age, relinquished command of the Newcastle force and on December 18th left for army headquarters in London. Westminster then consolidated all British forces in the north into a single army commanded by General Henry Hawley, who had served under Cumberland in Flanders.

On January 6th, Charles and his army arrived at Stirling. Two days later the town surrendered; however, Stirling Castle remained in British hands. The Jacobite army that arrived from Glasgow was reinforced with about 4,000 troops recruited and trained by Strathallan. Charles' field force now totaled approximately 9,000 men. The same day that Charles arrived at Stirling, Hawley and his troops from Newcastle reached Edinburgh. One week later Hawley sent 7,000 regulars west to relieve Stirling Castle. On January 17, 1746, the Jacobites attacked the British at Falkirk Muir. The British retreated and incurred about 500 casualties (killed, wounded, and captured). The Jacobites captured the British artillery and incurred about 50 dead and 80 wounded. Westminster and the British public viewed Falkirk as a defeat and as a result, the government replaced Hawley with Cumberland as Commander-in-Chief, Scotland.

After Falkirk, Charles besieged Stirling Castle even though he lacked the heavy artillery that could reduce its walls. On January 29th, with the castle still in British hands, Lord George and most of the clan chiefs asked Charles to abandon the siege and withdraw farther north. Casualties, desertions, and sickness had reduced the Jacobite army to about 7,000 effectives and British reinforcements were arriving in Scotland daily. On February 1st, Charles withdrew north to secure the western Highlands. He believed that there his army could hold out until French assistance arrived. The Jacobites captured Ruthven Barracks, then advanced on Inverness. Loudon, the British commander at Inverness, evacuated the town but left a garrison at its fort. On February 20th, Fort George surrendered and three weeks later Charles sent about 1,000 men to take Forts Augustus and William. Fort Augustus fell but Fort William held and on April 4th, the Jacobite force, now greatly reduced, withdrew towards Inverness.

As the British army increased in strength, the Jacobite army, with no economic base and starved of supplies from France by the Royal Navy, waned. By early April, it was down to 5,000 troops. Charles decided to engage the British before his army disintegrated. On April 16, 1746, near the village of Culloden, the Jacobite army met Cumberland's nearly 9,000 regulars and was quickly defeated. British superiority in artillery and numbers, higher morale, and better organization, all combined to give Cumberland a quick and decisive victory. The Jacobites lost about 1,800 killed, 600 taken prisoner, and 1,100 gone missing. The missing were mostly old men, boys, and reluctant soldiers that were deployed in the army's rear.

They had fled when the front line soldiers retreated due to heavy losses. British casualties totaled about 600 (killed and wounded).

Charles sent what was left of his army south to Ruthven Barracks while he and his staff withdrew towards Fort Augustus. With his army decimated and a French invasion unlikely, Charles abandoned the campaign. On April 20th, the 1,500 Highlanders at Ruthven Barracks received their last order from Charles: "Let every man seek his safety in the best way he can." The few soldiers with French Army credentials surrendered to the British and hoped for prisoner-of-war status.[72] The rest of Charles' men, fearing the scaffold for treason, went into hiding.[73] Many of the Jacobite notables made for the coasts to await rescue by French ships, as did Charles.[74] On September 20th, the Pretender's son left Scotland aboard a French naval vessel.

After Culloden, the British, with a heavy hand, began to subdue the Highlands. Cumberland let loose his troops to engage in "the hunting of fugitives, the burning of villages, and the destruction of crops."[75] For these depredations, Cumberland quickly acquired the epithet of "the butcher." His ruthless methods proved effective as the Highlands was flattened into submission. After The '45, the British Army fought no organized force at home until the 1916 Easter Rising in Dublin.

In Aid of Civil Authority, Eighteenth Century England [76]

Eighteenth century England was plagued with tumults and riots by "mobs and crowds." The disturbances were sparked and fueled by political partisanship, religious intolerance, and

---

[72] The British government did not accord prisoner-of-war status to French soldiers who were born in Ireland or Great Britain and treated the mercenaries as rebels.

[73] Many former soldiers, as well as Highlanders who never took up arms, were killed by Cumberland's troops in the pacification of the Highlands. Numerous others were imprisoned as suspected rebels. In early summer, thirty-four captured Jacobite notables, including four peers, were indicted for treason. The peers were convicted by the House of Lords and sentenced to death. Three were beheaded while one had his sentence commuted to imprisonment. The other Jacobite leaders, including Hamilton and Townley who surrendered at Carlisle, were convicted by English courts. They were sentenced to death by hanging, the corpse to be mutilated and beheaded.

In 1747, Parliament, through an Act of Pardon, gave amnesty to all but the key participants in the attempt to restore the Stuart Dynasty. Expressly excluded from the pardon were persons of the Clan MacGregor, persons convicted of treason prior to June 15, 1747, persons who committed treasonous acts prior to July 20, 1745, and traitors who were "beyond the Seas" between July 20, 1745 and June 15, 1747. 20 Geo. 2, c. 52.

[74] John Murray, Charles' Secretary of State, was captured and turned King's evidence. He was released from prison in 1747 and pardoned a year later. John O'Sullivan, Charles' Quartermaster and Adjutant-General, escaped to France. Of the four Jacobites who served as lieutenant-generals, Strathallan died at Culloden, Lord Elcho escaped to France, Lord George Murray escaped to Germany, and the Duke of Perth died of fever aboard ship while bound for France. All four forfeited their estates to the Crown.

[75] Fortescue, *A History of the British Army*, Vol. 2, 146.

[76] Source: Stevenson, *Popular Disturbances in England 1700-1870*, 17-142.

economic grievances. Mob violence gave rise to legislation such as the Riot Act of 1715 and the "Spitalfields Act" of 1773. In many cases the British Army was used to restore order, protect property of employers and landowners, and guard prisoners (from mobs intending either to kill them or free them).

The politically motivated disturbances that occurred early in the first half of the century were by Whig and Tory supporters who used violence to both intimidate office-holders and suppress their opponents' vote. Disturbances from 1745 onward generally had an economic characteristic. In the countryside, violence was triggered by the Enclosure Acts, while in London disturbances arose from trade union-like actions. Additionally, during economic downturns employees often attacked establishments of their employers' rivals to maintain job security. Food riots, which resulted from shortages, perceived shortages, and rising prices, erupted during 1709-10, 1727-29, 1739-40, 1756-57, 1766-68, 1772-73, 1783, 1795-96, and 1799-1801.[77]

The most serious mob violence of the eighteenth century was that of the Gordon Riots of 1780. No civil disorder since, in England, has led to greater bloodshed or more widespread destruction of property.[78] George Gordon (son of the 3rd Duke of Gordon), a former Member of Parliament, in 1779 organized the English Protestant Association, which sought repeal of the Catholic Relief Act of 1778. He modeled the organization on the Protestant Association formed a year earlier in Scotland. There its members instigated anti-Catholic riots in Edinburgh, Glasgow, and many small towns, though no loss of life was reported. Lord Gordon's organization would prove to be deadly. On June 2, 1780, he assembled in London 60,000 angry supporters to coerce a Parliamentary repeal of the Relief Act. On the following, day mobs attacked Catholic-owned property and churches. Troops were called out to maintain order and several Gordonites were arrested. Such violent acts occurred sporadically for several days as anti-Catholic mobs streamed into London. By June 6th, several hundred hostile Gordonites had gathered in front of Lambeth Palace (guarded by troops), and multiple thousands surrounded the Houses of Parliament. The government called out the Foot and Horse Guards to provide passages for Members to enter and exit the legislative buildings. The violence culminated on June 7th when throughout the metropolitan area mobs attacked houses, shops, and other premises belonging to Catholics, as well as public buildings. When the distillery of Thomas Langdale was attacked its 120,000 gallons of gin fueled a fire that destroyed 20 houses. Pillaging and assaults continued throughout the night while troops fired into crowds that refused to disperse. The Bank of England building came under attack, but soldiers held off the rioters. The riot had international implications as anti-Catholic mobs destroyed the chapels of the Piedmontese and Bavarian embassies. By June 9th, 10,000 army troops, along with several regiments of militia, had restored order. The riots left 285 dead and caused at least £70,000 of damage to private property and £30,000 of damage to public buildings. Gordon was among the 450

---

[77] Ibid., 91.

[78] Ibid., 76.

persons arrested. Of those tried and convicted, 25 were hanged. Gordon, charged with high treason, was acquitted at trial.[79]

Throughout the eighteenth century the Customs and Excise used the army to assist in the collection of duties and to guard contraband seized from smugglers. Smuggling was a popular occupation in coastal regions as tariffs were high and consequently, profits from the illegal practice were large. In addition to revenue collection, troops were used to suppress the smuggling-related practice of "wrecking." Wrecking was the illegal seizure of goods from coastal shipwrecks. This practice was particularly widespread and well organized in Cornwall.[80]

The Nine Years' War, 1692-1697 [81]

The later-named Nine Years' War (a term that encompasses the Glorious Revolution and the Williamite Wars in England, Scotland, and Ireland), was between a multi-state alliance and the France of Louis XIV. After pacification of Ireland in 1691, King William turned his attention to European affairs. He convinced Parliament to finance a 70,000-man army which would allow England to aid the Dutch and Germans then battling the French in Flanders. William personally commanded the force Parliament sent to the Continent. In early 1692, the field forces of the "British" Army totaled 54,930 deployed as follows:[82]

England	17,133	Flanders	20,859
Ireland	12,960	Colonies	1,160
Scotland	2,818		

---

[79] Lord Gordon went on to lead an unusual life. In 1786 he was excommunicated from the Church of England for reasons unrelated to the London riots. That same year Gordon converted to Judaism and took the name Israel Abraham Gordon. His conversion was unrelated to his Anglican excommunication. In 1787, while living in London, Gordon befriended Joseph Balsamo, the notorious charlatan "Cagliostro." On December 7, 1787, while living in the Jewish quarter of Birmingham, he was arrested for criminal libel against the French Queen, the French Ambassador to Great Britain, and the British judiciary. The libel arose from a pamphlet Gordon had previously written and published. He was convicted and sentenced to five-years' imprisonment. While in Newgate Prison, Gordon observed all practices of Orthodox Judaism and gained a reputation as a pious man. After expiration of his sentence in January 1793, he appeared in court for his release, which was contingent on financial guarantees of his future good conduct. For Gordon, the hearing got off to a bad start as he insulted the judge by refusal, on religious grounds, to remove his hat. Gordon presented as his would-be guarantors two poor, Jewish immigrants from Poland. The judge rejected them for lack of financial means and returned Gordon to Newgate. He died there later that year of typhoid fever. Solomons, "Lord George Gordon's Conversion to Judaism."

[80] Cornish smuggling and wrecking gave rise to the Gilbert & Sullivan operetta, *The Pirates of Penance*.

[81] Fortescue, *A History of the British Army*, Vol. 1, 351-79.

[82] Walton, *History of the British Standing Army*, 499-502.

National Wars of the Eighteenth Century

During the eighteenth century the British Army fought in five national wars that encompassed 40 years: War of the Spanish Succession (1701-1714), War of the Austrian Succession (1740-1748), Seven Years' War (1756-1763), French/Spanish/Dutch Wars (1778-1783), and War of the French Revolution (1793-1802). The British Army fought on four continents: Europe, North America (including the Caribbean), Africa, and Asia (the Indian sub-continent). In all those wars the British fought the French, in four wars the British fought the Spanish, and in three the British were allied with the Dutch.

It was during the War of the Spanish Succession that Gibraltar became a British possession. On July 21, 1704, an Anglo-Dutch fleet, commanded by British Admiral George Rooke, landed 1,800 sailors and marines on the sandy isthmus of the Gibraltar peninsula.[83] The ground force, supported by gunfire from the ships, defeated the Spanish garrison in the fortified town nestled against "The Rock." In the three-day battle, the British suffered 60 killed and 216 wounded. The Anglo-Dutch fleet then departed, but Rooke left behind the British marine contingent and placed a German general, Prince George-Louis of Hesse-d'Armstadt, in command of the new, temporary British possession. The German Prince was in effect the first British governor of Gibraltar.[84] The units of marines that first garrisoned Gibraltar would evolve into line regiments of the British Army (4th, 31st, and 32nd Foot). "The intrinsic value of the Rock in those days was small, and its value as a military position was little understood in England; but it was at any rate a capture and very soon it became a centre of sentiment."[85]

The Napoleonic Wars, 1803-1815

For nearly 13 years the British Army, sometimes on its own, most often allied with armies of Continental states, fought the armies led by the Corsica-born, Napoleon Bonaparte, First Consul of the French Republic, later Emperor of France. During this period, the British Army increased in size from about 140,000 men in 1802, to a peak strength of about 285,000 in 1813. British combat casualties were 14,055 killed in action and a further 61,319 wounded. Total army deaths; however, greatly exceeded those who fell in combat. Deaths from all causes during the 13 years of war totaled 137,218, approximately 10,000 per year.[86] At the time, the War Department did not identify separately "deaths from wounds and injuries." Presumably, many of the wounded died of causes related to their wounds, such as disease contracted while hospitalized. Most of the deaths were not causally related to combat and resulted from the poor living conditions of soldiers on campaign (poor field sanitation, malnourishment, and exposure to the elements).

---

[83] A much larger force than the "4 drunken English sailors" that Molly Bloom thought took Gibraltar from the Spanish. U (Gabler) Penelope 18:756.

[84] Drinkwater, *A History of the Siege of Gibraltar*, 9-10.

[85] Fortescue, *A History of the British Army*, Vol. 1, 448.

[86] Hodge, "On the Mortality arising from Military Operations." Hodge relies on data in returns to Parliament prepared by the various military departments.

During the Napoleonic Wars, 267,000 men joined the British Army. Such number, added to the starting strength of 140,000, indicates 407,000 men in total served from 1803 through 1815. Accordingly, the total death rate was 33.7%; the killed-in-action rate was 3.5% and the wounded-in-action rate was 15.1%.[87] Assuming one-third of the wounded died of their injuries, then deaths from illness and disease were about four times that of deaths from combat.

Imperial Campaigns, 1744-1853 [88]

During the 110-years prior to the Crimean War, the British Army fought abroad to either maintain, defend, or expand British possessions in Asia, Africa, North America, and New Zealand. In Africa, British troops fought in the western region known as "the Gold Coast" and the southern region that became the dominions of Natal and The Cape of Good Hope. On the North American mainland, British soldiers fought American rebels to keep the "13 Colonies" British and protect loyalist Upper and Lower Canada. In the Far East and Pacific, British troops served in furtherance of the government's imperial policies in China and Southeast Asia, and in support of British settlers in New Zealand. What would become by far the area of the British Army's greatest imperial endeavors was the sub-continent of India.

The first "British" territorial possessions in India were the private property of the British East India Company. The EIC acquired such territory with soldiers recruited locally from unemployed European adventurers and Indian mercenaries. The earliest experiences of the British Army in India took place in the 1750s. The British government sent troops to assist the Company's efforts to displace the rival French East India Company Over the years the British government curtailed both the political and commercial powers of the EIC. With respect to governance, by the early nineteenth century the Company was simply an agent of Westminster. In 1813, by statute, the Company's trading activities were limited to the Chinese opium trade and the export of tea.[89] Twenty years later, Parliament stripped the Company of its remaining commercial powers.[90] From 1833 to 1858, the EIC was simply Westminster's paid agent for governance of the British East Indies.

Excluding campaigns that were adjuncts to European wars (such as the Seven Years' War), following are the major imperial campaigns and wars in which the regulars of the British Army fought:

Carnatic Wars	1744-1763	India, East, intermittent
Mysore Wars	1766-1799	India, South, intermittent
1st Maratha War	1775-1782	India, Central
American War of Independence	1775-1783	North America
2nd Maratha War	1800-1805	India, Central

---

[87] Ibid.

[88] Source: Haythornthwaite, *The Colonial Wars Source Book*.

[89] East India Company Act, 1813, 53 Geo. 3, c. 155.

[90] Government of India Act, 1833, 3 & 4 Will. 4, c. 85.

The War of 1812	1812-1815	North America
3rd Maratha War	1817-1818	India, Central
1st Ashanti War	1823-1831	Western Africa
1st Burmese War	1824-1826	Burma
1st Opium War	1839-1842	China, Hong Kong
1st Afghan War	1839-1842	Afghanistan
The Sind Campaign	1843	India, West
1st Sikh War	1845-1846	India, West
1st New Zealand War	1845-1846	New Zealand
7th Xhosa Frontier War	1846-1847	Southern Africa
2nd Sikh War	1848-1849	India, West
8th Xhosa Frontier War	1851-1852	Southern Africa
2nd Burmese War	1852-1853	Burma

The wars and campaigns fought against non-Europeans in Asia and Africa were not always a matter of a modern, well-armed, and organized army against a technologically backward, unorganized collection of irregulars. For example, the Sikh army was a large, professional, European-style force and armed comparably to its Crown and Company opponents. The Sikh artillery gunners were arguably the finest in the world.

In Aid of Civil Authority, Nineteenth Century England [91]

A soldier's life at home in the nineteenth century was not confined to training and garrison duties. Like in the previous century, regular troops were called upon by the civil authority in England. During the first 70 years of the nineteenth century magistrates called for troops to suppress disturbances and riots, protect private property, and break-up peaceful assemblies they deemed seditious. After 1870, there were few popular disturbances and those that did occur were usually handled by police. Following are summaries of state use of regular troops to maintain domestic order in England.

The Burdett Disturbances, April 10, 1810

On April 6, 1810, Parliament voted the arrest of Sir Francis Burdett, radical MP for Westminster, for "breach of privilege" which was essentially criminal libel. Hundreds of Burdett's supporters assembled to prevent his seizure by bailiffs. To effectuate the arrest, and to crush what the government perceived as a nascent rebellion, Parliament dispatched thousands of troops to London.

---

[91] Source: Stevenson, *Popular Disturbances in England*, 151-300.

Luddite Attacks, 1811-1812

In the Midlands and the North, especially Yorkshire, troops were deployed to protect machinery from vandalism.

Corn Law Disturbances and Protests, 1815-1846

Magistrates, on numerous occasions, called out troops in response to demonstrations and riots sparked by high food prices. The Corn Laws were tariff and embargo laws to protect British agriculture. For, example wheat imports were prohibited when the price was below 80s. per quarter (eight bushels).

Post-Napoleonic War Popular Actions, 1816 – 1820

When war spending ended, the economy fell into recession. Some industries suffered an economic depression resulting in wage cuts and widespread unemployment. The resulting mass meetings, assemblies, and riots were viewed by the establishment "as part of a gigantic plot to promote a revolution."[92] Magistrates called for troops to protect property and break-up meetings they deemed seditious. The most infamous use of troops in aid of civil authority was the "Peterloo Massacre" of August 16, 1819. In Manchester, soldiers, along with some Yeomen, attacked a crowd of 60,000 peaceful protesters and within 15 minutes 15 civilians were killed and another 400 wounded. Troops were brought in by the magistrates because to them, the large demonstration "bore the appearance of an insurrection."[93]s

Parliamentary Reform Protests, 1830-1832

During this period soldiers broke-up meetings and quelled disturbances and outright riots. The bloodiest incident occurred in Bristol on October 24, 1831. Magistrates called out troops to disperse demonstrators who gathered to protest the Lords' rejection of the Reform Bill passed by Commons. Soldiers killed 12 civilians and wounded 94.

Protests Against the Poor Law Act of 1834

Troops broke up demonstrations against the law's provision that ended cash benefits for unemployed, able-bodied men ("Out-Door" benefits, that is payments to those "outside" the workhouse door).

Chartist Demonstrations, 1835-1845

Troops assisted bailiffs in the arrest of thousands who demonstrated in support of electoral reform. In the early years, arrests for "riot" averaged 2,000

---

[92] Ibid., 208.

[93] Ibid., 214.

annually. Such arrests increased and peaked in 1843 at 5,700. Mass arrests ended after 1845 but street protests continued until passage of the Reform Act, 1867.

Army Disorganization [94]

From 1661 through 1853, while the Crown had the allegiance of tens of thousands of red-coated soldiers, and Parliament appropriated millions of pounds for their equipage and upkeep, there was no corporate body that was "the British Army." In the early nineteenth century, the ground forces of the United Kingdom were governed by thirteen departments consisting of uniformed and non-uniformed personnel headed by five ministers (three of Cabinet), two senior generals, and six boards and commissions. "Any cooperation between these departments was largely dependent upon such customs and traditions as had developed over the years, and the whims and caprices of the departmental heads and their clerks."[95] Though the army at home was dependent on the navy to carry it to the theatre of war, there was no peacetime joint planning by admirals and generals. What little communication there was between the sea and ground forces took place between two civilians, the Under-Secretary of State for War and the Secretary of the Admiralty.[96] The uniformed force itself was simply a collection of independent regiments, each with its own traditions, to be hastily assembled into brigades and divisions by an expeditionary force commander, whenever the government sent it off to war. By 1837, control of the Army was streamlined somewhat and there were eleven departments, boards, and offices responsible for military matters.

The three cabinet ministers with military responsibilities were the Chancellor of the Exchequer, the Secretary of State at Home, and the Secretary of State for War and the Colonies. The Treasury, through its Commissariat Department, was responsible for feeding the army and for all army transport. It did not employ butchers and bakers, nor did it possess any horses and wagons. Its uniformed civilian officers, assisted by clerks, simply contracted for the required ration of meat and bread, and arranged with private carters and shipping firms for the movement of army supplies from Ordnance stores to the regiments. This system operated both at home and abroad, in peace and war. The Home Secretary supervised the militia, yeomanry, and volunteer force through the County Lord Lieutenants. The Secretary of State for War and the Colonies, through the War Department, ordered troop movements and was effectively commander-in-chief for the army garrisons in the colonies. He had no control over British Army units in India as they were commanded by

---

[94] Sources: Thomson, *The Military Forces & Institutions of Great Britain and Ireland*; Wheeler, *The War Office Past and Present*; Clode, *The Military Forces of the Crown*, chaps. 7, 11, 12, 19-21, 23, 25-26; Fortescue, *A History of the British Army*, Vol. 11, 454; Moyse-Bartlett, "The British Army in 1850."

[95] Farwell, *For Queen and Country*, 18.

[96] Ibid., 19.

the East India Company which executed orders received from the government's Control Board for India.

The other ministers with Army responsibilities were the Paymaster General, whose office paid the troops and handled all other cash disbursements, and the Secretary-at-War. The latter headed the War Office (separate and apart from the War Department) which had both civilian and military officers. Its main components were the departments of the Adjutant-General (personnel and general administrative matters), Quarter-Master General (logistics), and Judge Advocate-General (legal). The Adjutant and Quarter-Master Generals; however, were subordinate to the Commander-in-Chief (whose duties were assumed by the Secretary-at-War when that military position was vacant). The War Office also "owned" all military facilities other than fortifications and was responsible for all financial matters.

The two generals at the top of the military pecking order were the Commander-in-Chief whose office was Horse Guards, and the Master-General of the Ordnance, who chaired the Board of Ordnance which was a state ministry. Horse Guards administered the infantry and cavalry (recruitment of soldiers, commissioning of officers, training, and regulation). The Board of Ordnance administered the artillery and engineers, owned the fortifications at home, purchased weapons, munitions, and all other military equipment, manufactured some weapons in its arsenals, and stored all army goods in its warehouses.

Independent commissions and boards involved with military matters were the Board of General Officers for inspection of clothing, the Army Medical Board, and the Commissioners of Chelsea and Kilmainham Hospitals (army pensioners).

The system's workings are illustrated by how an infantry regiment would receive blankets for its soldiers. Parliament, as part of its annual army estimate vote, would appropriate funds for the purchase of blankets. The Board of Ordnance then contracted with manufacturers for delivery of the blankets to a Stores Department warehouse. (This, and other purchases, would in due course be audited by the War Office.) When an infantry regiment was running short of blankets, its quartermaster, through the regiment's commander, sent a request to the War Office. Upon approval the War Office forwarded the request to the Commissariat. The Commissariat verified that the Stores Department had blankets to spare, then arranged for their shipment. The contracted delivery firm picked up the blankets at the warehouse and delivered them to the barracks. There the regimental quartermaster first checked that the quantity delivered was correct. If satisfied that the shipment was in order, he asked the regimental sergeant-major for soldiers to off-load the blankets and carry them into the regiment's storeroom. Additional paperwork was then done at the regiment, War Office, Commissariat, and Board of Ordnance to confirm delivery of the blankets.

"Thus the system of administration and command, founded on political mistrust, went creaking on its way. Developed from small beginnings in an era of protracted conflict, subject to complex checks and dubious balances, it was just workable in time of peace."[97] In 1854, the army, the government, and the British public, would find out how workable the system was in time of war.

---

[97] Moyse-Bartlett, "The British Army in 1850.".

Chapter Bibliography

Brady, Ciaran. "The captains' games: army and society in Elizabethan Ireland." In *A Military History of Ireland* edited by Thomas Bartlett and Keith Jeffrey. Cambridge: Cambridge Univ. Press, 1996.

Childs, John. *The Army of Charles II*. London: Routledge & Kegan, 1976.

Clode, Charles. *The Military Forces of the Crown*. London: Murray, 1869.

Drinkwater, John. *A History of the Siege of Gibraltar*. London: Murray, 1905.

Dunlop, M. A., ed. *Ireland under the Commonwealth being a selection of documents relating to the government of Ireland from 1651 to 1659*. Manchester: Univ. of Manch. Press, 1913.

Ellis, Stephen G. "The Tudors and the origins of the modern Irish state." In *A Military History of Ireland* edited by Thomas Bartlett and Keith Jeffrey. Cambridge: Cambridge Univ. Press, 1996.

Farwell, Byron. *For Queen and Country*, London: Penguin, 1981.

Ferguson, Kenneth Patrick. "The Army in Ireland from the Restoration to Act of Union." PhD Thesis, Trinity College (Dublin), 1980.

Fortescue, J. W. *A History of the British Army*. Vol. 1, London: MacMillan, 1899.

——— *A History of the British Army*. Vol. 11, London: MacMillan, 1923.

Grose, Francis. *Military Antiquities*. Vol. 1, London: Hooper, 1786.

Guy, Alan J. "The Irish Military Establishment, 1660-1776." In *A Military History of Ireland,* edited by Thomas Bartlett and Keith Jeffrey. Cambridge: Cambridge Univ. Press, 1996.

Hallam, Henry. *The Constitutional History of England*. Vol. 2, London: Murray, 1855.

Hamilton, F. W. *The Origin and History of the First or Grenadier Guards*. Vol. 1, London: Murray, 1874).

Haythornthwaite, Philip J. *The Colonial Wars Source Book*. London: Arms & Armour, 1995.

Hodge, William Barwick. "On the Mortality arising from Military Operations (Concluded)." *The Assurance Magazine and Journal of the Institute of Actuaries* 7 (April 1858): 275-85.

Horse Guards. *Historical Record of the Life Guards*. London: HMSO, 1835.

Jenkyns, Henry. "History of the Military Forces of the Crown." In *Manual of Military Law*. London: HMSO, 1907.

Johnston, S. H. F. "The Scots Army in the Reign of Anne." *Transactions of the Royal Historical Society* 3 (December 1953): 1-21.

Miller, John. "The Earl of Tyrconnell and James II's Irish Police." *Historical Journal* 20, no. 4 (December 1977): 803-24.

Moyse-Bartlett, Hubert. "The British Army in 1850." *Journal of the Society for Army Historical Research* 52, no. 212 (Winter 1974): 221-37.

Murray, Alice Effie. *History of the Commercial and Financial Relations between England and Ireland.* London: P.S. King, 1903.

Packe, Edmund. *The Royal Regiment of Horse Guards.* London: Parker, Furnivall, & Parker, 1847.

Roper, Michael. *The Records of the War Office and Related Departments.* London: PRO, 1998.

Solomons, Israel. "Lord George Gordon's Conversion to Judaism." *Transactions of the Jewish Historical Society of England* 7 (1911-1914): 222-71.

Stevenson, John. *Popular Disturbances in England 1700-1870.* New York: Longman 1979.

Thomson, H. Byerly. *The Military Forces & Institutions of Great Britain and Ireland.* London: Smith, Elder, 1855.

Walton, Clifford. *History of the British Standing Army.* London: Harrison, 1894.

Wheeler, Owen. *The War Office Past and Present.* London: Metheun, 1914

Sources for Rebellions and Risings at Home: The '45

Blaikie, Walter Biggar. *Itinerary of Prince Charles Edward Stuart.* Edinburgh: Scottish History Society, 1897.

Duffy, Christopher. "The Jacobite Wars, 1708-46." In *A Military History of Scotland* edited by Edward M. Spiers, Jeremy A. Craig, and Matthew J. Strickland. Edinburgh: Univ. of Edinburgh Press, 2012.

Fortescue, J. W. *A History of the British Army.* Vol. 2, London: MacMillan, 1910, 124-48.

Tomasson, Katherine and Francis Buist. *Battles of the '45.* New York: MacMillan, 1962.

Army Estimates for 1745, *Journals of the House of Commons*, Vol. 24, 702-04.

Chapter 3
The Crimean War

The only regimental affiliation Joyce attributed to Major Brian Tweedy was that of the Royal Dublin Fusiliers. The Dublins evolved from two of the nine European-manned infantry regiments of the British East India Company, regiments which served only in the Company's sphere-of-interest.[1] Accordingly, young Tweedy must have enlisted into the EIC Army, not the British Army, and would not have participated in the Crimean War. Joyce likely associated Tweedy with the 2nd Battalion, Royal Dublin Fusiliers as it was the only Irish battalion that saw no active service during the last 50 years of the nineteenth century.[2] Tweedy only experienced combat vicariously through books and newspapers.[3]

Prelude to War

The spark to the tinderbox that ignited war between the Russians and the Turks in 1853, were the keys to the main door of the Church of the Nativity in Bethlehem, then situated within the Ottoman Empire.[4] Over the centuries an understanding developed among the various Christian denominations as to ownership of the principal shrines of the Holy Land and sectarian usage rights thereto. The resulting customs were codified in 1757 through a *firman* (decree) of Sultan Osman III. As no authoritative writing of this decree survived into the nineteenth century, each denomination (Latin, Greek, and Armenian), claimed legal support for its demands as to the Holy Places. Principal rights to the Church of the Nativity were, and are still, held by three bodies, the Franciscans (Catholic), and both the Greek and Armenian Patriarchates of Jerusalem (Orthodox). Minor rights are held by the Syriac Patriarchate of Antioch and the Coptic Orthodox Patriarchate of Jerusalem (both Orthodox). In 1852, the keys to the Church's main door were held by monks of the Greek Patriarchate.

---

[1] Mainwaring, *Crown and Company*. In addition to India, EIC troops served in Africa, Aden, the Persian Gulf area, Afghanistan, Southeast Asia, and China.

[2] From the end of the 2nd Sikh War in 1849 to the outbreak of the 2nd Boer War in October 1899, 2/Royal Dublin Fusiliers (and its antecedents), had only garrison assignments (India, Gibraltar, Egypt, and at home). The battalion did; however, furnish 200 volunteers for the Persian Expedition in 1856. Ibid.

[3] Russo-Turkish War of 1877-78, "Hozier's *History of the Russo-Turkish War*" U (Gabler) Ithaca 17:1385, 1416; Zulu War of 1878-79 and the Gordon Relief Expedition to the Sudan of 1884-85, "captain Groves and father talking about Rorke's drift and Plevna and sir Garnet Wolseley and Gordon at Khartoum" U (Gabler) Penelope 18:690-91.

[4] Royle, *Crimea*, 15.

In 1850, French Emperor Napoleon III presented the Ottoman Foreign Office ("the Porte") a demand for "restoration" of the Franciscan's rights to the Holy Places in Ottoman Palestine. These rights included possession of the keys to the Church of the Nativity.[5] The French ambassador argued that the Franciscans' rights were guaranteed under a 1740 Franco-Ottoman treaty. In February 1852 Sultan Abdulmecid acceded to the French demand. Shortly thereafter, Czar Nicholas I of Russia, who viewed this agreement an affront to the Russian Empire and all people of the Orthodox Faith, convinced the Sultan to back out of his recent accord with the French. The Czar argued that a later Russo-Ottoman treaty of 1757 made Russia protector of Christianity in the Ottoman Empire and gave the Orthodox church supremacy with respect to the Holy Places.[6]

It was now Napoleon III who was affronted by the Sultan. That summer, to express his displeasure with the Sultan's action (and to pose a thinly-veiled threat), he sent a steam-powered warship through the Dardanelles in contravention of a treaty to which France was a signatory.[7] Sultan Abdulmecid understood what the French implied by this show of strength and in December 1852 awarded the now troublesome church keys to the Franciscan monks in Bethlehem.

The Czar, insulted by the Sultan and threatened by the French, made plans to bolster Russia's standing in the region and restore its pre-eminence at the Porte. This he would accomplish by the military occupation of the Romanian Principalities of Moldavia and Wallachia, autonomous, tribute-paying suzerainties of the Ottoman Empire.[8] In March 1853, the Czar's ambassador made demands upon the Porte to give the Orthodox Churches control of the Holy Places in Palestine, grant the Russian Monarchy exclusive jurisdiction over such churches within the Ottoman Empire, and ensure "privileges" for the Sultan's Orthodox Christian subjects. By the end of June, Russia had in place on the Moldavian border two army corps, approximately 70,000 troops, commanded by Prince Mikhail Gorchakov.[9] This was the standing army of the southwestern military district, headquartered in Kiev.

In May, the French government dispatched a squadron of ships from its Mediterranean Fleet to Besika Bay, the Aegean entrance to the Dardanelles. The squadron consisted of five

---

[5] Ibid., 9.

[6] Royle, *Crimea*, 18-19.

[7] Article IV of the London Straits Convention of 1841 prohibited foreign warships from passing through the Dardanelles and Bosporus in times of peace. James Shotwell, "A Short History of the Question of Constantinople and the Straits;" Royle, *Crimea*, 19-20.

[8] Kinglake, *Invasion of the Crimea*, Vol. 1, 193-94. Note: Serbia was the third autonomous, European Ottoman Principality. In 1863, Wallachia and Moldavia united and in 1881, after another Russo-Turkish war, the autonomous principality became the independent Kingdom of Romania. A year later, Serbia proclaimed itself an independent kingdom.

[9] Royle, *Crimea*, 20, 40, 47-48, 51. The strength of a Russian army corps in peacetime was 35,000. Kinglake, *Invasion of the Crimea*, Vol. 1, 55.

sailed battleships (two with steam-powered paddles) and five steam-powered escort vessels (15-20 guns each).[10]

In the Spring of 1853, the British Mediterranean "Fleet" consisted of five sailed battleships, four large steamers (about 40 guns each), and six steam-powered escort vessels (15-20 guns each). The home port was Malta, but some vessels would moor temporarily at Gibraltar.[11] At that time the French Mediterranean Fleet, based primarily at Toulon, was much larger. It had fourteen sailed battleships (two with steam-powered paddles), six sailed escorts, six steam-powered escorts (10-20 guns), and seven steam-powered gunboats (2-8 guns).[12] On June 2, 1853, the British government ordered the Mediterranean Fleet to send six ships to join the French squadron at Besika Bay. The British flotilla arrived on June 13th.[13]

On July 2, 1853, with the Russian demands of March still unmet, Gorchakov's army crossed the River Prut into Moldavia and began occupation of the Romanian Principalities.[14] By the end of October, this army, through reinforcements, totaled 88,000 men.[15] For the remainder of the summer, French and British diplomats strived in vain to convince the Russians to withdraw from the Balkans and avert war. Meanwhile, the British public, antagonistic to the Czar, clamored for decisive government action to protect Turkey from Russian imperial designs.[16]

At the beginning of October, the French and British governments ordered their Besika Bay naval forces to enter the Dardanelles and proceed to Constantinople.[17] On October 4, 1853, with Anglo-French support, the Ottoman Empire declared war on Imperial Russia.[18]

Early Engagements of the Russo-Turkish War, 1853-1856

The Sultan began the war with an army of 300,000 men (123,000 of the standing army, 177,000 reservists and irregulars). About half of the troops were in the Balkans and a majority

---

[10] Lambert, *The Crimean War*, 74. Most steamers (equivalent to sailed frigates) typically had 15-20 guns. Lesser steam-powered vessels, gunboats, had 4 to 10 guns. There were also large steam-powered vessels with 40 to 60 guns. de Bazancourt, *L'Expédition de Crimée, La Marine Française,* vol. 1, 399-403.

[11] Lambert, *The Crimean War*, 50-51.

[12] Dodd, *Pictorial History of the Russian War*, 565-66.

[13] Lambert, *The Crimean War,* 74; Royle, *Crimea,* 53.

[14] Kinglake, *Invasion of the Crimea*, Vol. 1, 205.

[15] Badem, *The Ottoman Crimean War*, 106.

[16] The British people were very sympathetic to Kossuth, former leader of the short-lived independent Hungarian state, then exiled in London. They never forgave Russia for its role in the forced restoration of Hungary to the Austrian Empire, as well as its oppression of the Poles. Royle, *Crimea*, 45-46, 57.

[17] Ibid., 78.

[18] Badem, *The Ottoman Crimean War*, 99.

of those were of the standing army. The army rank and file were nearly all Muslim young men from the provinces who were conscripted at age 20. Conscripts spent five years with the colors and seven years in the first-line reserve. Military service; however, was not universal for Muslims. Those resident in Constantinople were exempt from conscription and those in the provinces could purchase exemption. Accordingly, the army was manned with poor, village youths. Non-Muslims were ineligible for military service and paid a special military tax. Overall, the army was poorly armed, paid, clothed, and fed. Many of these shortcomings were due to the routine embezzlement of funds and conversion of army goods (uniforms, rations, and personal equipment) by the army's officers. Despite these limitations, the Balkan force was up to Russian Army standards (which for a European army were low). The best troops were the career volunteers from Albania, Bosnia, and Bulgaria.[19]

The army's numbers were supplemented with contingents provided by Muslim rulers of two Mediterranean states. Abbas Pasha, the effectively independent governor of Ottoman Egypt, sent 23,000 troops and Ahmed Bey, ruler of Tunis, sent 7,000.[20]

General Omar Pasha commanded the Balkan Army. He began life as Mihaylo Latas, son of a Croatian army officer in the Austrian Imperial Army. Latas was one of many European mercenary officers in the Ottoman Army. Latas converted to Islam, changed his name, and rose in rank through success in suppression of revolts (Bosnian, Arab, and Kurd) and marriage to a Turkish heiress.[21]

The Ottoman Navy consisted of 68 warships, 4 of which were sailed battleships. The smaller ships were 46 sailed escorts, 6 steam-powered escorts (15-20 guns), and 12 steam-powered gunboats (<10 guns). This establishment was augmented by 9 ships from Egypt: 3 sailed battleships and 6 escort vessels of which 2 were steamers. The navy was manned mostly with conscripts and accordingly lacked sufficient numbers of experienced sailors. At the outbreak of hostilities, the ships were in fair to good condition.

In January 1853, the Russian Army had approximately 750,000 standing imperial troops, 250,000 irregulars (nearly all Cossack and mostly cavalry), and 265,000 reservists.[22] In late 1853, the British Ambassador to Russia estimated that the Czar had 645,000 troops, including reservists, available for operations against the Ottoman Empire. Forces near the Ottoman frontier were as follows: 120,000 in the Balkans, 60,000 in the Bessarabia-Black Sea area, 25,000 in the Trans-Caucasus centered on Kutais, and 80,000 in the Caucasus.[23]

---

[19] Royle, *Crimea*, 108-09; Badem, *The Ottoman Crimean War*, 50-52, 103, 110, 292, 296; Uyar, *A Military History of the Ottomans*, 159-61.

[20] Badem, *The Ottoman Crimean War*, 81, 112, 155.

[21] Royle, *Crimea*, 92, 108; Badem, *The Ottoman Crimean War*, 49-50, 102.

[22] de Todleben, *Defense de Sebastopol*, Vol 1, appx. 4.

[23] Dodd, *Pictorial History of the Russian War*, 22-23. Dodd relies on a report on the Russian Army dated October 27, 1853, from Ambassador Seymour to the British Foreign Minister, George Villiers (4th Earl of Clarendon). Seymour's troop numbers appear derived from the war-strength establishment for a Russian army corps (60,000 men organized in three infantry divisions, one cavalry division, and one artillery division). Curtiss, *The Russian Army under Nicholas I*, 108-09.

The Crimean War

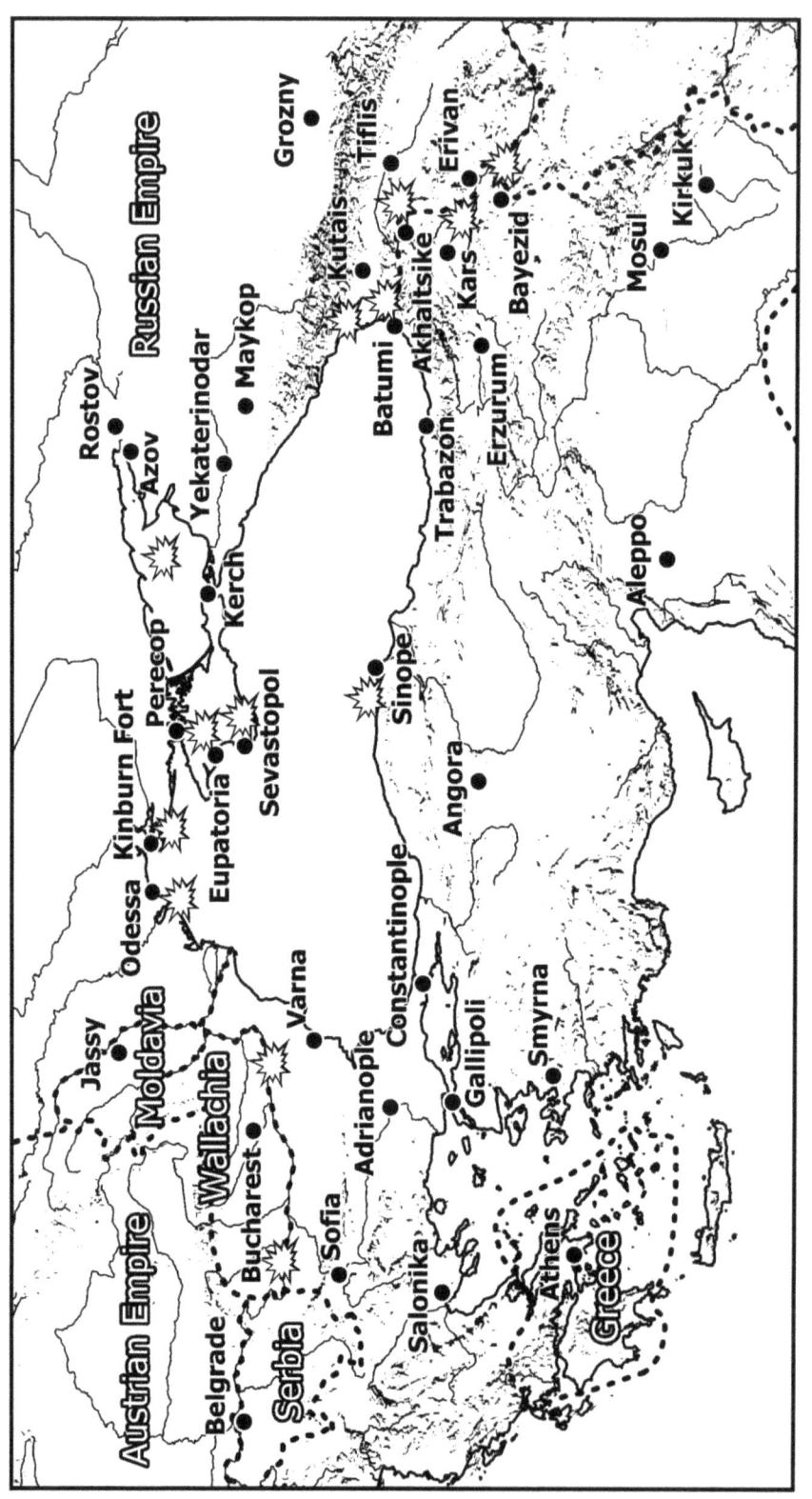

Russo-Turkish (Crimean) War, 1854-1856
Black Sea Theater

The army was manned by conscripted serfs and graduates of "cantonist" military schools. Attendance at such schools was compulsory for children born to serving soldiers. Additionally, ethnic and religious minority male children, as young as age nine for Jewish boys, plus orphans and delinquents, were drafted into such schools on a selective basis. At the outbreak of the Crimean War, 378,000 boys were enrolled in cantonist schools, where discipline was harsh and conditions barbarous. Upon attainment of age eighteen, cantonists entered the army along with the drafts of serfs. The total military service obligation was 25 years: 20 years with the colors and 5 years with the reserves. Conscripts with good records; however, were released to the reserves after 15 years service.[24] Veterans did not receive pensions, but many were employed by the para-military Internal Defense Corps. Most, but not all, disabled soldiers entered state-run old soldiers' homes or received civil service jobs with undemanding duties.[25]

The Russian Army lacked facilities to house all its soldiers. Accordingly, soldiers not posted to fortresses or urban barracks, about 72% of the total, were quartered with civilians. The state provided rations to village communes for sustenance of the quartered soldiers, but such foodstuffs rarely reached the householders. Soldiers who lacked the money to purchase meals were generally malnourished. The dispersed soldiers were assembled into encampments for four to five months of the year where they received training, took part in field exercises, and labored on state infrastructure projects. Training was parade ground drill and unrealistic field exercises as participants simply followed detailed scripts. A soldier's life was miserable and the peace-time, annual mortality rate was 3.7%, three times higher than the rate for civilian males 20 to 40 years of age. Accordingly, about half of each annual intake would die of non-combat causes during a 20-year term of service.[26] Disease, accidents, and combat took a severe toll on the Russian soldier. Only about 20% of conscripts completed their service intact and in reasonably good health.[27]

The quality of Russian infantry and cavalry was substandard for European armies. Soldiers were armed with muzzle-loading, smooth-bore muskets, many of which had flintlock firing mechanisms. Accordingly, Russian soldiers, under ideal conditions, could fire no more than one aimed shot per minute. Additionally, the effective range of Russian infantry fire was half that of the British infantry's range.[28] Russian generals were indifferent to the use of firearms. Army doctrine established the bayonet as the principal infantry weapon, while the sword and lance were the principal cavalry weapons. The quality of Russian engineering services and artillery was on par with that of other European armies.[29]

---

[24] Curtiss, *The Russian Army under Nicholas I*, 233-252; Wirtschafter, "Military Service and the Russian Social Order.".

[25] Curtiss, *The Russian Army under Nicholas I*, 254.

[26] Wirtschafter, "Military Service and the Russian Social Order;" Curtiss, *The Russian Army under Nicholas I,* 118-19, 246-47, 250.

[27] From 1826 to 1850 the army inducted, on average, 80,000 men annually. Each year about 15,000 soldiers entered the reserve. Curtiss, *The Russian Army under Nicholas I,* 234, 253.

[28] Ibid., 120-38.

[29] Ibid., 144-49.

The Crimean War

The Russian Imperial Navy, manned by conscripted serfs with a cadre of mercenary, foreign seamen, consisted of two fairly modern fleets and three small, independent squadrons. The Baltic Fleet had its primary base at Kronstadt, a fortified island at the eastern extremity of the Bay of Finland 30 kilometers offshore from St. Petersburg. Its main armament was 26 sailed battleships (3 with steam-powered paddles), 2 large steam-powered escorts (40 guns), 11 small steam-powered escorts (10 guns), and 15 steam-powered gunboats (2-3 guns). Including unarmed vessels, it totaled 288 ships and was manned by just under 50,000 personnel (ship and shore establishments).[30] The Black Sea Fleet had its primary base at Sevastopol, a fortified town on the Crimean Peninsula. Its main armament was 14 sailed battleships, 4 steam-powered escorts (15-20 guns), and 23 steam-powered gunboats (4 guns). Including unarmed vessels, it totaled 136 ships and was manned by 35,000 personnel.[31]

There were lightly gunned squadrons on the White Sea (Archangel) and the Caspian Sea with 34 sailing vessels (total 66 guns), and 30 sailing vessels (total 49 guns) respectively. The navy also maintained a small Far East Flotilla at the port of Petropaulovsk on the Kamchatka Peninsula. It consisted of 8 small sailing ships (total of 30 guns) and 800 personnel. Total personnel for these three lesser establishments was about 5,000.[32]

On October 28, 1853, 12,000 Turkish troops commanded by General Ismail Pasha, crossed the Danube at the western extremity of Wallachia's border with Ottoman Bulgaria and threatened the Russian army's right flank.[33] In response, Gorchakov reoriented his forces along the Danube and on January 4, 1854 attacked Ismail Pasha's troops dug in at Kalafa. The Ottoman line held and after four days of fighting the Russians withdrew to Bucharest, terminus of their supply line from Warsaw. After this defeat, the Czar sacked Gorchakov and placed the Balkan force under direct command of his superior, the 72-year-old Count Paskevitch. Paskevitch was commander of Russian forces in the west and Viceroy of Poland. St. Petersburg sent reinforcements to the Balkans and directed Paskevitch to prepare for an advance on Adrianople.[34]

Concurrent with the crossing of the Danube, the Ottoman Anatolian Army, at the other end of the empire, advanced into the Russian Caucasus. This army, two-thirds of which consisted of reservists and irregulars, totaled about 85,000 troops of which 15,000 held the fortified Black Sea port of Batumi. It was commanded by Abdi Pasha, an Anatolian Turk who had studied in Vienna for five years. The chief-of-staff was an illiterate Kurd, Ahmed Pasha, whom Abdi Pasha held in contempt. Due to low literacy rates among native Ottoman officers, most of the Anatolian Army's staff officers were foreign mercenaries (who did not have to convert to Islam).

---

[30] de Todleben, *Defense de Sebastopol,* Vol 1, appx. 8.

[31] Ibid.

[32] Ibid.

[33] Badem, *The Ottoman Crimean War*, 107.

[34] Kinglake, *Invasion of the Crimea,* Vol. 2, 38-45; Royle, *Crimea*, 92; Badem, *The Ottoman Crimean War*, 104.

On October 28th Ottoman troops at Batumi took the opposing Russian position of Fort St. Nicholas. The Russians, fearful that the Anatolian Army would receive naval support, and not sure whether they could control the simmering insurgency in the area, withdrew from the Mingrelian coast of the Black Sea.[35]

Farther south, the Anatolian Army advanced towards Tiflis and Erivan, the capitals of Russian Georgia and Armenia, respectively. After penetration of Russian territory, the advances stalled and the Russian counter-attacks were successful. On November 14, 1853, the Russians took the Ottoman fortified frontier town of Akhaltsike. By the end of the month the Anatolian Army, greatly weakened by casualties, had withdrawn to the fortresses of Batumi, Kars, and Bayezid.[36] Allied with the Anatolian Army were the irregulars of Ali Shamyl who waged guerilla war against the Russians in Caucasian Dagestan and Chechnya, territories which were seized by Russia earlier in the century. Shamyl claimed 20,000 part-time fighters and could at times field a force of 10,000 to 15,000 men.[37]

To bolster his flagging Anatolian Army, the Sultan dispatched by sea, reinforcements and additional supplies from Constantinople. These troops and provisions were carried by two transport vessels and twelve small warships (Egyptian and Turkish), of which two were steam-powered. The convoy's destination was Batumi, but it anchored at Sinope to shelter from a developing storm. The Ottoman vessels were spotted by a Russian division of three sailed escorts commanded by Admiral Nakhimov. Nakhimov informed his superior, Admiral Kornilov of the Black Sea Fleet, of this opportunity to interdict the Ottoman supply line. Kornilov quickly dispatched from Sevastopol six sailed battleships, followed soon by three steamers. On November 30, 1853 the Russian squadron attacked the Ottoman ships anchored at Sinope. Within the battle was the first combat in history between steamed warships. The Russian naval guns, firing explosive shells which the Ottoman Navy lacked, devastated Sinope's defenders. Nakhimov's squadron sank all the Ottoman vessels, destroyed the harbor and its protective fort, killed and wounded about 1,500 Ottoman soldiers and sailors, and took about 200 hundred prisoners, including the wounded commander of the Turkish flotilla. The Russian dead numbered about 35. News of the battle stunned the British government, but the French were unperturbed.[38]

On March 20th, the British Ambassador in St. Petersburg, at the behest of both the British and French governments, presented the Russian government with a formal demand that it begin withdrawal from the Balkans within six days. On March 28th, two days after the ultimatum expired, with no Russian troop movement or diplomatic response, France and the United Kingdom declared war on Russia and sent their ships at Constantinople into the Black Sea.[39] The first Anglo-French engagement with the Russians took place on April 22nd

---

[35] Badem, *The Ottoman Crimean War*, 154.

[36] Royle, *Crimea*, 92-93; Badem, *The Ottoman Crimean War*, 145-48, 166-70; Uyar, *A Military History of the Ottomans*, 161; Curtiss, *The Russian Army under Nicholas I*, 316.

[37] Badem, *The Ottoman Crimean War*, 149-52; Curtiss, *The Russian Army under Nicholas I*, 318-19.

[38] Royle, *Crimea*, 93-95; Badem, *The Ottoman Crimean War*, 117-23; de Bazancourt, *L'Expédition de Crimée*, xxx.

[39] Royle, *Crimea*, 126-27.

The Crimean War

when six British and three French ships bombarded the Russian port city of Odessa.[40] This marked the commencement of the Allied campaign to take control of the Black Sea.

Shortly after the Anglo-French declaration of war, Paskevitch, after having shored up his army's right flank, sent a force of 60,000 troops south from Bucharest. Its first objective on the route to Adrianople was the Ottoman Danubean fortress of Silistra. The siege of Silistra began in May 1854. Repeated Russian assaults against the troops of General Musa Pasha, the Ottoman garrison commander, were unsuccessful and the advance stalled.[41]

French and British "Boots on the Ground"

On February 8, 1854, the British government decided to send 10,000 soldiers to Malta as the core for a possible expeditionary force to protect Constantinople. The French also took steps to send ground forces to the Balkans. The British appointed Master-General of the Ordnance Fitzroy Somerset, 9th son of the Duke of Beaufort and recently created 1st Baron Raglan, as commander of their forming army. The French appointed General and War Minister Leroy de Saint-Arnaud as commander of their force.[42] Neither commander would survive the upcoming war. Saint-Arnaud, terminally ill with stomach cancer, would die in September. Lord Raglan's death from dysentery would follow nine months later.

Lord Raglan, age 66, had combat experience as a junior officer in the Napoleonic Wars in which he also served as Wellington's military secretary. As an army commander's aide, he had some exposure to the logistical problems posed by a large force in the field. At the Battle of Waterloo, he lost an arm. Raglan also spoke French fluently.[43]

Saint-Arnaud, age 55, gained extensive combat experience through 15 years of intermittent, colonial warfare in North Africa. In Summer 1851, after completion of an extraordinarily successful campaign, he was promoted to Major-General. At the time, President Louis Napoleon sought a royalist officer to serve as Minister of War and partake in his scheme to restore the Napoleonic monarchy. His military aide recommended Saint-Arnaud, and in October the General became Minister of War. While holding this office he played an important role in the December coup by which Charles-Louis Napoleon Bonaparte became Emperor Napoleon III. Saint-Arnaud also oversaw the bloody suppression of the Parisian uprising in defense of the Republic.[44]

In April 1853, the British government directed the various administrative components of the army to prepare and deploy to the Balkans a force of 10 cavalry and 25 infantry regiments

---

[40] Ibid., 149.

[41] Kinglake, *Invasion of the Crimea*, Vol. 2, 199-216.

[42] Royle, *Crimea*, 112-13.

[43] Ibid.

[44] Kinglake, *Invasion of the Crimea*, Vol. 1, 217, 240-43, 264-65; de Bazancourt, *L'Expedition de Crimée*, Vol. 1, 12-13.

with an appropriate number of artillery batteries and support personnel.[45] At the start of 1854, the British Army totaled 146,755 men maintained at an annual cost of £9,600,000. The British taxpayer had to support only 114,383 of its soldiers as the 32,372 troops in India were paid for by the East India Company. At home were 8,690 cavalry, 50,694 infantry, 10,245 artillery, and 1,218 engineers, a total of 70,847 regulars to both defend the Isles and man an expeditionary force.[46]

In addition to the uniformed personnel there were civilian employees of the Board of Ordnance, Treasury Department, War Department, War Office, and the Medical Department, many uniformed, all with designated, equivalent army rank. The £9.6 million army appropriation for 1853 allocated £2.2 million for ordnance, commissariat, and other support services.[47] Backing up the army were the part-time militia and the volunteer force. The militia had 35,813 men, maintained at an annual cost of £250,000, while the volunteers totaled 14,721 men (largely self-funded with supplementary grants through the Home Office).[48] By law, the two auxiliary forces could be employed only within the United Kingdom. In all, the United Kingdom had 121,381 "trained" ground troops at home.

The number of men available for home defense in 1853 was nearly the same as in 1819, the first year after full demobilization from the Napoleonic Wars. In 1819, the Army had 122,918 troops of which 61,142 were at home. This peacetime army (excluding 21,723 soldiers in India) was maintained at an annual cost of £10,000,000. Backing up the army was the 61,190-man volunteer force.[49] Therefore, 122,332 men were available then for home defense, less than one percent more than in 1853. Though the nominal amount the British spent per soldier (outside of India) declined from £99 in 1819 to £84 in 1853, real spending had increased. The early nineteenth century was a deflationary period for the United Kingdom and from 1819 to 1853 prices had declined 30%.[50] Accordingly, real per capita spending on the British Army increased about 20% between the Napoleonic and Crimean Wars.

From a collection of separate infantry and cavalry regiments, plus artillery batteries and engineer companies, the unwieldy British Army administrative apparatus had to crank out a 25,000 to 30,000 man "Eastern Army" for deployment in an inhospitable landscape.[51] When

---

[45] *Return concerning the late Army of the East*, 1857 Sess. 1, H.C. Accounts & Papers, No. 42.

[46] War Office, *Return of the Number of Officers and Men, 1800-1858*, 1859 Sess. 2, H.C. Accounts & Papers, No. 88.

[47] Ibid.

[48] War Office: *Return of the Total Expenses incurred on account of the Embodied Militia*, 1860, H.C. Accounts & Papers, No. 380; *Return of the Number of Officers and Men, 1800-1858*, 1859 Sess. 2, H.C. Accounts & Papers, No. 88.

[49] War Office, *Return of the Number of Officers and Men, 1800-1858*, 1859 Sess. 2, H.C. Accounts & Papers, No. 88.

[50] Composite Price Index from 1750, Office for National Statistics (UK), www.ons.gov.uk/economy/inflationandpriceindices.

[51] The British contingent is styled Eastern Army to distinguish it from the French Army of the East (*Armée de l'Orient*).

The Crimean War

war was declared the Eastern Army consisted of 8,600 men stationed in Malta (from the troop deployment ordered in February). In early April, these troops moved to the Turkish Straits and the British command established its headquarters at Scutari, a suburb of Constantinople on the Asian side of the Bosphorus. Over the next two months a further 14,700 men arrived from England.[52] As was the custom of the times, the Eastern Army was accompanied by some officers' wives, who paid their own way, and a portion of the "on-the-strength" enlisted men's wives, whose passage and rations were provided by the state.[53]

As the British moved 23,000 troops to the Asian edge of the Balkans, the French sent a similar number of men. At the end of May an Anglo-French expeditionary force of 47,000 men, organized into seven divisions (four British and three French), was bivouacked along the Turkish Straits from Gallipoli to Constantinople.[54] Beginning June 11, 1854, the Allied force began a sea-borne move to Varna, a port town on the Black Sea in Ottoman Bulgaria.[55] As the troops moved north, the British established a logistical base at Scutari staffed with about 1,000 men. During the movement of the Anglo-French army to Varna, the British bolstered its contingent with an additional 5,000 men from home. The British Eastern Army, including detachments in the Turkish Straits, had reached its authorized strength of 28,000 men and 60 artillery guns.[56]

The Eastern Army was untried and ill-prepared for campaigning. Only 6 of the 25 infantry regiments had engaged in combat within the last 30 years. Of the five infantry division commanders (Raglan formed a fifth division after arrival of the June reinforcements) only two had combat experience in command of a multi-battalion force, and only one was under age 60. The cavalry commander had no combat experience, although at 53 he was of an age appropriate for his position. Supply of rations and transport of all materiel was administered by the 66-year-old former head of the Treasury's Commissariat Department, James Filder who held equivalent army rank of Brigadier-General. He was assisted by four subordinate officers and hastily hired temporary clerks, assistant store-keepers, and interpreters. This

---

[52] *Report of the Commissioners appointed to inquire into the Regulations Affecting the Sanitary Condition of the Army*, 1857-58, [2318], at 524. Hereafter cited *Sanitary Report*. Royle, *Crimea*, 140-41, 145.

[53] Families of enlisted men who married with their commanding officer's consent were placed on the regimental establishment. The state provided such families with housing, food at cost, and job opportunities (washing and mending soldiers' clothes, cooking for the soldiers, servants to officers' families). Family housing in the 1850s was typically a screened off portion of a barracks room ("corner accommodation"), or a room separate from the single men, but shared by several families. Wives of soldiers who married without consent, and their children, received nothing from the state. Approximately 5 to 7% of the soldiers sent to the Crimea had on-the-strength wives. The War Office would only authorize a portion of the wives to go on campaign with their husbands, and such wives were selected by lot. It is likely that 750 to 1,250 military wives (women with officer husbands and women on-the-strength) were with the Eastern Army. Trustram, *Women of the Regiment*; Rappaport, *No Place for Ladies*.

[54] *Sanitary Report*, at 524; Royle, *Crimea*, 140-41; Russell, *The British Expedition to the Crimea*, 49, 56, 63.

[55] Kinglake, *Invasion of the Crimea*, Vol. 2, 198-99.

[56] *Sanitary Report*, at 524; Kinglake, *Invasion of the Crimea*, Vol. 2, 348.

Commissariat detachment was utterly unprepared to support the 27,000 men in and around Varna.[57]

Throughout 1853 and early 1854, the government of the Hapsburg Austrian Empire observed with keen interest, the diplomatic and military maneuvering regarding Constantinople. Emperor Franz Joseph was greatly concerned that the Czar might tilt the Balkan balance of power to his favor at the expense of the Austrians'. At first, Franz Joseph was loath to turn against his Russian ally whose army in 1849 brought the breakaway Hungarian state back into the Hapsburg fold. As events in the Balkans unfolded, Franz Joseph became antagonistic to Russia. On June 2nd, with the support of Prussia, Austria demanded that Russia withdraw from Wallachia and Moldavia. Czar Nicholas was enraged by this "betrayal" but he had to consider the demand with a cool head as the Austrian Army had mobilized into position along the northern Wallachian frontier. With all the other major European powers against him, Czar Nicholas capitulated and ordered his Balkan army back to Russia. Two weeks later the Porte agreed to joint Austrian-Ottoman occupation of Wallachia and Moldavia to protect the principalities from future Russian aggression. By the end of July, the Russians were out of the Balkans.[58]

With Ottoman Europe no longer threatened by a Russian army, and the Russian Black Sea Fleet confined to its ports by the Allied combined fleet, the Anglo-French expeditionary force lost its *raison d'etre*. The war between the Ottoman and Russian Empires had become a border clash in the Caucasus, far from both British and French areas of geopolitical interest. But Anglo-French armed support on the ground for "the sick man of Europe" did not end in the Summer of 1854.

Invasion of Crimea

National policies required the Anglo-French armed forces to inflict a significant punishment on the Czar to deter future Russian aggression in the Black Sea region. The Allied priority now changed from the defense of Constantinople to an attack on Sevastopol. Though British and French ships had complete control of the Black Sea, Allied planners viewed the large Russian naval facility as a long-term threat to the political *status quo* in the region. Its seizure and destruction would weaken Russian regional strength for years to come and make clear to Czar Nicholas that future action against the Sultan's empire, without Great Power consent, would be an undertaking of great peril. Targeting of Sevastopol by the Allies was not a new development. After the naval battle at Sinope the previous year, British government and military leaders viewed Sevastopol as a high priority objective. In March, James Graham, First Lord of the Admiralty, began promoting within the government a British attack on the Russian base, while in April, Henry Pelham-Clinton (Duke of

---

[57] Spiers, *The Army and Society,* 99-100; Russell, *The British Expedition to the Crimea,* 6, 21. By the end of the year the Commissariat would have 31 officers attached to the Eastern Army. *Hart's New Annual Army List, 1855.*

[58] Royle, *Crimea,* 65, 170.

Newcastle), Secretary of State for War and the Colonies, directed Raglan to determine Russian strength in the Crimea and evaluate the vulnerability of Sevastopol.[59]

All-in-all, the British government viewed their combined Black Sea fleet (Ottoman, French, and British vessels) and the Anglo-French Expeditionary Force, as offensive weapons forged to thwart Russian expansion. The British also noted that Allied troops placed in Crimea could be supplied easily from Constantinople now that the Russian Fleet was confined to its bases. Sevastopol had become an irresistible target. France too was eager to attack Sevastopol. Napoleon III saw an opportunity for a cheap victory in that he believed the European navies could simply bombard the Russian fortress into submission. On June 28th, the British government, supported by the French, ordered Raglan to prepare to seize Sevastopol and to execute his plan unless he should decide, based on new intelligence, that the mission "… could not be undertaken with a reasonable prospect of success." After receipt of this order Raglan and his staff met with their French counterparts, and the British and French naval staffs, to plan the amphibious assault. The Ottoman commanders were uninvited and accordingly took no part in the planning.[60]

While the Allied governments formulated grand strategy, severe problems developed for the troops at Varna. In June, cholera had broken out amongst the French and by the end of July had spread to the other armies. By August, British deaths from disease averaged 15 to 16 per day (an annualized mortality rate of 20%). On August 10th a major fire destroyed a quarter of the town of Varna and consumed a significant amount of army stores. By the end of the month, cholera had spread to the ships anchored in Varna Bay. Compounding the problem of cholera was that a significant portion of the troops suffered from diarrhea. When September began, about 5,000 of the 27,000 British soldiers were unfit for duty due to illness (a number equal to the June reinforcements). The French and Ottoman contingents experienced similar health problems. The Allied army in Varna was wasting away.[61] The Allied commanders were unperturbed by the sorry state of their army. Afterall, disease, exposure to the elements, and malnourishment were always the causes of most wartime casualties. In the undeveloped Balkans, with little modern 1850s infrastructure, an ineffectives rate of 19% was not an extraordinary occurrence. Besides, they would soon leave the pestilence of Bulgaria for glory in Crimea.

On August 20th, the final plans for the invasion were in place. The invasion force, 24,000 French, 23,000 British (including 1,000 cavalry troopers with their horses), and 5,000 Turkish and Egyptian, would depart on September 2nd. Plans for loading the ships were made and their captains were told to prepare to sail northeast; however, a landing site at the destination had not been decided upon. Incredibly, that decision was to be made after the amphibious force departed Varna.[62]

The Allies began their departure as planned. On September 5th, Raglan, in accordance with a prior Allied arrangement, went to Saint-Arnaud's headquarter to thrash out the

---

[59] Ibid., 183; Badem, *The Ottoman Crimean War*, 195, 245.

[60] Royle, *Crimea*, 183-86.

[61] Russell, *The British Expedition to the Crimea*, 57-65.

[62] Royle, *Crimea*, 192-94.

campaign's final details only to discover that the French commander had already departed for Crimea. The staffs then set for September 8th a meeting at sea to take place on Saint-Arnaud's command ship. This meeting took place as scheduled; however, the two ground force commanders did not attend. The terminally-ill Saint-Arnaud was prostrate in bed and the one-armed Raglan could not safely make the transfer from his ship to Saint-Arnaud's. Their subordinates met and eventually agreed that they lacked sufficient information about Crimea's terrain, as well as the strength and disposition of the Russian forces, to make an informed decision as to a landing site. Accordingly, they postponed that decision pending a naval reconnaissance. The next day Raglan's command vessel, with French staff officers on board, cruised Crimea's west coast from Sevastopol north to the minor port of Eupatoria. The senior officers present now agreed with Raglan's previous conclusion that Sevastopol was too strong to be taken directly by sea. With concurrence of the French, Raglan selected a landing site at Kalamita Bay approximately midway between Eupatoria and Sevastopol. The plan was that once fully ashore, the Allied Expeditionary Force would quickly march south past Sevastopol and secure supply bases. The British would utilize the minor port of Balaklava. The French, after crossing the Chernaya River, would wheel westward to the inlets of Kamiesh and Kazach where they would establish their supply bases. Should Sevastopol prove weaker than originally thought, the Allies retained the option of a direct attack shortly after landing.[63]

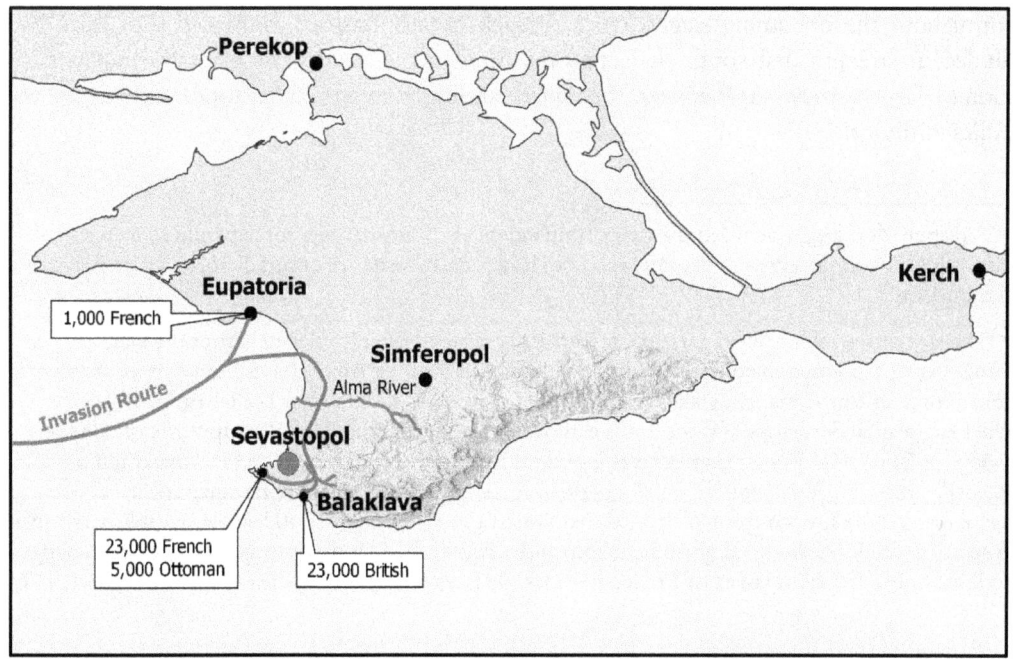

Invasion of Crimea, September 14 – October 14, 1854

---

[63] Ibid., 194-98.

On September 13th seaborne Allied troops occupied undefended Eupatoria. Raglan and Saint-Arnaud planned to hold it as a backup supply base in case the Russians blocked their southern advance.[64] The expeditionary force began to land early the next morning. It took four days for it to disembark and form up for the advance to Sevastopol. While assembling, the Allies were harassed occasionally by Russian cavalry patrols. On September 19, 1854 51,000 French, British, and Ottoman troops (and the French and British wives), with 128 artillery guns, marched south. At most, 1,000 troops remained at Eupatoria.[65] The advancing army was covered, in part, by the guns of the Allied warships that proceeded down the coast in tandem.[66]

Awaiting the Allies were the 69,000 soldiers and sailors commanded by Prince Alexander Menshikov, Commander-in-Chief, Crimea. Subordinate to Menshikov were Admiral Kornilov and General Khomoutov. Kornilov commanded the 12,000 sailors who manned and supported the ships anchored in Sevastopol harbor. Khomoutov commanded 12,000 soldiers spread throughout central and eastern Crimea. Menshikov had personal command of the Sevastopol fortress troops and their supporting mobile ground forces, which together consisted of 39,000 soldiers and 6,000 naval infantry. Artillery for the defense of the port was 88 field guns, 205 fortress guns (145 facing south, 60 facing north), and 530 guns of the anchored warships.[67]

The Russian high command had anticipated an assault on Sevastopol as early as March 1853 and accordingly began improvement of its defenses. In the summer of that year, Menshikov stepped up the pace of such work.[68] This engineering work would continue throughout the upcoming siege. After Cossack cavalry patrols confirmed the Allies had landed in strength just south of Eupatoria, Menshikov decided to place the bulk of his Crimea force somewhere between the Allied landing zone and Sevastopol and engage the Allies in the field.[69]

---

[64] A dispute developed between the two commanders as to how large a force would remain at Eupatoria. A comprise was apparently reached; however, records are contradictory as to the size of the garrison.

[65] The size of the initial Allied garrison at Eupatoria remains open to debate. After the war, Kinglake, who accompanied the Eastern Army, claimed that the Allies did not leave more than a token force in Eupatoria. Kinglake, *Invasion of the Crimea,* vol. 2, 348. Von Todleben, the Russian chief engineer at Sevastopol, wrote in his campaign history that the Allies left upwards of 3,000 troops in Eupatoria. His assessment was based on Russian reconnaissance (which was often faulty). Russell, *General von Todleben's History of the Defence of Sevastopol,* 32. Baron de Bazancourt, who accompanied the French troops, claimed that at first the Allies planned to leave a sizeable garrison at Eupatoria (2,000 Turks, an English battalion, and a French battalion); however, orders were changed to leave only a few companies of French marines. de Bazancourt, *L'Expédition de Crimée,* Vol. 1, 167, 177.

[66] Royle, *Crimea*, 198-211.

[67] Russell, *General von Todleben's History of the Defence of Sevastopol,* 26-27, 70-71; de Todleben, *Defense de Sebastopol,* Vol. 1, appx. 17; Royle, *Crimea,* 189.

[68] Russell, *General von Todleben's History of the Defence of Sevastopol,* 25-27, 70-73; Royle, *Crimea,* 242.

[69] Royle, *Crimea,* 242-43.

Menshikov could have sat tight in the fortified port while Khomoutov, whom Menshikov could have reinforced with part of the Sevastopol garrison, disrupted the Allied attempt to establish supply bases and form protective lines. That is pretty much what the Allied commanders expected of him. Menshikov's decision to engage the Allies before they reached Sevastopol was probably influenced strongly by his vague knowledge of Allied strength. Russian intelligence could be no more precise than to estimate the size of the Allied force as between 50,000 and 100,000 men (both onshore and afloat). Should the upper end of the estimate be correct, then the amphibious force would be strong enough to encircle Sevastopol and Kornilov's fleet. With the supply line through Simferopol to Perecop then cut, the fall of the fortified base would be just a matter of time.[70]

Menshikov marched north with an army of 33,600 troops (28,300 from Sevastopol and 5,300 from Khomoutov's field force, mostly cavalry), and dug in on the heights on the south bank of the Alma River. The army included 3,600 cavalry and 96 artillery guns. Menshikov left 17,300 troops in Sevastopol. Of this reduced garrison, 6,600 were untrained naval infantry formed from the crews of Russian ships sunk to block the harbor's entrance. Eight battalions of mostly army reservists accounted for another 7,300 of the garrison's troops. The balance was of the standing army.[71] Menshikov planned to stop the Allied force 40 kilometers from its reserve base of Eupatoria. If his Alma line held, the Allies would have to either retreat to Eupatoria and dig in for the winter, or board their ships and quit the campaign.

On September 20, 1854 the Allied Expeditionary Force attacked the Russians along the Alma River. For the British, and French, it was their first land battle with a European power since Waterloo, nearly 40 years earlier. The Allies over-powered the Russians and forced their withdrawal. In reports to their respective governments, Menshikov admitted to 5,700 total casualties, Saint-Arnaud 1,300, and Raglan 1,500.[72] The actual losses were much higher. The dead alone, those killed in action and those who succumbed to their wounds, was about 5,000 in total for all combatants.[73] With medical care at this stage of the war inadequate for all sides (non-existent for the Ottoman troops), nearly half the wounded died.[74] Accordingly, the aggregate combat casualty rate for the Battle of Alma, dead and wounded, was about 10 to 11% of the troops deployed. Menshikov withdrew most of his force eastward, the balance he sent to Sevastopol. He planned to await reinforcements for his field force, then attack the Allies' exposed eastern flank as they besieged Sevastopol.[75]

The Allies continued their advance and according to plan, the British entered the undefended port town of Balaklava, southeast of Sevastopol, where they established a supply

---

[70] Ibid., 188-89, 206.

[71] de Todleben, *Defense de Sebastopol*, Vol 1, appxs. 10, 13.

[72] Report from Saint-Arnaud to Napoleon III, September 21, 1854. de Bazancourt, *L'Expédition de Crimée*, Vol. 1, 358-61; Russell, *General von Todleben's History of the Defence of Sevastopol*, 64.

[73] Royle, *Crimea*, 231.

[74] Ibid., 246-60.

[75] Ibid., 262.

The Crimean War

base. The French established their forward supply depots at the bays of Kamiesh and Kazach, west of Sevastopol. The French would supply the Ottoman contingent. The Allies formed a siege line opposite the Russian fortifications and prepared defensive positions to guard their exposed eastern flank. By mid-October allied ships had transported a total of 67,000 troops to Crimea: 33,000 British, 27,000 French, and 7,000 Ottoman. This was the force with which the Allies hoped to take Sevastopol.[76]

Twice the Russians attacked the Allied eastern flank, first near Balaklava in the south (October 25th), then near Inkerman in the north (November 5th). The Allied line held. British casualties at Balaklava totaled 500, including 107 killed in the notorious "Charge of the Light Brigade" while the Russians incurred losses of 1,700.[77] The much bloodier Battle of Inkerman resulted in about 14,500 casualties amongst all sides: 11,000 Russian, 2,500 British, and 1,000 French.[78]

The Siege of Sevastopol then began. It lasted until September 9, 1855 when the Russians abandoned the town and established new defensive positions on the northern bank of the harbor and the Chernaya River. During the siege, the Kingdom of Piedmont and Sardinia declared war on Russia and in January 1855 sent 15,000 Italian troops to join the British, French, Turkish, and Egyptian soldiers in the trenches before Sevastopol. The King, Victor Emanuel II of the Savoy Dynasty, hoped by such measure to receive French support for his quest to become sovereign of a unified Italian state.[79]

Misery and Death at Sevastopol

When the Allies departed Varna, they did not leave cholera behind them in Bulgaria. During the voyage to Crimea, 150 British soldiers died of the disease and a further 300 became too ill to land on shore. By the time the Allied force established its position opposite Sevastopol, deaths from cholera had equaled the number of combat deaths from the Battle of Alma.[80]

Cholera was not the only health problem that the soldiers had to deal with upon arrival in Crimea. Exposure to the elements soon had to be endured by the Eastern Army. The British Army in the field slept in large, heavy, multi-man tents that required animal transport for movement. These tents were not landed at Kalamita Bay until the day before the army

---

[76] de Bazancourt, *L'Expédition de Crimée*, Vol. 1, 193.

[77] Russell, *The British Expedition to the Crimea*, 161-62.

[78] Royle, *Crimea*, 289-90.

[79] Ibid., 324. The House of Savoy would prove to be the longest-reigning European dynasty. It was established in 1003 when a mercenary captain, known as Humbert the White-Handed, was created Umberto I, Count of Savoy, by his employer, Holy Roman Emperor Conrad II. The dynasty reigned first as counts, then dukes, and finally kings until 1946, when Italy voted 54% to 46% in favor of a republic. See, Robert Katz, *The Fall of the House of Savoy* (New York: MacMillan, 1971).

[80] Russell, *The British Expedition to the Crimea*, 89, 139.

marched south. For three days the British slept in the rain, while the French slept in their individual "scraps of tents." For the advance to Sevastopol, the British tents were put back on the ships as there was insufficient land transport to move them with the advancing soldiers. The British soldiers again slept in the open until their base was established at Balaklava and the tents, once again, were unloaded from the ships.[81]

The British soldiers during their first months in Crimea were weakened physically by inadequate rations, overwork, and exposure to the elements. Food supply in Varna had been adequate as there was plenty available in Bulgaria. Additionally, with Varna's large port and serviceable roads to the encampments, it was a simple matter for the Commissariat to provide food purchased in Constantinople, Salonika, and Smyrna. By law, the Commissariat only provided the basic meat and bread ration, for which the soldiers were charged the regulation 3.5d. per day. At Lord Raglan's request, the Commissariat in Varna also supplied coffee, sugar, and rice to the soldiers at cost, 1d. per day.[82] Vegetables and other food items were obtained locally by each regiment, the cost of which was charged to the soldiers. Problems feeding the troops developed in Crimea after depletion of the edible stores carried with the amphibious force. Deficiencies in both quantity and quality of food were due to lack of local supply and transport problems. Inability to purchase food in Crimea was a serious problem as the British Army, even on campaign, sourced its food supply locally. The Commissariat obtained food from Constantinople but its delivery from the ships at Balaklava to the regiments in the field was hampered by poor roads (dirt tracks that turned to mud in the Autumn rainy season) and an insufficient number of draft animals (most of the mules, horses, and oxen brought from Varna had died from disease, over-exertion, and insufficient forage).[83] Despite continual arrivals of draft animals, their number didn't increase as 100 died each week.[84] On January 8, 1855, the Eastern Army's Quartermaster-General complained to the Commissary-General that "... almost half this army is employed in fatigue parties, carrying up provisions, warm clothing, blankets, planking and every description of stores, which service should properly be performed by the Commissariat Department."[85] Such provisions and stores were man-handled with great difficulty and these fatigue assignments took a great physical toll on the soldiers as it sometimes took twelve hours to bring supplies up the muddy tracks from the harbor to the British lines.[86]

---

[81] Ibid., 89-92.

[82] *Report of the Commission of Inquiry into the Supplies of the British Army in the Crimea*, 1856, [422-1], at 45-46. Hereafter cited *McNeill-Tulloch Report*.

[83] By January 16, 1855, the entire transport establishment for the Eastern Army of 31,000 troops consisted of 125 carts, and 345 effective draft animals (including 12 camels). This works out to 2 carts with 5 draft animals per 560 men (the approximate strength of a regiment in Crimea). *McNeill-Tulloch Report*, at 66.

[84] Russell, *The British Expedition to the Crimea*, 201.

[85] *McNeill-Tulloch Report, Appendix*, at 63.

[86] *McNeill-Tulloch Report*, at 16.

The Crimean War

Food for the troops was inadequate in both quantity and quality. For the Winter of 1854-55, the British troops subsisted on what were essentially emergency sea rations: salted, dried meat and dry biscuits (a substitute for soft bread). No portable ovens were brought to Crimea so there was no fresh bread, and the Commissariat was unable to provide enough fresh meat to meet the regulation requirement of one pound per day for each soldier. For the five months ended March 31, 1855, fresh meat deliveries accounted for only one-third the daily requirement. The shortfall was usually, but not always, made up for with salted, dried meat shipped from Britain.[87] Also, in Crimea, the Commissariat did not provide the ration supplement of coffee, sugar, and rice as it did at Varna. Neither did it provide vegetables and other food items; such provisions were not in the Commissariat's statutory remit.[88]

A further hindrance to prompt delivery of provisions to the front, was the attitude of the Commissariat officers. As guardians of the public purse, they were sticklers for regulatory compliance. Additionally, they were trained to follow established procedures and not to act on their initiative. They refused to move outside the Treasury regulations, many of which dated from the Napoleonic Wars.[89] A seemingly mindless adherence to rules resulted in tragic absurdities. For example, the British Queen's son, Prince Albert, wrote the following to the newly appointed War Minister:[90]

> "It so happens that one of the Crimean Relief Societies sent out a whole shipful of vegetables. On its arrival at Constantinople, the man in charge of it reported himself to the Commissary (I believe Smith, reported to be our best), who was delighted to hear of the arrival of provisions; when he saw the list, however, and found they were vegetables, he declined purchasing 'as the Commissariat had no power to purchase vegetables'!!"

William Howard Russell, *The Times* war correspondent, presented in his book on the Crimea campaign a similar episode. He recounted the following ludicrous exchange between a surgeon and a commissary officer concerning the surgeon's request for two or three heating stoves for the wards of a hospital ship:[91]

Surgeon: Three of my men died last night from choleraic symptoms, brought on from the extreme cold, and I fear more will follow.

---

[87] Ibid., *Appendix*, at 55.

[88] *McNeill-Tulloch Report*, at 8.

[89] Funnell, *Accounting at War*, 17-23.

[90] Letter from Prince Albert to Lord Panmure, February 10, 1855. Douglas, *The Panmure Papers*, 54. Fox Maule Ramsay (2nd Baron Panmure), was appointed Secretary of State for War February 8, 1855.

[91] Russell, *The British Expedition to the Crimea*, 208.

Commissary:	Oh, you must make your requisition in due form, send it up to head-quarters, and get it signed properly, and returned, and then I will let you have the stoves.
Surgeon:	But my men may die the meantime.
Commissary:	I can't help that; I must have the requisition.
Surgeon:	It is my firm belief that there are men now in a dangerous state whom another night's cold will certainly kill.
Commissary:	I really can do nothing; I must have a requisition properly signed before I can give one of these stoves away.
Surgeon:	For God's sake, then, lend me some; I'll be responsible for their safety.
Commissary:	I really can do nothing of the kind.

Early in 1854, Westminster expected that British Army intervention in the war between the Russian and Ottoman Empires would be limited to a single, short, colonial-style campaign. War planning by the generals proceeded under such assumption. Accordingly, the British had dispatched their soldiers with only summer uniforms, the regulation one blanket per man, lightweight tents, and no heating stoves with fuel supply.[92]

In September, the Eastern Army had encamped at Balaklava unprepared for the soon to arrive winter. Crimean winters are not particularly severe. At an average temperature of 2.3°C (36.1°F) they're only slightly colder than those of London 3.8°C (38.9°F).[93] But such temperatures mandate heating of quarters and wearing of heavy clothing. At the beginning of December, only 22% of the troops had received an extra blanket. It would not be until mid-January 1855 before all the troops were so supplied. In mid-December only 8% of the soldiers had received overcoats. By the end of March, that figure would rise to only 28%.[94]

At the end of October, simply from lack of adequate food and shelter, plus overwork, of the 35,600 British troops of the Eastern Army, only about 17,000 troops opposite Sevastopol were fit for service.[95] Greatly compounding the Allies' problem of survival was the massive storm of November 14th. It sank 21 supply ships in the Crimean ports including a British

---

[92] Royle, *Crimea*, 113, 139-40, 298; *McNeill-Tulloch Report*, at 23-25. Stoves began arriving in December in which month 656 were delivered. By the end of January, 2,011 stoves had been received at Balaklava. Return of Receipts, *McNeill-Tulloch Report*, at 94-96.

[93] *McNeill-Tulloch Report, Appendix,* at 187.

[94] Ibid., at 89.

[95] Ibid., at 151.

transport that had arrived with winter uniforms and boots. The high winds not only sank ships but damaged or destroyed most of the British Army's thin-canvassed, summer tents.[96]

Both officers and other ranks went without adequate shelter until the arrival of enough wood planking for scratch construction of huts. Planking arrived in eleven shipments from November 25th through March 13th. Sufficient wood was received to construct huts to house 10,000 men and 3,500 horses. In November, Lord Raglan ordered enough assembleable huts to house 1,500 men, and the first of these arrived from England on December 25th. Further deliveries were received at Balaklava throughout the winter. The hut components were large and heavy; it took 250 to 300 men to carry them from the docks to the encampments. Accordingly, Raglan's staff decided to assemble the huts near the harbor and use them as warehouses, offices, and housing for the contracted civilian laborers (who began to arrive in December).[97]

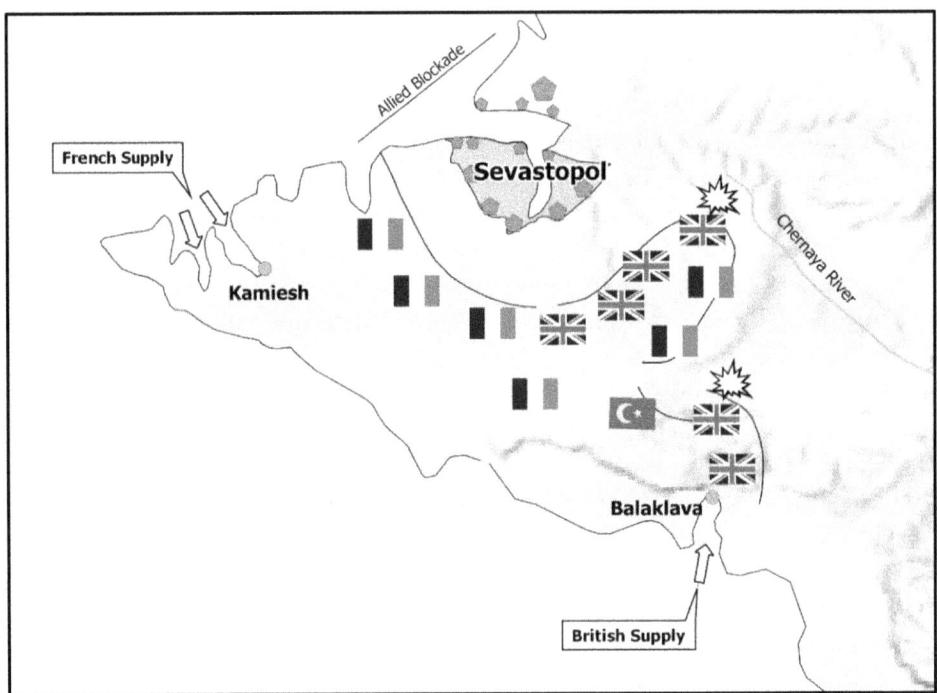

Sevastopol: Allied Positions on November 1, 1854

The Eastern Army denuded the countryside of firewood and by December the problem of insufficient fuel for heating and cooking developed. Charcoal and firewood were received at the harbor of Balaklava but there was no means to transport it efficiently to the

---

[96] Royle, *Crimea*, 266-67.

[97] *McNeill-Tulloch Report*, at 33-36.

encampments. Regiments received whatever firewood and charcoal their troops could carry from the harbor.[98]

Wounded and ill British soldiers in Crimea received immediate medical care at aid stations termed "regimental hospitals." The serious cases were evacuated to army hospitals established in and around Constantinople. The largest such hospital was at the British theatre headquarters in Scutari. The Scutari Army Hospital became infamous through the writings of Thomas Chenery, *The Times* correspondent in Constantinople, and Florence Nightingale, leader of a newly-formed, civilian nursing team that arrived there November 4, 1854. The British public first learned of the egregious state of army medical care through Chenery's news account of October 12, 1854.[99]

> "Not only are there not sufficient surgeons – that, it might be urged, was unavoidable – not only are there no dressers and nurses – that might be a defect of system for which no one is to blame – but what will be said when it is known that there is not even linen to make bandages for the wounded? The greatest commiseration prevails for the unhappy inmates of Scutari, and every family is giving sheets and old garments to supply their want. But, why could not this clearly foreseen event have been supplied?"

Despite engagement of civilian physicians and surgeons, the British Army was woefully short of medical officers in both hospitals and regimental establishments. As for support staff, the British Army had no professional nurses, trained surgical assistants, or paraprofessional medical attendants and emergency technicians. The wounded and ill were simply carried by the regimental musicians to the regimental aid stations where they were treated by surgeons assisted by ordinary soldiers assigned to care for the casualties. Accommodation at field and rear area hospitals was rudimentary and disgusting. Not even the most basic sanitary precautions were taken regarding disposal of corpses and human waste. Additionally, there were insufficient quantities of medicines, surgical materials, bandages, and clothing for the patients.[100] The War Minister, Lord Newcastle, in response to numerous alarming reports from army officers and newspaper correspondents, on October 24, 1854, appointed three commissioners to investigate "the state of the hospitals of the British Army in the Crimea and Scutari."[101]

By mid-December it became apparent to the British soldiers, their officers in the field, and most importantly the war correspondents of the British press, that the army's support apparatus was unable to adequately feed, clothe, shelter, and provide medical care for a large field force deployed in a hostile environment.

---

[98] Ibid., 11, 16; *Evidence, McNeill-Tulloch Report*, at 95.

[99] Royle, *Crimea*, 246-47.

[100] Ibid., 246-56.

[101] The appointees were two physicians, Alexander Cummings and Thomas Spence, and a lawyer, Peter Benson Maxwell. Their findings would be reported to the War Office on February 23, 1855. *Sanitary Report*.

Frances (Fanny) Isabella Duberly was one of the officers' wives who went to Crimea with her husband, Henry Duberly, Paymaster of the 8th Regiment of Light Dragoons - The King's Royal Irish Hussars. She kept a journal, published in 1855, which became a many time reprinted, classic account of the Eastern Army in Varna and Crimea. Her entry for Sunday, December 3, 1854, provides a vivid description of the British encampment at Balaklava in the Winter of 1854-55.[102]

> "If any body should ever wish to erect a 'Model Balaklava' in England, I will tell him the ingredients necessary. Take a village of ruined houses and hovels in the extremest state of all imaginable dirt; allow the rain to pour into and outside them, until the whole place is a swamp of filth ancle-deep; catch about, on an average, 1000 sick Turks with the plague, and cram them into the houses indiscriminately; kill about 100 a-day, and bury them so as to be scarcely covered with earth, leaving them to rot at leisure – taking care to keep up the supply. On to one part of the beach drive all the exhausted *bât* ponies, dying bullocks, and worn-out camels, and leave them to die of starvation. They will generally do so in about three days, when they will soon begin to rot, and smell accordingly. Collect together from the water of the harbour all the offal of the animals slaughtered for the use of the occupants of above 100 ships, to say nothing of the inhabitants of the town, – which, together with an occasional floating human body, whole or in parts, and the driftwood of the wrecks, pretty well covers the water – and stew them all up together in a narrow harbour, and you will have a tolerable imitation of the real essence of Balaklava."

Throughout the latter months of 1854, British newspapers ran stories of the appalling conditions endured by soldiers of the Eastern Army. Leading the press in this regard was *The Times*. Not only did it publish the uncensored reports of its Crimean correspondent, William Howard Russell, but it printed letters from officers at the front that confirmed Russell's findings. The middle class public was outraged over the government's management of the war. In January 1855, John Arthur Roebuck, a backbench, maverick Member of Parliament, called for a committee to investigate the conduct of the war. The government quickly condemned Roebuck and opposed formation of such a committee. On January 28, 1855, Commons approved Roebuck's call for a special committee by vote of 305 to 148. The government then resigned. Its replacement was headed by Henry John Temple (Viscount Palmerston) with Fox Maule-Ramsay (Baron Panmure) as Secretary of State for War.

By mid-winter, British deaths from disease had become appallingly high. In January 1855 nearly 10% of all troops in the theatre died from causes other than wounds and injury

---

[102] Duberly, *Journal Kept During the Russian War*, 143-45.

(annualized rate of 114%). As shown in the following table, the probability that a British soldier could survive for a year in Crimea had become no better than 25%.[103]

Table 2.
Annualized Death Rate by Cause, British Eastern Army
(Percent of Average Monthly Strength)

Month	Army Status	Wounds & Injury	Other Causes
July, 1854	At Varna	--	16
August	At Varna	--	34
September	Move to Crimea	3	34
October	Battles in Crimea	5	25
November	Recovery from Battles	12	39
December	Siege of Sevastopol	4	68
January, 1855	Siege of Sevastopol	3	114

Source: See note 104.

For the French soldier, data indicates his survival chances were about twice as good as that of his British counterpart. Of the 89,855 French troops that spent the Winter of 1854-55 in Crimea, 12.2% died; the comparable British mortality rate was 23.0%.[104] The French, with their army's *Intendance Militaire*, (logistics and medical organization of uniformed civilians subject to military discipline) were much better prepared than the British to deal with casualties. In the field, each French army division of about 8,000 men was provided with 46 doctors, 104 professional nurses, and 7 administrative officers. Additionally, each division had a field evacuation establishment of trained medical aides and specially-built ambulance wagons.[105] Furthermore, unlike the British, the French had assembled wooden barracks prior to the onset of winter.[106]

For the Ottoman soldier, a posting to Crimea was practically a death sentence. The Sultan's army made almost no arrangements for medical care of its soldiers.[107]

On February 19, 1855 Lord Panmure dispatched two Scotsmen, John McNeill and Colonel Alexander Murray Tulloch to investigate "the whole arrangement and management of the Commissariat Department" at Crimea and Constantinople. McNeill, a retired British East India Company surgeon, was Chairman of the Board of Supervision for administration

---

[103] Derived by averaging the annualized total death rate for September 1854 (37%), with that for January 1855 (117%). *Sanitary Report*, at 524.

[104] McDonald, "Florence Nightingale, Statistics and the Crimean War." McDonald relied on data compiled by Jean-Charles Chenu and published in *De la mortalité dans l'armée et des moyens d'économiser la vie humaine, extraits des statistiques médico-chirurgicales,* (Paris: Hachette, 1870).

[105] Royle, *Crimea*, 251, 258.

[106] Curtiss, *The Army of Nicholas I*, 338.

[107] Ibid., 260; Russell, *Expedition to the Crimea*, 145, 192, 201.

of the Scottish Poor Law Act, while Tulloch, an officer on half-pay, was the War Office's Military Superintendent of Outpensioners (retired soldiers not resident in the army's nursing homes in London and Dublin, the Royal "Hospitals" at Chelsea and Kilmainham).[108]

In June 1855, McNeill and Tulloch reported their findings to Parliament. With respect to the mortality rates of the Eastern Army, they wrote

> "… it seems to be clearly established that this excessive mortality is not to be attributed to anything peculiarly unfavorable in the climate, but to overwork, exposure to wet and cold, improper food, insufficient clothing during part of the winter, and insufficient shelter from inclement weather."[109]

This was official confirmation of the public's belief that the army's administration could not provide properly for its soldiers.

During 1855 the new British government made several administrative and operational changes to the army. Palmerston abolished the position of Secretary-at-War and merged its bureaucracy into the War Department under the Secretary of State for War. His government moved the Commissariat from the Treasury to the War Office and its officers were given some, but not much, discretionary power.[110] The War Minister, Panmure, created in Crimea the Land Transport Corps (LTC); a military organization commanded by a colonel and responsible for movement of all army supplies. Its junior officers were enlisted men newly commissioned as cornets (lieutenants) and quartermasters of brigades (captains). Its senior officers were combatant officers from the British and EIC armies.[111] The actual handling of the goods was done by civilian employees, a measure that freed thousands of British soldiers to fight the Russians.[112] To keep watch over Raglan, the government sent out from England Lieutenant-General James Simpson to fill the new position of Chief-of-Staff, Eastern Army. Though subordinate to Raglan, Simpson was required to report regularly to Westminster.[113] Raglan died of dysentery on June 28th and the government made Simpson his replacement.[114]

The army cleaned up the hospitals, provided better nursing care, and established a regimen of field sanitation.[115] The government also purchased and sent to Constantinople, a

---

[108] Royle, *Crimea*, 332; *Hart's New Annual Army List, 1856*; Blackden, "The Board of Supervision and the Scottish Parochial Medical Service."

[109] *McNeill-Tulloch Report*, at 3.

[110] Royle, *Crimea*, 326-34.

[111] *Hart's New Annual Army List, 1856*.

[112] Royle, *Crimea*, 404-05.

[113] Ibid., 331.

[114] Simpson did not last long in command. Shortly after his appointment, he developed clinical depression and in November was convinced by Panmure to resign. To replace Simpson, the government selected Major-General William Codrington, then one of the Eastern Army's divisional commanders. Ibid., 402-05.

[115] *Sanitary Report*, at xxxi.

modern, assembleable hospital designed by the renowned English engineer, Isambard Kingdom Brunel.[116] To solve the problem of transport from Balaklava harbor to the troop positions overlooking Sevastopol, the British government funded construction of a railroad. It became operational in March 1855.[117] With the railroad, experience gained by the logistics officers, and simply the passage of time, the supply bottleneck ended.[118]

Improvements in logistics, sanitation, and healthcare paid big dividends for the British soldiers of the Eastern Army. During its second winter in Crimea, the army's overall mortality rate was 2.5%, about one-tenth of the 23.0% mortality rate for the previous winter.[119] That winter's worst month for casualties was November 1855 when the annualized non-traumatic death rate was 7%. This represented a startling improvement over the Winter of 1854-55. The worst month of that winter was January 1855 for which the annualized, non-traumatic death rate was 114%.[120]

During the Allies' second winter in Crimea the mortality rate of the French soared. Of the 106,634 French soldiers of the Army of the East, 19.9% perished, a rate nearly twice as high as that for the first winter.[121] Unlike the British, the French had made no significant changes to their army's way of doing business. The French leadership simply saw no need to do so. There was no public outcry for change as the public did not know the conditions of the French Army of the East. (Throughout the war Napoleon III maintained strict censorship of all press reports on the French Army.)[122] Troop losses were of no great concern to French ministers and senior generals as conscription produced enough young men to provide the necessary replacements.

The Wider War

The "Crimean" War, during which the British public and government focused their attention first on the Balkans, then on the death-trap of Sevastopol, encompassed several engagements and campaigns elsewhere. Battles were fought throughout the Black Sea region, on the Baltic Sea, in the Caucasus and on the Kamchatka Peninsula in the Russian Far East. The British Army was not involved in the Baltic Sea and Russian Far East operations, nor was it deployed to the Caucasus.

---

[116] Royle, *Crimea*, 257.

[117] Ibid.

[118] Russell, *Expedition to the Crimea*, 237-40.

[119] McDonald, "Florence Nightingale, Statistics and the Crimean War."

[120] *Sanitary Report*, at 524.

[121] McDonald, "Florence Nightingale, Statistics and the Crimean War."

[122] Royle, *Crimea*, 333.

The Crimean War

The Battle of Eupatoria [123]

In December 1854, the Sultan agreed to an Anglo-French request for him to transfer troops from the quiet Balkan theatre to Allied-held Eupatoria, north of Sevastopol. A large force there would pose a major threat to the rear of the Russian forces at Sevastopol. Additionally, such force, if reinforced with French and British troops, would give the Allies a viable option to attack Perecop at the land entrance to the Crimean Peninsula. Should Perecop fall to the Allies, the Russians in Crimea, cut off from supply, would have to surrender.

The Ottoman troop movement took three months to complete. By the end of February 1855, the Ottoman Eupatoria force, commanded by Omar Pasha (the former Balkan commander) totaled 35,000 men, of which about one-third were Egyptian.

By early January, the Ottoman troop arrivals in Eupatoria had been observed by the Russian cavalry that screened Eupatoria. Menshikov reported this dangerous development to Czar Nicholas. The Czar, who believed the Allies planned to march on Perecop, ordered Menshikov to launch a pre-emptive attack. The battle took place on February 17, 1855. The attacking Russian force of 19,000 men was repulsed by Omar Pasha's army with minor assistance from the recently reinforced, but still small, French garrison, and major assistance from the guns of three Allied steamers anchored nearby. Russian casualties totaled about 800, while Ottoman casualties, mostly Egyptian, 400, and French casualties 13.

After the unsuccessful Russian attack, the Czar replaced Menshikov, Commander-in-Chief Crimea, with Mikhail Gorchakov, whom he had sacked previously as the Balkan theatre commander. This change in command was one of the last official acts of Nicholas I. He died of pneumonia on March 2, 1855. Nicholas was succeeded by Alexander II.

Allied Seizure of Kerch and the Sea of Azov Naval Bombardments [124]

In April 1855, the Allied command decided to seize Kerch, located at the eastern extremity of Crimea. They assembled an amphibious force of 15,000 troops (7,000 French, 5,000 Ottoman, 3,000 British), which arrived at the Kerch Straits on May 24th. There, the Allies discovered that the Russian Kerch garrison of 9,000 troops had withdrawn to the west, while the small Russian naval force normally anchored there, had fled into the Sea of Azov. After disembarkment of the soldiers, the Anglo-French naval force followed the Russian naval force into the Sea of Azov where it engaged and destroyed the Russian Kerch flotilla.

For about three weeks, Allied warships cruised the Sea of Azov, bombarding many minor, Russian ports, and sinking numerous, small, merchant vessels. The Allied force of 14 steam-powered escort vessels lacked the firepower to undertake a bombardment of the fortified

---

[123] Sources: Badem, *The Ottoman Crimean War*, 278-80, Royle, *Crimea*, 335-36, Russell, *General von Todleben's History of the Defence of Sevastopol*, 247-48.

[124] Source: Kinglake, *The Invasion of the Crimea*, Vol. 9, 38-86.

Russian port of Azov, which guarded the approach to the large city of Rostov, upstream on the Don River.[125]

In early June, Allied warships withdrew from the Sea of Azov and the eastern Black Sea taking with them most of the ground troops then at Kerch. The Allies left behind a garrison of the entire Ottoman contingent (5,000 men) and 1,000 of the French soldiers.

Allied Seizure of Kinburn Spit [126]

On Oct 8, 1855 an Anglo-French amphibious force of 90 vessels (warships, transports, and tenders), carrying 9,000 ground troops, assembled some distance from the confluence of the Dnepr and Southern Bug Rivers at the Black Sea. There, a narrow strip of land runs into the sea from the east bank of the Dnepr to form a gulf. This strip of land is the Kinburn Spit and on it the Russians had a fort which the Allies intended to seize.

The amphibious force was more powerful than necessary to subdue the Russians. The task force included ten steam-powered battleships and for the French three, newly built, ironclad, steam-powered escorts. Participants in the upcoming engagement would witness the first combat of armored, steam-powered, warships.

The bombardment began on October 14th and the next day the ground troops landed. On October 17th, the Russian commander surrendered the fortress and its 1,400 troops. The defenders suffered 45 dead and 130 wounded.

Resolution of The War and British Army Reorganization

Russian Evacuation of Sevastopol

On August 16, 1855 the anti-climax of the Crimean War began, but at the time few, if any, were aware of it. By mid-summer, the Russians were losing the war of attrition at Sevastopol. Troop replacements were hard to come by and the Russian engineers had fallen behind in repairs to fortifications damaged by Allied artillery fire. Gorchakov decided to go on the offensive to break the siege. The Russian attack failed, and Gorchakov's troops took 8,000 casualties; the Allies suffered only 2,000 dead and wounded.[127]

With the Russians now seriously weakened, the Allied commanders ordered "the big push" to take Sevastopol. On August 17th the Allies began the most intensive bombardment of Sevastopol to date. The shelling continued until September 8th when the Allied infantry assaulted the Russian fortifications. The British objective was the bastion known as the Great Redan, located at the southeastern approach to Sevastopol. The French objective was the Malakov, the bastion that controlled the eastern approach to the town. The British attack

---

[125] Slade, *Turkey and the Crimean War*, 394.

[126] Source: Clowes, *The Royal Navy,* Vol. 6, 470-72.

[127] Royle, *Crimea*, 407-09.

The Crimean War

failed, but the French took the Malakov and breached Sevastopol's defensive ring. At the end of the day, the Russians counted 13,000 dead and wounded, the Allies 10,000.[128]

The Russian position in Sevastopol was now untenable. The following day Gorchakov ordered a withdrawal to the fortified northern bank of the harbor. That night, the defenders of Sevastopol evacuated the town over a previously constructed pontoon bridge. The only troops left behind were the worst of the hospital cases. On September 9, 1855 Allied troops entered Sevastopol.[129] Quiet descended on the Crimea as the combatants on both sides, from the generals down to the privates, sensed that the war was effectively over. As the weeks passed with no official news of an armistice, anxiety set in among the troops. This feeling of uneasiness heightened when on January 29, 1856, the Russians shelled the Allied positions in an intensive bombardment. This artillery action proved to be the last warlike act in the theatre. On February 28, 1856, in Paris, representative of the belligerents signed an armistice agreement.

Cost of the War

The war was expensive, both in terms of lives and money. The number of Russian and Ottoman casualties remains open to debate. Estimates for Russian war dead, from all causes on all fronts, range from 450,000 to 630,000.[130] In Crimea, the Russians incurred about 100,000 combat casualties (killed, wounded, and missing).[131] Far more died there of disease. For the Ottoman Army, war dead estimates range from 35,000 to 175,000. The wide range is because the Ottoman War Ministry simply did not keep such records.[132] The western allies maintained and published what historians accept as accurate casualty records. The number of military deaths, shown in Table 3 on the following page, was far greater than expected by the French and British back in 1853.[133]

---

[128] Ibid, 409-13.

[129] Ibid., 414-15.

[130] The 450,000 figure is from a published account by the Medical Department, Russian Ministry of War. John Shelton, *Russia's Crimean War* (Durham, NC: Duke Univ. Press, 1979), 470. Curtiss considers it accurate as the number coincides with his estimate based on army starting strength, enlistments, and ending strength. The 630,000 number is the estimate of Dr. Jean-Charles Chenu of the French military medical school. Chenu, *Rapport au Conseil de Santé des Armées*.

[131] Curtiss, *The Russian Army under Nicholas I*, 359.

[132] Badem, *The Ottoman Crimean War*, 5-6. Dr. Chenu estimated 35,000 deaths for the Balkans and Crimea and did not address the Caucasus Front. Chenu, *Rapport au Conseil de Santé des Armées*.

[133] Combat deaths are killed-in-action plus subsequent death from wounds. British: War Office, *Return of the Total Number of Officers and Men in the Army who have been Killed in the Crimea*, 1857, H.C. Accounts & Papers, No. 57. French and Sardinian-Piedmontese: Chenu, *Rapport au Conseil de Santé des Armées*, 564, 579, 614, 617.

Table 3.
Deaths from All Causes, Western Allies
(Black Sea Theatre of War)

	Total Troops	Combat Deaths	Other Deaths	Total Deaths	Mortality Rate, %
France	309,000	18,941	76,674	95,615	30.9
U.K.	97,000	4,774	16,041	20,815	21.5
Sardinia-Piedmont	19,000	38	2,156	2,194	11.6
Total	425,000	23,753	94,871	118,62	27.9

Source: See note 133.

French wounded totaled 39,868 (12.9% of participants) of which 31,157 survived. British wounded totaled 11,867 (12.2% of participants) of which 9,848 survived.[134] Accordingly, for the wounded French the survival rate was 78%, for the wounded British the survival rate was 83%.

In Table 3, "Other Deaths" consists almost entirely of men who died of disease. Note that this category accounts for 80% of all deaths. The ratio of four non-combat deaths for every one combat death is about what the British Army experienced in the Napoleonic Wars.[135] The very high proportion of Other Deaths for the Savoyard troops is a result of their arrival in Crimea after the battles of the Alma, Balaklava, and Inkerman had been fought. Also, as untested troops, the Allied command employed few of them in the final assault on Sevastopol.

The financial cost of the war for the United Kingdom was enormous. For the peacetime year 1853, the combined army and navy appropriation was £15,859,202.[136] The demands of war increased military, naval and other expenditures markedly. For 1854, the additional expense was £8.7 million, for 1855, £32.5 million, and for 1856, £25.1 million.[137]

The Treaty of Paris

On March 30, 1856 representatives of the major European powers signed the treaty which officially ended the war. The treaty's 34 articles essentially called for a return to the *status quo ante bellum*. A notable exception was that warships were banned from the Black Sea, which rendered redundant the remnants of Russia's Black Sea Fleet.[138] As for the keys

---

[134] Ibid.

[135] This work, The Napoleonic Wars, 1803-1815 in "History of the British Army, Cromwell to 1853," *supra*.

[136] *Navy Estimates*, 1852-53, H.C. Accounts & Papers, No. 73; *Army Estimates*, 1859 Sess. 2, H.C. Accounts & Papers, No. 88.

[137] Beaulieu, "Crimean War, Chapter II."

[138] Royle, *Crimea*, 482.

to the main door of the Church of the Nativity, the Franciscans were compelled to turn them over to the Greek Orthodox monks.[139]

British Army Reorganization

During the Crimean War advocates of British Army reform became more vocal and garnered further support in Parliament. The main issues were the fractured nature of army control, army dependency on civilians for its transport and supply, purchase of commissions and officer promotion in the cavalry and infantry, inability of other ranks to obtain commissions, promotion by seniority, officer staff appointments resulting from family influence, an aged officer corps, and the general constitution of the army as the private domain of the privileged classes. Early in the war some reform took place, mostly regarding logistics, but by mid-1855, the reformers lost what Parliamentary support they had had.[140] Changes to army administration made by Parliament and the government were as follows:[141]

> The Colonial Office became a separate ministry. The domain of the former Secretary of State for War and the Colonies was limited to military matters.
>
> The duties of Secretary-at-War and Secretary of State for War merged, and the War Department took over the functions of the War Office.
>
> The Board of Ordnance became the Ordnance Branch of the War Department and the administration of the artillery and engineers was placed under the Commander-in-Chief.
>
> The Commissariat moved from Treasury to the War Department and acquired its own transport service (Land Transport Corps, later renamed The Military Train).
>
> Administration of the militia moved from the Home Office to the War Department.
>
> The Army Medical Department, previously an independent state agency, became a component of the War Department.

Command and administration of the army became bifurcated with the Secretary of State for War responsible for support and the civilian-oriented departments, and the Commander-in-Chief, practically autonomous, responsible for administration of uniformed personnel and fortifications. The bureaucracy for the civil side of the army was soon labelled the "War Office" while the bureaucracy for the uniformed side became known as "Horse Guards." By the end of 1857, the principal administrative components of the army were as follows:[142]

---

[139] Cust, *The Status Quo in the Holy Places*, 9.

[140] Spiers, *The Army and Society*, 108-17.

[141] Ibid.; Wheeler, *The War Office Past and Present*, 173-79.

[142] *Hart's New Annual Army List, 1858*.

Secretary of State for War (War Office)

Ordnance Branch
Commissariat Branch
Inspector-General, Militia

Judicial Department
Military Store Department
Barracks Department
Medical Department

Commander-in-Chief (Horse Guards)

Inspector-General of Fortifications
Adjutant-General
Adjutant-General, Royal Artillery

Quartermaster-General
Inspector-General, Cavalry
Inspector-General, Infantry

Chapter Bibliography

Adye, John. *A Review of the Crimean War.* London: Hurst & Blackett, 1860.

Badem, Candan. *The Ottoman Crimean War.* Boston: Brill, 1970.

de Bazancourt, César Lecat. *L'Expédition de Crimée.* Paris: Amyot, 1856.

——— *L'Expédition de Crimée, La Marine Française.* Paris: Amyot, 1858.

Beaulieu, Paul LeRoy. "Crimean War, Chapter II." *Advocate of Peace*, New Series, 1, no. 8 (August 1869): 117-21.

Blackden, Stephanie. "The Board of Supervision and the Scottish Parochial Medical Service." *Medical History* 30, no. 2 (April 1986): 145-72.

Chenu, Jean-Charles, *Rapport au Conseil de Santé des Armées sur les résultats du service médico-chirurgical pendant la campagne d'Orient en 1854-56.* Paris: 1865.

Clowes, Wm. Laird. *The Royal Navy.* Vol. 6, London: Sampson, Low, Marston, 1901.

Cust, L.G.A. with Abdullah Kardus. *The Status Quo in the Holy Places, Report to the High Commissioner of Palestine.* London: HMSO, 1929.

Dodd, George. *Pictorial History of the Russian War.* London: Chambers, 1856.

Douglas, George and George Dalhousie Ramsay, eds. *The Panmure Papers.* London: Hodder & Stoughton, 1908.

Duberly, Frances Isabella. *Journal Kept During the Russian War.* London: Longman, Brown, 1855.

Funnell, Warwick and Michele Chwastiak. *Accounting at War.* New York: Routledge, 2015.

Kinglake, A.W. *Invasion of the Crimea*, 6th ed. Vol. 1, London: Blackwood, 1877.

——— *Invasion of the Crimea*, 6th ed. Vol. 2, London: Blackwood, 1877.

——— *Invasion of the Crimea*, 6th ed. Vol. 9, London: Blackwood, 1901.

Lambert, Andrew, *The Crimean War: British Grand Strategy against Russia.* London: Routledge, 2011.

McDonald, Lynn. "Florence Nightingale, Statistics and the Crimean War." *Journal of the Royal Statistical Society: Series A*, part 3 (2014): 569-86.

Rappaport, Helen. *No Place for Ladies: The Untold Story of Women in the Crimean War.* London: Aurum, 2008.

Royle, Trevor. *Crimea.* New York: Palgrave MacMillan, 2000.

Russell, William Howard. *The British Expedition to the Crimea.* London: Routledge, 1877.

——— *General von Todleben's History of the Defence of Sevastopol.* London: Tinsley, 1865.

Shotwell, James. "A Short History of the Question of Constantinople and the Straits." *International Conciliation*, no. 180 (November 1922). Reprinted in *International Conciliation,* Vol. 8 (Buffalo: Hein, 1997).

Slade, Adolphus. *Turkey and the Crimean War*. London: Smith, Elder, 1867.

Spiers, Edward M. *The Army and Society 1815-1914*. London: Longman, 1980.

Trustram, Myna. *Women of the Regiment*. Cambridge: Cambridge Univ. Press, 1984.

Uyar, Mesut and Edward J. Erickson. *A Military History of the Ottomans*. Santa Barbara, CA: ABC-CLIO, 2009.

Wheeler, Owen. *The War Office Past and Present*. London: Metheun, 1914.

Wirtschafter, Elise Kimerling. "Military Service and the Russian Social Order, 1649-1861." In *Fighting for a Living,* edited by Erik-Jan Zürcher. Amsterdam: Amsterdam Univ. Press, 2013.

Chapter 4
Late Victorian Military Campaigns and Army Reform

The post-Crimean British Army was kept busy with two major wars and a series of colonial campaigns in Asia and Africa. Near the close of the century, in June 1899, the British Army was a splendid looking, peculiar little force governed by uniformed and civilian aristocrats and led by gentlemen officers. It fought well in distant foreign lands (usually against poorly armed, unorganized, and technologically backward opponents), and operated under an archaic, public school, code of conduct. Six months later, the British public learned that its well-drilled army that looked so impressive on parade in its scarlet tunics, was an anachronism. Beginning in October 1899, on the *veld* of southern Africa, it took nearly three years for the entire home-based, professional army (plus tens of thousands of mobilized auxiliaries and volunteers from the dominions), to subdue a motley, enemy army of amateurs in civilian garb, that did not drill or salute, and never fielded at one time more than 35,000 armed men.

The Indian Mutiny

On the heels of the Crimean War came a sequence of violent events in India that the British termed "The Indian Mutiny." In the north of the subcontinent soldiers of the East India Company mutinied, civilians rebelled, displaced ruling families reclaimed their authority, and some dependent native rulers renounced their allegiance to the British. Part and parcel of the uprising and its suppression were the killing of prisoners and the massacre of civilians by all parties. The mutiny and its aftermath gave rise on both sides to mistrust and hard-feelings that lasted until partition and independence of British India in 1947.

British East India Company Administration of India [1]

In 1857, the United Kingdom governed India through the British East India Company, which at the time was no longer a mercantile organization. The Government of India Act, 1833 had stripped the Company of its remaining trading powers and made it the Crown's agent for the governance of India, production of tea, and the production and export of opium.[2] The statute also made the United Kingdom guarantor of the Company's annual dividend, fixed at £10.5 per share. As the market value of EIC shares in 1833 was about £200, the guaranteed return to investors was 5.25%.[3] For most of the nineteenth century the

---

[1] Sources: Reid, *Commerce and Conquest*, 161-65, 178-201, 223-29, 243-49; Heathcote, *The Indian Army*, 14-16, 155-57.

[2] 3 & 4 Will. 4, c. 77.

[3] Robins, *The Corporation that Changed the World*, 31.

interest rate on perpetual, but callable, Treasury securities (Consols) was 3.0%. Accordingly, Parliament gave the EIC shareholders a state-guaranteed return higher than that for direct, state obligations. The guaranteed dividend was in effect the profit on the Company's management fee to administer British India. The 1833 statute also granted an open-ended option to the British government to buy out the EIC's shareholders at £200 per share.

The EIC collected taxes and maintained armed forces, a well-paid civil service, and a judiciary. For the fiscal year ended April 30, 1857, the EIC had Indian receipts of £22.1 million (including £3.7 million from Crown opium sales) and expenditures of £23.1 million.[4] Total dividends paid to shareholders of £630,000 represented 2.7% of Indian revenues.[5] For administrative purposes, the Company had divided its governing apparatus into three "presidencies." The largest was the Bengal Presidency, headquartered in Calcutta. It had jurisdiction over northern India from the Khyber Pass east to the Bay of Bengal. The other presidencies were Bombay in the west-central part of India and Madras in the east-central and southern parts of India. The Governor of Bengal was also Governor-General of India and had authority over the governors of Bombay and Madras. Under the letter of the law, the EIC's board nominated persons for the governorships and the British government either rejected or accepted the candidates. In practice, governorships, like all senior EIC positions, were government appointments. Entry-level positions in the civil establishment were filled by graduates of the Company's civil college in Hertfordshire, England. Admission to the college was by competitive examination.

The British government's India policy was set by the Board of Commissioners for the Affairs of India. This body, commonly called the "Control Board," consisted of several government ministers and was headed by a President who was in effect the India Minister. The Control Board also had authority over the Company's policy-making body, the "Court of Directors." One-third of the directors were appointees of the British government; two-thirds were elected by shareholders. The Court of Directors set internal administrative policies (mostly personnel matters) subject to Control Board approval.

The army of the EIC was actually three armies; one for each presidency. Each presidency army had a commander-in-chief and the CINC Bengal Army was also the Commander-in-Chief India. Accordingly, the commanders-in-chief of the Bombay and Madras armies were subordinate to both their presidency's governor and the CINC Bengal Army. Senior army positions were effectively government appointments. The EIC armies had two types of component units, Native and European. Native regiments were staffed with Indian enlisted personnel and European officers.[6] European regiments were staffed entirely with Europeans. Nearly all European soldiers and officers were British; there were a few North Americans and Continental Europeans (predominantly in the enlisted ranks). Officers were graduates of the Company's military college in Addiscombe, Surrey, England or recipients of direct

---

[4] EIC, *Accounts respecting the Territorial Revenues and Disbursements of the East India Company for the Year 1855/56*, 1857 Sess. 2, H.C. Accounts & Papers, No. 135.

[5] EIC, *Home Accounts of the East India Company*, 1857 Sess. 2, H.C. Accounts & Papers, No. 110.

[6] The EIC Armies had native officer ranks, but such ranks were subordinate to British officer and NCO ranks. Native officers were effectively the senior NCOs of native troops.

commissions. Admission to the military college was by competitive examination. Enlisted men were recruited by the EIC pursuant to War Office regulations and the enlistment process was supervised by the British Army. The recruit training depot was at Warley in Essex, England.

The British government bolstered the EIC armies with British Army regiments rotated through India. From 1852 through 1856 the British Army maintained on average 25,000 troops in India. In the immediate period preceding the Crimean War, British Army strength averaged 145,000 men.[7] Accordingly, about 17% of the British Army garrisoned India. British regiments in India were in the EIC command structure and the cost to maintain them was borne by the Company (*i.e.*, Indian taxpayers). British military personnel in India received the same pay and allowances that EIC European military personnel received. Table 4 shows the troop strength of all forces in India as of April 30, 1857. Table 5 shows the number of regiments into which the force was organized. Note that for table cells with two numbers the first is the number of regular regiments, the second the number of irregular regiments.

Table 4.
British and EIC Troops in India, 1857

Army	British	European	Native	Pct.	Total
Bengal	15,795	8,571	135,767	84.8	160,133
Bombay	4,718	5,712	45,213	81.3	55,643
Madras	3,750	6,975	51,244	82.7	61,970
Total	24,263	21,259	232,224	83.6	277,746

Source: EIC, *Returns Relating to the Armies in India*, 1857-58, H.C. Accounts & Papers, No. 201.

Table 5.
British and EIC Regiments in India, 1857

Army	Infantry British	Infantry European	Infantry Native	Cavalry British	Cavalry Native	Artillery Euro.	Artillery Native
Bengal	15	3	73 + 41	2	10 + 27	9	3
Bombay	4	3	26 + 10	1	3 + 6	3	2
Madras	3	3	42	1	8	5	1
Total	22	9	141 + 51	4	21 + 33	17	6

Sources: EIC, *Returns Relating to the Armies in India*, 1857-58, H.C. Accounts & Papers, No. 201; *East-India Register and Army List 1857*.

---

[7] War Office, *A Return of the Number of Officers and Men from 1800 to 1858*, 1859 Sess. 2, H.C. Accounts and Papers No. 88; EIC, *Returns Relating to the Armies in India*, 1857-58, H.C. Accounts & Papers, No. 201.

As is readily apparent from the tables on the previous page, Indians made up the bulk of the Company's soldiers and the Bengal Army was by far the largest of the presidency forces. Indians accounted for 83.6% of the EIC's troop strength and native regiments accounted for 82.9% of the combat units in India. Clearly, the British could only control India if they had the loyalty of a substantial number of the EIC's native soldiers.

The Company's regiments were not confined to the Indian sub-continent as the EIC's original charter granted trading authority over all lands east of the Cape of Good Hope to the Straits of Magellan. Company troops had fought in Africa, the Indian Ocean islands, the Arabian Peninsula, Persia, Afghanistan, Southeast Asia, and China. At the time of the Mutiny, 17 native regiments of the EIC were outside of India. There were 13 regiments in Burma, 1 in Aden, 1 in Malaya, and 2 in Persia.[8] Table 5 shows the number of regiments in India as of March 31, 1857.

Mutiny & Rebellion [9]

In 1857, British India consisted of territory under direct EIC control and protectorate Native States. Native State rulers had a free hand in internal affairs so long as their actions did not incur the displeasure of the EIC. The EIC monitored the Native States through "Residents" who functioned as the Company's emissaries as well as advisors to the rulers. Native States had their own armies, often with EIC officers attached. Many of these states had devolved from provinces and tributary states of former empires. For example, when the Maratha Empire displaced the Mughal Empire in the late seventeenth century, the former large Mughal vassal states of Oudh (Awadh), Hyderabad, and Behar-Orissa became independent kingdoms. Behar-Orissa was annexed by the EIC soon after its independence from the Mughal Emperor.

The Maratha Empire, also known as the Confederation of Rajput, was the principal Indian power of the eighteenth century. The Rajputs were high caste Hindu, military clans. Throughout the eighteenth century, the EIC encroached on Maratha territory and fought two wars against the Maratha Empire. After the 3rd Maratha War in 1818, the British dismantled the empire. The EIC granted a life pension to the last Maratha Peshwa, Baji Rao II. After Baji Rao's death, his adopted son, Nana Sahib, claimed the imperial title and asked the EIC for a pension. The British denied his request. Nana Sahib and Tantia Tope, a former minister in the Peshwa's court, were two leaders of the 1857 rebellion. Many of the Maratha Empire's former soldiers joined the EIC's Bengal Army. Throughout the first half of the nineteenth century the Bengal Army recruited heavily among the Rajputs, especially in Oudh.

Much of the disaffection within the Bengal Army in 1857 was due to the actions of Governor-General James Brown-Ramsay (1st Marquess of Dalhousie), in office January 12,

---

[8] *East-India Register and Army List, 1857*.

[9] Sources: Wagner, *The Great Fear of 1867*; Pati, *The 1857 Rebellion* – Majumdar, R.C., "The Character of the Outbreak of 1857," Chaudhuri, S. B., "Theories on the Indian Mutiny," Sen, S. N., "Reflections on the Mutiny;" Haythornthwaite, *The Colonial Wars Source Book*, 98-109.

1848 through February 28, 1856. Dalhousie established the Doctrine of Lapse whereby the EIC proclaimed its intent to annex any protected state where the ruler misgoverns egregiously or dies with no male issue. During Dalhousie's administration, there were two controversial, major annexations in northern India, both of which were approved by the British government.

> Nagpore State, population 4.6 million: Annexed by the EIC in 1853 when Raja Raghoji III Bosale died without male issue.

> The Kingdom of Oudh, population 5.0 million: Annexed by the EIC in late 1856 for misrule by Nawab Wajid Ali. The Company pensioned off Wajid Ali and exiled him to Calcutta.

Together, these states were the family home of many of the Bengal Army's native soldiers, both Hindu Rajputs and Muslims. From May 1, 1851 through December 31, 1856, the EIC added to its domain about 146,000 square miles of territory and 10.5 million inhabitants.[10]

Table 6 shows the population of British India and the extent of territory of each presidency and the protectorates. Note that the EIC exercised direct authority over 71.2% of British India's 180 million inhabitants.

Table 6.
Population and Territorial Proportion of EIC Jurisdictions, 1857

Jurisdiction	Population	Percent of Population	Percent of Area
Bengal	97,763,472	52.3	39.2
Bombay	11,790,942	6.5	9.0
Madras	22,437,297	12.4	9.0
Native States	48,376,247	28.8	42.8
British India	180,367,948	100.0	100.

Source: EIC, *A Return of the Area and Population of each Division of each Presidency in India,* 1857 Sess. 2, H.C. Accounts & Papers, No. 215.

Causes of the military and civil turbulence in India during 1857 are numerous and scholars still debate their relative importance. Some of the causes of mutiny and revolt which nearly all historians mention follows:[11]

> Fear among highly caste-conscious Hindus and revivalist Muslims of the Bengal Army that the British had implemented a plan to "pollute them" reli-

---

[10] EIC, *Statement of the Territories and Tributaries in India acquired since the 1ˢᵗ day of May 1851*, 1857-58, H.C. Accounts & Papers, No. 201.

[11] Wagner, *The Great Fear of 1857,* 21-22.

giously which would place them outside their caste or faith. Of immediate concern were paper powder cartridges for new rifles that Indian soldiers and civilians believed were greased with cow and pig fat. There were also fears within the ranks that British-supplied flour to the regimental cookhouses contained the ground bones of cows and pigs. Indians viewed such pollution as the first phase of a British plot to Christianize the EIC's soldiers and the civilian population.

Resentment of British rule by the populace and leaders of native states annexed recently by the EIC.

EIC changes to the system of land tenure which resulted in significant transfers of farmland from traditional, rural aristocrats to town and city merchants.

Lack of British respect for Indian religious sensitivities and customs.

Increased activity by Christian missionaries whom the populace viewed as instruments of the EIC.

Complete social segregation between the British and the Indians.

Local rebellions against the British usually followed a common pattern. First, a regiment in a garrison town mutinied, often at the instigation of the townspeople. The soldiers, joined by civilians, then attacked the EIC officers and their families, EIC administrators, and Indian "collaborators." The surviving British, carrying tales of atrocities committed by Indians, fled to a town garrisoned by British or EIC European troops. With the British gone, the insurgents seized arms and munitions, looted Company and private property, and set fire to the buildings of civil administration. The collapse of EIC administration in the garrison town touched off rioting in nearby towns, and the rioters frequently killed European inhabitants. Rioters also attacked Indian merchants who profited from Company rule. To restore order in a chaotic district, local aristocrats, relying on their prestige and armed retainers, took charge. Some sided with the uprising and were supported by the mutinous troops, others simply restored order pending re-imposition of British control. There were skirmishes between aristocratic landlords if there was disagreement as to who would control, and collect taxes from within, a given area.

Approximately 55% of the Bengal native regiments actively mutinied, that is attacked their British officers or deserted *en masse* with their weapons. Another 15% of the Bengal regiments were disarmed or disbanded by the British as a precautionary measure. Only 30% of the Bengal native regiments remained "loyal" to the EIC. The British deemed a regiment loyal if most of its troops obeyed orders or fought to suppress rebellion. About 85,000 Bengal soldiers mutinied, 25,000 were disarmed or dismissed, and 25,000 remained loyal to the Company. Mutiny in the Bombay and Madras armies was negligible though many regiments exhibited more than the usual discontent. Table 7 shows the regimental extent of the mutiny.

Table 7.
EIC Native Units and the Mutiny

Army	Regiments in India	Actively Mutinied	Disarmed	Disbanded	Loyal to EIC
Bengal	154	84	20	2	48
Bombay	47	2	2		43
Madras	51			1	50
EIC	252	86	22	3	141

Sources: *East-India Register and Army List 1857*; India Office, *A Return of those Regiments in the Native Bengal Army, both Regular and Irregular, that remained Faithful to the British Government*, 1861, H.C. Accounts & Papers, No. 21; India Office, *A Return of the Name or Number of each Regiment and Regular and Irregular Corps in India which has Mutinied*, 1859 Sess. 1, H.C. Accounts & Papers, No. 133; Haythornthwaite, *Colonial Wars Source Book*, 108; David, "The Bengal army and the outbreak of the Indian mutiny," appx. 2.

The rulers of twelve Native States rebelled or were viewed by the British as having done so. Shorapur (a vassal state of Hyderabad), in the Madras Presidency, and Nargund, in the Bombay Presidency, were the only Native States outside of the Bengal Presidency's supervision that broke with the British.[12] Troops of several loyalist native rulers mutinied. The largest native state force that experienced widespread mutiny and rebellion was that of Gwalior (population 3.2 million). All its seven infantry and two cavalry regiments, plus artillery units, mutinied and fought the British.[13] The map on the following page shows the sites where EIC regiments mutinied. The smallest mutiny icon represents one regiment, the largest five regiments. Note that the bulk of the mutinied regiments were stationed generally around a line beginning in the northwest at Peshawar then running southeasterly to Allahabad.

Areas that the British authorities considered in rebellion did not uniformly break with the EIC. Uprisings took place in a checkerboard pattern where in each area some towns and districts rebelled while other towns, typically under Indian leadership, remained loyal to the EIC. The territory of the western part of the modern Indian State of Bihar and all the State of Uttar Pradesh had the highest density of rebellion and effectively became independent. In Delhi, mutinous troops proclaimed 82-year old Bahadur Shah emperor of a restored Mughal Empire. Bahadur Shah was a direct descendant of the last ruling Mughal Emperor and retained the title "King of Delhi." In Lucknow, Nana Sahib, adopted son of Wajid Ali, was proclaimed nawab of an independent Oudh. Oudh became a vassal state of the nominally revived, Mughal Empire. Provisional governments formed with power held not by the proclaimed monarchs, but by executive councils of military leaders and landed aristocrats. The new governments collected taxes, borrowed money, raised armies, and attempted to extend their areas of control.

---

[12] Schwartzberg, Joseph E., ed., *A Historical Atlas of South Asia*, 2nd impression (New York: Oxford Univ. Press, 1992), 62.

[13] *East-India Register and Army List, 1857*.

Late Victorian Military Campaigns and Army Reform

The Indian Mutiny, 1857-1858

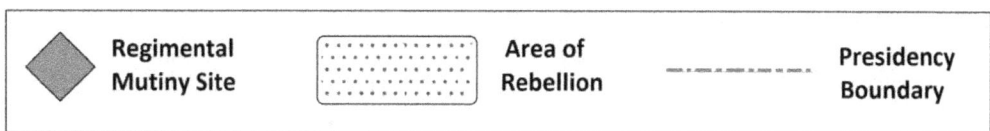

Sources: Schwartzberg, Joseph E., ed., *A Historical Atlas of South Asia*, 2nd impression (New York: Oxford Univ. Press, 1992); *Cruchley's New Map of India, Seat of the Mutinies* (London: 1857); sources for Table 7.

British Reinforcements and Restoration of Control

The British regained control of the Bengal Presidency area with Crown and European regiments, loyal native units, and new EIC formations that recruited Sikhs and Muslims of the northwestern provinces. Reinforcements from the United Kingdom began to arrive in India in late 1857. The War Office in total sent 68 additional regiments to India; 52 infantry, 8 cavalry, and 8 artillery. The EIC raised 3 new infantry regiments and 4 new cavalry regiments. By the end of 1858, the British had effectively suppressed the rebellion.[14]

By mid-February 1859, there were 114,489 British troops in India; 89,916 Crown and 24,573 former Company.[15] Like it had done during the Crimean War, the British government embodied militia units for home service to free regular units for deployment to the theatre of war. Militia and other auxiliary units were not sent to India.

The War Office did not tabulate British Army deaths attributable to the hostilities of 1857-1858. A return to Parliament of enlisted men deaths in India from 1856 through 1864 indicates an increase of 6,671 in such deaths for the two years after 1856.[16] Therefore, including officers, total British Army deaths in India indicated for the mutiny and rebellion is about 7,000. The return also indicates about 2,500 additional British military personnel were invalided home during the hostilities. For the years 1856, 1859, and 1860 the overall mortality rate for British enlisted men in India averaged 3.8%. For 1857 the mortality rate was 12.2%, for 1858 it was 9.1%. About 80% of British Army deaths in India during hostilities were caused by sunstroke and disease.[17] The EIC did not tabulate European deaths in the Bengal Army for the year 1857.[18] Accordingly, the number of EIC European military deaths during 1857 is unknown. For 1858, the mortality rate for such troops was 4.3%. As in Crimea, the army experienced widespread, significant logistics problems; however, such deficiencies were not reported in the press.[19] On October 2, 1857 the Commander-in-Chief of the British Army wrote to the War Minister that the army in India

> "... is deficient in *everything*. Not a spare set of harness in store, no shoes, no ammunition, no man able to make use of the beautiful machinery sent out to make Minie bullets. It is almost incredible, and yet from the first I feared it and told you so, but you thought otherwise, relying very naturally on the assurances of the Company."[20]

---

[14] EIC, *Returns Relating to the Armies in India,* 1857-58, H. C. Accounts & Papers, No. 201; India Office, *Return of the actual Strength, both of the Queen's and the East India Company's Forces in the Three Presidencies, and in the Punjab,* 1859 Sess. 2, H.C. Accounts & Papers, No. 64.

[15] India Office, *Return of the actual Strength, both of the Queen's and the East India Company's Forces in the Three Presidencies, and in the Punjab,* 1859 Sess. 2, H.C. Accounts & Papers, No. 64.

[16] War Office, *Return of the Mean Strength and of the Deaths and other Casualties among Non-Commissioned Officers and Men,* 1865, H.C. Accounts & Papers, No. 341.

[17] Spiers, *The Army and Society,* 135.

[18] India Office, *Return of the Mean Strength and of the Deaths and other Casualties among Non-Commissioned Officers and Men of the European Forces,* 1865, H.C. Accounts & Papers, No. 486.

[19] Spiers, *The Army and Society,* 134-35.

[20] Douglas and Ramsay, *The Panmure Papers,* 441.

Aftermath

The 1857 mutiny and rebellion prompted Parliament to strip the EIC of its role in Indian affairs. The Company lost its governance authority through the Government of India Act, 1858.[21] The duties of the President of the Control Board were assumed by a new cabinet minister, the Secretary of State for India. The Board's staff merged with the EIC's home establishment and became the India Office, and the EIC's administrative apparatus in India was transformed into the Government of India. The Company continued to exist but functioned only as tea trade agent for the Crown, provisioner of the shipping support facility at St. Helena in the South Atlantic, share registrar, and payor of share dividends funded by the Indian government.

The India Act relabeled the EIC military establishment as the Indian Army and subordinated it to the new Indian government. All European military personnel became soldiers of the Crown. Soldiers of the European regiments remained, temporarily, with their old units. Separate recruitment for European regiments continued through 1860. Between mid-1859 and mid-1862, the European regiments were disbanded or transferred to the British Army. The surviving nine European infantry regiments of the Indian Army became the British Army's 101st through 109th Regiments of Foot. The surviving three European cavalry regiments became the British Army's 19th through 21st Regiments of Light Dragoons. The European artillery was re-organized into 14 Royal Artillery battalions.[22]

To preclude future, wide-spread mutinies, British government policy now required that in India the ratio of native soldiers to British soldiers not exceed three to one. Accordingly, for the post-mutiny period of British rule in India, the War Office kept a substantial portion of the British Army on the sub-continent. For the five-years of 1863 through 1867, the British Army's strength in India averaged 70,000 men. During that same period total British Army strength averaged 220,000 men.[23]

In January 1857. there were 24,000 Crown troops and 16,000 Company regimental troops in India, for a total of 40,000 effectively British combat troops.[24] With 70,000 British troops stationed in India after full restoration of British control, the War Office had increased by 30,000 the number of combat troops stationed there compared to pre-mutiny levels. This mandated a 15% increase in total army strength for the years 1863 through 1867, which exacerbated the army's chronic recruitment problem.[25]

For 16 years the Indian Government paid the EIC shareholders the dividend of £10.5 per share guaranteed by the Government of India Act, 1833. In 1874, Parliament dissolved the

---

[21] 21 & 22 Vict., c. 106.

[22] General Orders by the Governor General of India in Council of 1861, No. 332, *The Calcutta Gazette*, Extraordinary, April 22, 1861.

[23] War Office, *Return of the Number of Officers and Men*, 1867-68, H.C. Accounts & Papers, No. 412.

[24] EIC, *Returns Relating to the Armies in India*, 1857-58, H.C. Accounts & Papers, No. 201.

[25] 30,000 extra troops in India plus 190,000 baseline troops give a total strength of 220,000. $30,000/190,000 = 0.158$, the increase in authorized strength.

Late Victorian Military Campaigns and Army Reform

East India Company.[26] Shareholders were offered £200 in British government bonds at 3% interest, or Indian government bonds at 4% interest, for each share of EIC stock. The British bonds were perpetual but callable and redeemable at will; the Indian bonds had a 14-year term.[27]

Imperial Campaigns of the Late Nineteenth Century

The army of which Rudyard Kipling wrote was known for the colonial campaigns that took place in the second half of the nineteenth century. Following is a list of such campaigns and minor wars.[28]

2nd Opium War	1856-1860	China, North and South
Maori Wars	1856-1880	New Zealand, intermittent
2nd Ashanti War	1863-1864	Africa, West
Bhutan War	1864-1865	India, North
Abyssinia Expedition	1867-1868	Africa, East
3rd Ashanti War	1873-1874	Africa, West
9th Xhosa Frontier War	1877-1878	Africa, South
2nd Afghan War	1878-1880	Afghanistan
Zulu War	1879	Africa, South
1st Boer War	1880-1881	Africa, South
Egyptian Expedition	1882	Africa, North
3rd Burma War	1885	Burma
4th Ashanti War	1894	Africa, West
Sudan Expedition	1898-1899	Africa, East

The Abyssinia and Sudan Expeditions warrant examination as they were short in duration, low in financial cost, and resulted in few British casualties.

The Abyssinia Expedition [29]

The campaign of the Anglo-Indian, Abyssinian Expeditionary Force was atypical of Britain's nineteenth century foreign military endeavors in that the British Army (more correctly, the Indian Army) was charged with a humanitarian rescue mission. For the British,

---

[26] East India Stock Dividend Redemption Act, 36 & 37 Vict., c. 17.

[27] India Office Notice, *The Economist* 32 (January 10, 1874): 58.

[28] Source: Haythornthwaite, *The Colonial Wars Source Book*.

[29] Sources: Holland, *Record of the Expedition to Abyssinia*; Haythornthwaite, *The Colonial Wars Source Book*, 159-61.

the war had no political purpose other than to deter future seizures of British diplomats by foreign governments. Because the Abyssinian Expeditionary Force achieved its objective relatively quickly and inexpensively, and with almost no British casualties, its campaign became in the public's mind, the ideal of the late-Victorian little war.

In 1864, Tewodros II, Emperor of Abyssinia, after being insulted by Britain (through the negligence of the Foreign Office) and fearful of British support for his Muslim enemies, imprisoned the British envoy, Captain Charles Cameron. Tewodros, a fanatical Christian, wanted European support for a military campaign to detach Jerusalem from the Ottoman Empire. His letter to Queen Victoria asking for such support was never answered. In addition to the seizure of the British diplomat, Tewodros imprisoned all Europeans resident in his capital, the fortress town of Magdala (now Amba Mariam, Ethiopia). Apparently, he sought either to hold them as hostages to obtain British aid for his proposed crusade, or as prisoners punished for the insolence of their governments. The Foreign Office tried by diplomatic means to free the captives, but Tewodros would not relent. Accordingly, in June 1867, the British took steps to send a rescue force from India.

In July 1867, the Foreign Office arranged with the Khedive of Egypt for British troops to land near Massawa on the Red Sea coast. That same month, Westminster notified Calcutta to assemble a rescue force from the Bombay Army and to place it under command of LTG Robert Napier. On August 12th, having received no sign from Tewodros that he would release the captives, Westminster ordered Napier to proceed with the expedition to Abyssinia. In September 1867, Royal Engineers arrived in East Africa to select a landing site and survey a route into the interior for the expected expeditionary force. Early in October, Commissariat officers arrived in East Africa and Aden to arrange for the purchase of draft animals, food, and other supplies, and to hire muleteers and porters. By October, the engineers had chosen Zula at Annesley Bay for the landing site and at the end of that month troops from India disembarked there to secure the area. The soldiers were followed by workmen to construct piers and other facilities for what would be the expeditionary force's supply base. Construction also began on a light railway from Zula to the interior. In December 1867, the Anglo-Indian relief force began to arrive and assemble at Zula and points inland.

Bombay sent a force of 3,638 British and 9,883 native Indian troops along with 7,116 official "followers" (the force's support component). Also sent was an initial shipment of 3,604 draft animals. In early March 1868, Napier's force moved out of its assembly area and headed south for Magdala. As the army advanced, Napier left strewn behind detachments to protect his supply line to the sea.

On April 10th, Tewodros' army of 6,000 to 7,000 irregulars attacked the spearhead of Napier's force on the Arogi Plateau about 50 kilometers from Magdala. The defenders were the 1,900 men of the 1st Brigade, a unit whose critical components were two artillery batteries (one of mountain guns, the other of rocket launchers) and a 473-man battalion of the British Army's 4th Regiment of Foot. The attackers were decimated. Napier estimated the Emperor's army suffered 700 dead and 1,200 wounded. Anglo-Indian casualties were 2 dead and 18 wounded. Tewodros withdrew his now demoralized army to the fortress of Magdala.

Tewodros' opponents, and there were many, now smelled blood. Subordinate chiefs, many former enemies of Tewodros with old scores to settle, withdrew support for their Emperor. Muslim tribes which had mobilized after the Anglo-Indian Expeditionary Force landed, moved in for the kill. Realizing that his reign had come to an end, Tewodros released his remaining soldiers from their obligations to him. All but 400 left Magdala, and they prepared, half-heartedly, for the British assault.

On April 14, 1868, the Anglo-Indian force attacked Magdala and quickly breached its defenses. Tewodros, rather than be taken prisoner, shot himself. Cameron and the civilian prisoners were freed. With his mission accomplished, Napier marched his army back to Zula where it embarked for India.

For the Anglo-British force, both combat casualties and total deaths were surprisingly light. As in earlier wars, the killer of soldiers on campaign was not the armed enemy; it was disease. Of the total 334 dead, only 2 were killed in action. The annualized mortality rate for non-combat deaths was 8.4% for the 9,900 Indians and 3.8% for the 3,600 British (actual attrition compounded for a full year). In the first year of the Crimean War the annualized, mortality rate for the British was 43.1%, in the second year 8.9%.[30] Casualties in addition to the 2 British killed-in-action were 13 British wounded and 15 Indian wounded. All the wounded survived.

After the Abyssinia campaign, Queen Victoria created Napier Baron of Magdala. Eighteen years later the British government appointed Napier Governor and Commander-in-Chief of Gibraltar. He would serve in that position from June 1876 through December 1882. Down the chain of command from the Commander-in-Chief, Gibraltar, would be the *Ulysses* character Major Brian Tweedy, then serving on the Rock and living with his daughter, Molly.

Sudan Expedition, 1898-1899 [31]

Another fairly inexpensive and successful colonial campaign with lopsided casualty figures, was the Sudan Expedition through which Britain restored its indirect control over the Sudan. The climax of this British campaign was the battle of Omdurman fought on September 2, 1898. There, the bulk of the imperial force of about 8,000 British plus 17,000 Egyptian and Sudanese troops, all commanded by General Herbert Kitchener, engaged possibly as many as 50,000 irregulars of the religiously inspired rebel leader, Abdullah al-Taashi, successor to the Mahdi. The battle opened with a charge over an open plain by 16,000 of al-Taashi's troops. Many were quickly mowed down by British machine guns and rapid-fire artillery. The survivors pulled back leaving behind 4,000 dead and dying. Fighting continued throughout the day in a somewhat similar manner. By evening, the rebel army had incurred casualties of 13,000 killed, 12,000 wounded and 5,000 taken prisoner. Kitchener's force suffered 47 killed and 382 wounded.

---

[30] *Report of the Commissioners appointed to inquire into the Regulations Affecting the Sanitary Condition of the Army*, 1857-58, [2318], at 524.

[31] Source: Haythornthwaite, *The Colonial Wars Source Book*, 216-22.

The 2nd Boer War, 1899-1902

The Dutch Cape Colony

In 1652, at a large natural harbor in the southwesterly corner of the South African coast, the *Vereenigde Oostindiche Compagnie* (Dutch East India Company) established a naval support facility for its East Indies trade. The harbor was formed in part by a peninsula, at the base of which the VOC established a settlement, *Kaapstad* (Cape Town). Dutch traders, ships chandlers, tradesmen, their employees and slaves, moved in to service the vessels going to, and coming from, the Spice Islands. The colony soon attracted potential farmers and graziers. The surrounding area had fairly good soil, some forests, was well-watered with rivers., and received enough rainfall to support crop farming. The first settlers acquired the croplands near Cape Town. Subsequent Dutch, German and French Huguenot farmers and graziers had to move farther out to obtain land. VOC officials termed these settlers *Trekboers*, Dutch for "wandering farmers."[32] The land was not there just for the asking; the area was inhabited, though sparsely, by the San and Khoi-Khoi. The Dutch referred to the pastoral Khoi-Khoi as *Hottentots* and termed the generally more distantly located hunter-gatherers, the San, *Bosjesmen* (Bushmen). At the time the Dutch arrived, the San numbered at most 20,000 while there may have been as many as 100,000 Khoi-Khoi.[33] The indigenous population was soon decimated by European diseases to which they had little biological resistance. Smallpox was especially deadly for the Khoi-Khoi who were in closer contact with the Dutch than the San.[34]

The Dutch purchased, or seized, farmland from the coastal natives and the *Trekboers* pushed inland to obtain pasturage. The Boer campaign of expansion into San territory was particularly brutal as they slaughtered many of the inhabitants and took children as slaves termed "apprentices." Such apprentices were bound to their masters until age 18. Boer depredations spurred numerous San to migrate northward. The Europeans established large farms and ranches, hired Khoi-Khoi as domestic servants and ranch hands, and brought in slaves purchased in West Africa to work crops. More Dutch immigrants arrived, mainly from the lower economic strata of society, and they fueled the settlement's territorial growth.[35] Over about 150 years, the VOC colony expanded along the coast 700 kilometers east to the Fish River and 400 kilometers north to the Buffels River. It reached into the interior in depths ranging from 50 to 300 kilometers, well past the natural border of the barrier mountains some 175 kilometers from the coast.[36] This territory, the size of Great Britain and Ireland combined, came to be known to Europeans as the Dutch Cape Colony.

As the Boers migrated ever eastward from Cape Town, they came to lands claimed by the native, mixed-farming Xhosa, who also sought to increase their land-holdings. In the 1770s

---

[32] Meredith, *Diamonds, Gold, and Wars*, 1-10.

[33] Davenport, *South Africa*, 6-8.

[34] Thompson, *A History of South Africa*, 38-39.

[35] Ibid., 31-51; Davenport, *South Africa*, 21-40.

[36] Eric A. Walter, *Historical Atlas of South Africa* (Oxford: Oxford Univ. Press, 1922).

the two groups of farmers and ranchers met head to head. Eventually, each group tried to penetrate the other's territory and in 1779, "war" erupted between the Boers and the Xhosa. The Boers, armed by the VOC, organized informal militia units which they called *commandos*. The *commandos* included Khoi-Khoi, both free employees and slave-apprentices.[37] For nearly two years there were cattle raids, skirmishes between small bodies of men, killings of women and children, and punitive expeditions. The area of hostilities gained the name "Frontier." The 2nd Frontier War took place from 1789 through 1793. That war ended when all parties accepted reluctantly, the Fish River as the eastern boundary of the Cape Colony.[38]

In 1793, the VOC conducted a rough census which showed the Cape Colony had 13,830 white colonists and 14,747 slaves. The VOC did not estimate the number of indigenous inhabitants. Cape Town had 4,155 white residents and the western district had 4,640 white farmers. The arid northern and northeastern districts had 1,925 Boers while the Frontier had about 3,100 Boers.[39]

British South Africa and the Boer Republics

In the 1790s the VOC failed financially and in 1799 its charter lapsed. Control of the company's territories, including Cape Colony, devolved to the Dutch state. In 1806, during the Napoleonic Wars, the British seized Cape Colony from the Dutch Republic of Batavia, at the time allied with France. Through the Anglo-Dutch Treaty of 1814, the Dutch state, then the Kingdom of the United Netherlands, ceded Cape Colony to the United Kingdom for £6 million.[40] When the British took formal possession, the colony's population was about 75,000 of whom 25,000 were of European descent (overwhelmingly Dutch) and 10,000 to 20,000 were slaves from West Africa. The balance of about 35,000 consisted of Khoi-Khoi, San, and Basters.[41] Baster is an Anglicization of the Boer term for the children of indigenous and West African women and European men.[42] The British came to categorize, socially and politically, the inhabitants of South Africa into four groups. Afrikaners were the Dutch descendants that inhabited the towns and farms of the western part of the colony. Boers were the *trekboer* descendants who lived on ranches in the north and east. The Cape Coloured were the San, Khoi-Khoi, descendants of the West African slaves, and the Basters, whom the British termed Griqua. The Bantu-speakers of South Africa, called *Kaffirs* by the Dutch, the British labelled Africans.[43]

---

[37] Thompson, *A History of South Africa*, 49.

[38] Haythornthwaite, *The Colonial Wars Sourcebook*, 176.

[39] Thompson, *A History of South Africa*, 35-36, 47.

[40] Davenport, *South Africa*, 42-43.

[41] Meredith, *Diamonds, Gold, and War*, 1-2.

[42] *Bastaar*, from *bastaard*, the Dutch word for both children born out of wedlock and animals of mixed-breed.

[43] Davenport, *South Africa*, 33.

British rule did not sit well with the Dutch descendants, especially the Boers of the Frontier districts. The colonial government opposed the Boer's expansionist policy and imposed controls on employer treatment of native employees and slave-apprentices. The British colonial government made a dramatic change to the socio-economic order in 1828 when it granted "free people of colour" the right to own land and legal equality with whites. Additionally, all employment contracts, other than apprenticeships, were limited to a term of one year.[44] In 1834, the British government abolished slavery throughout the empire and changed the legal status of slaves to that of indentured servants with a four-year term of servitude. Most farmers and ranchers received little compensation for the loss of their human property.[45] Two years later, through the Cape of Good Hope Punishment Act, Westminster extended the Cape Colony's jurisdiction over British subjects north to the 25th Parallel (somewhat north of modern-day Pretoria).[46] While this 1836 act did not extend sovereignty, it made British subjects answerable to the Cape Colony courts for actions taken outside of the colony.

By the late 1830s, many of the Boers felt oppressed by British colonial government. They viewed British religious beliefs as un-Christian and British social mores as immoral. Furthermore, they disagreed strongly with British race-relations policy, especially with respect to land ownership and the grant of civil rights to the Cape Coloured. In general, the Boers became convinced the British would prevent fulfillment of their divine destiny: An independent Boer society that would operate in accordance with the principles of the Dutch Reformed Church.[47]

In 1837, the first disaffected Boers migrated out of Cape Colony. They were accompanied by their native apprentices, domestic servants, ranch hands, and livestock. The Boer emigrants called themselves *Voortrekkers,* pioneer migrants. By 1840, approximately 11,000 to 15,000 Boers and their retinues had left British territory in what became known in South Africa as "The Great Trek."[48] They had left Cape Colony "escaping from an alien government whose policies they had come to detest and hoping to find some Promised Land where they might make their own arrangements with one another, with their servants, and with the other inhabitants."[49]

The *Voortrekker's* destination was the sparsely inhabited territory to the far northeast of Cape Colony. In 1839, the Boers organized their first state, Natalia, in territory along the Indian Ocean coast south of the Tugela River. The inland *Voortrekker* lands north of the Vaal River would become in 1844, the Transvaal Republic, while the area between the Orange and Vaal Rivers would become in 1854 the Orange Free State. The Orange River was the

---

[44] Thompson, *A History of South Africa*, 60.

[45] Davenport, *South Africa*, 47.

[46] 6 & 7 Will. 4, c. 57.

[47] Davenport, *South Africa*, 51-53.

[48] Ibid.; Thompson, *A History of South Africa*, 67. About 6,000 to 8,000 Boers emigrated which represented 9-10% of the colony's white population.

[49] Thompson, *A History of South Africa*, 67.

Cape Colony's northern boundary.⁵⁰ The *Voortrekkers* believed they could reach an accommodation with the people located northeast of Cape Colony and acquire land by purchase. This proved not to be the case as the Africans, in general, did not want to part with their holdings. Also, Africans did not recognize private land ownership. South African chiefs and kings had long granted residency and grazing rights to migrants, but such grants were tenancies-at-will. Boers, when they received such rights, viewed their grants as conveyances of title. Accordingly, the new Boer states were established mostly by conquest.⁵¹

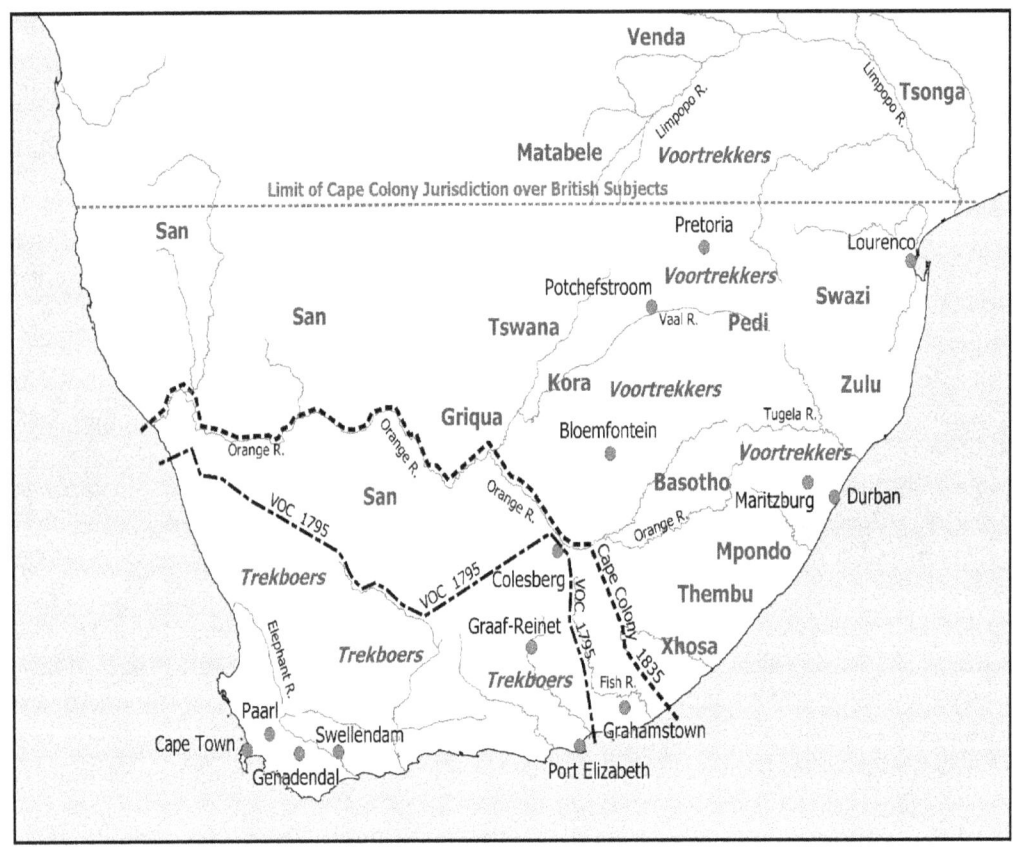

South Africa, 1795-1840

Sources: Davenport, Souuth Africa A Modern History; Thompson, South Africa.

In 1846, a Whig minority government in Parliament had displaced the Conservatives and changed imperial policy in South Africa. The new government directed South African colonial officials to protect African chiefs who felt threatened by the Boers and appointed a

---

⁵⁰ Davenport, *South Africa*, 80-84.

⁵¹ Davenport, *South Africa*, 77; Thompson, *A History of South Africa*, 71.

High Commissioner to look after British interests outside of Cape Colony and Natal. In 1848, High Commissioner Harry Smith declared that the Queen had paramount authority over all lands between the Orange and Vaal Rivers. In the southern part of that region, the *Voortrekkers* pledged allegiance to Smith. In response, the Boer leader Andries Pretorius and his followers attacked the British supporters. Smith in turn sent armed Griqua to aid the British-aligned Boers. Smith's forces were victorious, and he proclaimed a British "Orange River Sovereignty" between the Orange and the Vaal Rivers with its capital at Bloemfontein.[52]

In 1852, Westminster changed course regarding its *Voortrekker* policy. That year Britain recognized the independence of the Transvaal Republic. Two years later the British government renounced all claims to territory north of Cape Colony and dismantled the Orange River Sovereignty. In effect, Britain recognized the independence of the Boer republic of the Orange Free State.[53]

During the mid-nineteenth century Cape Colony expanded easterly and northernly through conquest and threat of war. To the north, the protected territories of the Griqua, San and other native people, were annexed. To the east, the Xhosa were conquered through a series of wars that ended in 1878. Natal expanded through the Zulu War of 1879. By 1880, the Boer Republics were surrounded by British territory and protectorates, and Portuguese East Africa.[54]

In December 1868, when the Liberals under Gladstone formed a new government, South African imperial policy changed again. The policy objective became formation of a federated South Africa that would include Cape Colony, Natal, the Orange Free State, and the Transvaal Republic. The federation would be self-governing and most importantly provide for its own defense.[55] The latter objective was due to the Liberals' desire to scale back expensive, British military commitments which then included British North America, Australia, and New Zealand. In furtherance of this objective, in 1872 the Cape Colony became the self-governing dominion of the Cape of Good Hope. Natal would attain dominion status in 1893. The Conservative government of Disraeli, which took office in 1874, continued the policy of federation. In September 1876, the Colonial Office decided to make the Transvaal a British possession. In January 1877, to effectuate control, the British government sent a small force of paramilitary Cape and Natal police to the Transvaal capital of Pretoria. The Transvaal government, then bitterly divided over internal issues, sullenly accepted the British takeover. Britain proclaimed the Transvaal a colony in April 1877.[56]

The Transvaal's status as a British colony, though self-governed by the Boers, lasted about four years. In 1880, Gladstone was returned to office and soon made clear that the Transvaal state would be incorporated into a federated, British South Africa. The Boers were united in

---

[52] Davenport, *South Africa*, 194-97.

[53] Ibid., 197-99.

[54] Haythornthwaite, *The Colonial Wars Source Book*, 179-93.

[55] Davenport, *South Africa*, 202.

[56] Ibid., 106, 120, 205-07.

opposition to federation and on December 13, 1880 the Transvaal declared independence. Boer *commandos* attacked and besieged the scattered, small British garrisons in the now former colony. In January 1881, British troops entered the Transvaal and fighting continued until March 6th when a truce was declared. The political situation was resolved by the Pretoria Convention of August 3, 1881, whereby the Transvaal became a self-governing, British dominion. In 1884, the British government recognized the Transvaal Republic, officially the South African Republic, as a sovereign state led by President Paul Kruger. The Boers had won their "First War of Independence."[57]

Though the British government had recognized the independence of the two Boer states, the long-term, strategic British objective remained federation. This goal was shared by the British residents of the Cape and Natal Dominions where they were a minority of the white population.

Gold, the Transvaal Republic, and the Path to War [58]

The discovery of large gold deposits in the Witwatersrand area of the Transvaal (the Rand), southwest of Johannesburg, made the Boer Republics tempting targets for British imperialists. In furtherance of expansionist aims, British politicians, both at home and in South Africa, adopted an aggressive stance toward the republics with respect to the plight of British immigrants in the Transvaal. These were nearly all men who arrived with the gold rush that began in 1886.

The Transvaal Boers termed the gold-rush immigrants *Uitlanders* (foreigners) who like the native peoples, had no political rights. Unlike the natives, many of the *Uitlanders* had substantial liquid assets and income from gold mining and ancillary businesses. Most *Uitlanders* were British; however, North Americans, Continental Europeans, Cape Afrikaners, and Cape Coloured accounted for a large minority of the fortune-hunters. The Boer's viewed the *Uitlanders* as fair game for high taxes, which they soon imposed. Adding to *Uitlander* grievances, high import duties and the monopolies granted by the Kruger regime (high monopoly profits meant high government business tax revenue) resulted in exorbitant prices for many goods they purchased. *Uitlanders*, who by 1896 made up most of Transvaal's white, adult male population, began to view inclusion into the British Empire as the cure for their political and financial disabilities. Their champion was Cecil Rhodes, the piratical, imperialistic, diamond and gold mining magnate who in 1895 was the Cape's Prime Minister.

In 1895, Rhodes conspired with several other gold barons to forcibly bring the Transvaal into the British Empire. Their plan was to mount a large raid into the Boer republic and instigate a simultaneous revolt of the *Uitlanders*. The conspirators expected the raiders and armed *Uitlanders* would quickly overwhelm the Transvaal police and easily seize control in Johannesburg and the Rand. Prime Minister Rhodes could then send Cape armed forces into the Transvaal to consolidate the insurgents' hold on the economic engine of the Boer state. Rhodes expected the Transvaal government, based in Pretoria, would then accept incorporation into the British Empire. The conspirators selected Leander Jameson,

---

[57] Haythornthwaite, *The Colonial Wars Source Book*, 193-95.

[58] Source: Pakenham, *The Boer War*, 1-115.

Administrator of the British South Africa Company, to lead the raid. The BSAC, of which Rhodes was a founding shareholder, was a British chartered company that controlled territory north of the Transvaal (later the British colonies of Northern and Southern Rhodesia). In 1895, Rhodes was the company's managing director. The raiders would mount their main attack from Pitsani in the Bechuana Protectorate and a subsidiary attack from nearby Mafeking in the Cape. They would have to ride 275 kilometers to reach Johannesburg.

Jameson failed to obtain support among the British *Uitlanders* and security measures for the planned coup were lax. Preparations for Jameson's invasion with about 600 men (400 BSAC police, 125 Cape volunteers, and 75 Cape Coloured horse-tenders and other non-combatants) were reported in the Kimberley press. On December 29, 1895 Jameson's force entered the Transvaal and soon encountered the mobilized Boer *commandos*. On January 2, 1896 Jameson, surrounded with no hope of relief, surrendered. The raiders suffered 16 dead and 49 wounded; the Boers 1 dead. Along with the captured raiders, the Transvaal Republic acquired the telegraph code book that Jameson carried. Telegrams sent by the conspirators showed conclusively Rhodes involvement in the failed coup.

Rhodes resigned as Prime Minister and the BSAC's board dismissed him as managing director. The Transvaal government handed over to the Cape authorities the raiders' officers. The British sent Jameson and his lieutenants to London where they were tried and convicted of violating the British Foreign Enlistment Act. The Court sentenced Jameson to 15-months imprisonment. To Westminster, the objective of a federated South Africa within the Empire now seemed nearly unattainable. The Transvaal Republic was by far the dominant economic power in South Africa and the Boers had again withstood British armed intervention. Kruger was rearming the Transvaal *commandos* and standing force with modern rifles and heavy artillery from Germany, rapid-fire artillery from France, and machine guns from Britain. Exacerbating the British imperial situation, Cape Afrikaners now felt solidarity with the Boers of the republics. It was conceivable that should there be a federated South Africa, it would be under Boer, not British, control.

Joseph Chamberlain, Colonial Secretary in the Conservative-Liberal Unionist government, and Alfred Milner, Cape Governor and High Commissioner for South Africa, made new plans to bring the Boers into the imperial fold. Chamberlain and Milner agreed that annexation of the Boer republics would require wide support of both the British public and the Afrikaners of the Cape. Chamberlain believed that over time the Boers would accept, resignedly, incorporation into the empire. Milner thought that time was working against British interests and he planned to provoke Kruger to declare war against the British dominions.

The idea that the Boers would attack the British Empire seems, on its face, ludicrous. The United Kingdom alone had a population of 32 million while the combined Boer population of the two republics was about 275,000. At the turn of the twentieth century, the Orange Free State had a population of about 270,000 of whom 90,000 were Boers. The balance was Cape Coloured and Africans. The Transvaal census of 1896 enumerated 245,397 white residents and estimated 622,500 other residents.[59] *Uitlanders*, including women and children,

---

[59] Bureau of Statistics, Department of the Treasury, *Colonial Administration 1800-1900*, (Washington, DC: GPO, 1903), 2831; Davenport, *South Africa*, 84-86.

totaled from 50,000 to 60,000 which indicated a Transvaal Boer population of about 185,000.[60]

Milner initiated a propaganda campaign that publicized Boer mistreatment of British and Cape Coloured *Uitlanders*. Milner also made progressively increasing demands upon Kruger to extend more rights to the *Uitlanders*, including the right to vote. These demands were made concurrently with the War Office's increase of the South African garrison. In December 1896, the British Army had in South Africa two cavalry regiments, four infantry battalions, and one artillery battery.[61] In May 1899, the regular army in South Africa consisted of two cavalry regiments, six infantry battalions, and four artillery batteries.[62] In August, the War Office sent a further two infantry battalions to South Africa.

Kruger accepted most of Milner's demands. *Uitlanders* with five-years residency would receive the right to vote and the Rand would be allocated 25% of the Transvaal's parliamentary seats. Milner's attempt to provoke Kruger to declare war appeared to have failed. The situation changed dramatically on September 8, 1899 when the British government announced the dispatch of 8,000 more troops to South Africa. Shortly thereafter Kruger met with Martinus Steyn, President of the Orange Free State. Kruger urged Steyn to support a joint attack on the British before their reinforcements arrived. Steyn took a wait-and-see position and did not support Kruger's plan for a pre-emptive seizure of the Natal port town of Durban. The situation changed again on September 22nd when the British Cabinet announced it would form in South Africa an Army Corps of three infantry divisions and one cavalry division, plus artillery and support troops. This would increase the number of British regulars in the Cape and Natal from about 11,000 to over 35,000. The Transvaal mobilized its commandos on September 28th and the Free State followed suit on October 2nd. On October 9th, Kruger presented an ultimatum to the British Agent in the Transvaal. The Boers demanded that the British land no more troops in South Africa, withdraw the troops that arrived after June 1st, and pull back its forces from the border areas. The British government did not accept Kruger's demands and the Transvaal government ordered the *commandos* to invade the British dominions.

Conventional Warfare in South Africa

The British government appointed General Redvers Buller commander of the South Africa Corps and doubled the number of troops, horses, and artillery guns it authorized previously for the Cape and Natal Dominions. During the buildup of strength, Buller made plans for a quick campaign of colonial conquest.[63]

---

[60] Most historians place the number of *Uitlander* men at 40,000 - 45,000. *Pakenham*, The Boer War, 38; Davenport, *South Africa*, 97; Thompson, *A History of South Africa*, 136.

[61] *Hart's New Annual Army List, 1897.*

[62] *Monthly Army List, May 1899.*

[63] Mobilization Actions, *Appendices to the Minutes of Evidence taken before the Royal Commission on the War in South Africa*, 1903, [Cd. 1792], no. 3. The report appendix is hereafter cited *Appendices, South Africa Report*.

As a precautionary measure, the War Office, on October 6, 1899, alerted the administrators of the approximately 79,000-man Army Reserve to prepare for a call to the colours. Twelve days later similar alerts were given for mobilization of the about 30,000 men of the Militia Reserve (the deceptively named additional reserve for the Regular Army).[64] At the time, about 78,000 British regular troops were in, or in transit to, South Africa. To some, mobilization of the reserves seemed unnecessary. John Dillon, Irish Nationalist MP, asked rhetorically in Commons why mobilization of the Militia Reserve was needed to defeat an amateur force drawn from a Boer population of at most 275,000.[65]

> "... 78,000 of the flower of the British Army, immense parks of artillery, and all the appurtenances of war. And yet the Government are not satisfied, and wish to call out the Militia Reserve! Anything more mean, cowardly, and disgusting I have not read of in history. It suggests the question, if such stupendous exertions are required for this South African expedition, where you have no organised military force opposed to you, what would the country do if called upon to fight a great European Power, say France, Germany, or Russia?"

On October 12, 1899 the Boers struck first while the British had only 22,104 regular troops in South Africa.[66] They soon surrounded the British garrisons near the Boer republics: Ladysmith in Natal; Kimberley and Mafeking in the Cape. Among those isolated in Kimberley was Cecil Rhodes. At Ladysmith, 13,496 imperial combatants were surrounded: 12,179 British Army soldiers, 280 Royal Navy gunners, and 1,037 dominion troops.[67]

In November and December, with reinforcements arriving daily, the British Army tried to relieve its besieged garrisons.[68] By mid-November, Buller had a force of three infantry divisions (consisting of ten brigades), a cavalry division, and several separate cavalry regiments. Buller assigned one infantry division to relieve Kimberley, another to hold the Orange River border area with the Free State, and the third to relieve Ladysmith. The cavalry division he dispersed throughout the Cape to defend against Boer raids and put down any Afrikaner uprisings. Buller left Cape Town to command personally British forces in Natal. As transport animals had not yet arrived from Britain, Buller's forces could not maneuver far from the rail lines that led to Kimberley and Ladysmith.[69] Accordingly, the British attacked, nearly frontally, the entrenched Boer infantry that was armed with machine guns and supported by artillery. The Boer trench lines held and the relief attempts failed. By the end of 1899, the British Army had incurred 5,700 casualties: 700 killed, 3,000 wounded, and

---

[64] War Office Précis of Events, *Appendices, South Africa Report*, no. 1.

[65] 77 *Parl. Deb.* (4th ser.) *(1899) 388-91.*

[66] Garrison in South Africa, *Appendices, South Africa Report*, no. 5; Total Strength on 1 October 1854, Ibid., no. 10.

[67] Maurice, *History of the War in South Africa,* vol. 2, appx. 2.

[68] Packenham, *The Boer War*, 197-251.

[69] Ibid., 169.

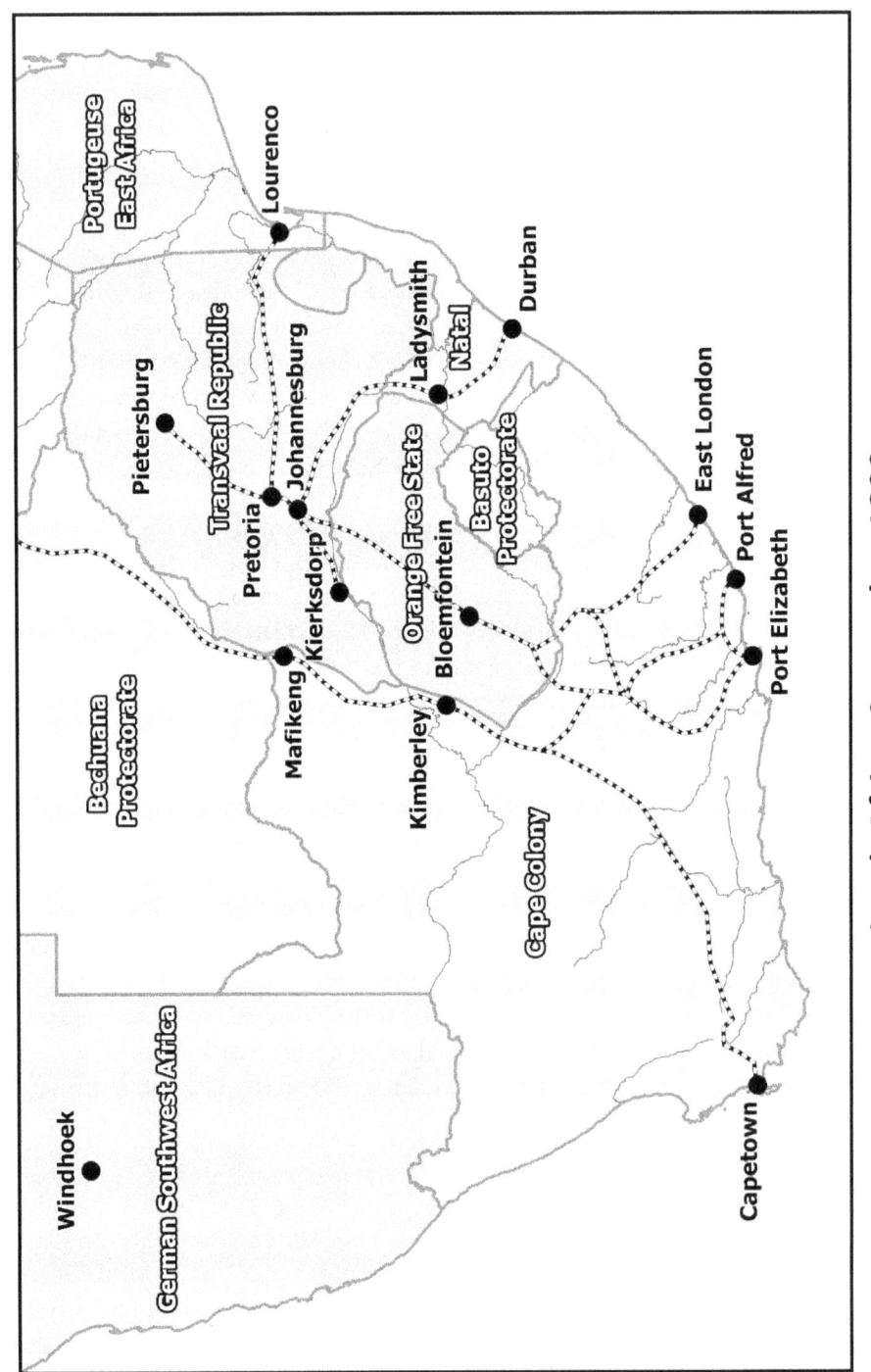

South Africa, September 1899

Source: Eric A. Walker, *Historical Atlas of South Africa* (Oxford: Oxford Univ. Press, 1922).

2,000 captured. Having memories of Magdala and Omdurman, British newspaper readers viewed the war news as reportage of something that simply should not have happened. The professionals of their army had been bested and humiliated by the amateurs of the Boer *commandos*.

The British government sacked Buller and replaced him with Field Marshal Frederick Roberts (Baron Roberts of Kandahar and the City of Waterford), a former East India Company officer who commanded the Irish Military District. Westminster allowed Buller to remain in South Africa and serve as commander in Natal, but subordinate to Roberts. Roberts soon concluded that the 84,016 regulars then in South Africa were insufficient to defeat what he and his lieutenants now viewed as a formidable foe.[70] With support of the army's Commander-in-Chief in London, Roberts requested more soldiers.

Though finding soldiers in peacetime to take the Sovereign's shilling was always a problem for British governments, wartime presented a different situation. During hostilities, Parliament offered short-term enlistments, paid large recruitment bounties, and awarded foreign service bonuses. During the early stages of the war, the personnel problem for the generals and the War Office's senior civilians was not recruitment but the "self-evident" constraints with regards to wartime manpower management:

> 1. Combat units should be of efficient size (for example, an infantry battalion should have at least 800 men).
>
> 2. Units should have supernumeraries of 10% to absorb three-months' "wastage" (combat, accident, and disease loses).
>
> 3. Soldiers on campaign should be "efficient," that is age 19, with six-months training, preferably at least one year of total military service, and meet certain physical requirements with respect to height, weight, and chest diameter.

Home infantry battalions, with respect to efficient soldiers, were usually significantly understrength. This was due to economy and the need to furnish replacements for battalions stationed abroad. Additionally, as the principal function of home battalions was instruction of recruits, many of their soldiers, as novices, did not meet the age and experience requirements for overseas service. For a home battalion to be deployed abroad, it would need to be filled out with reservists. For example, in October 1899, of the four Irish battalions at home and destined for South Africa, the 1st Battalion, Connaught Rangers had the most deployable soldiers, 477, while the 2nd Battalion, Royal Irish Rifles, had the least, 442. A total of 1,616 reservists were required for these four battalions to be brought up to deployment strength.[71] By the end of March 1900, almost all the Regular Army's deployable troops had been sent abroad.

---

[70] Total Strength on 1 December 1899, *Appendices, South Africa Report*, No. 10.

[71] Detail of Field Force, *Appendices, South Africa Report*, No. 3.

The first wave of reinforcements for South Africa totaled 47,081 men, of whom 20,589 were reservists. Subsequent reinforcements, through March 31, 1900, totaled 52,429 men of whom 22,368 were reservists.[72] Of the 103,052 regulars at home on October 1, 1899, within six months 57,366 went to South Africa as reinforcements, 8,533 went there as replacements, and 1,126 went to India as replacements.[73] In addition to troops raised at home, the British Army in South Africa received soldiers provided by the governments of Canada, Australia, and New Zealand. These troops were poorly trained militiamen and wartime volunteers, organized in their own units. During the war, the Dominions sent 31,347 soldiers to South Africa.[74]

In December 1899, with the number of available regulars at home shrinking fast, the government turned to the supposedly trained militiamen, yeomen, and Volunteers to offer their services for South Africa. Additionally, in January 1900, what the government presumed to be "almost" combat-ready civilians (horse enthusiasts with shooting experience) were recruited into new, mounted infantry formations, the Imperial Yeomanry, and sent off to war after a few months training.[75]

By April 1, 1900, the initial war-time recruitment drive had replenished the ranks at home; there were then 103,023 regulars in the United Kingdom.[76] The problem for the War Office was that few of those soldiers were deployable. The at-home total included 37,333 still immature soldiers (who were on the strength in the previous October), raw recruits, reservists who had been found unfit for foreign service, and casualties sent home from South Africa.[77]

At the end of September 1900, the army's manpower pool was exhausted. In addition to the reinforcements of regular units, volunteered militia units, and Imperial Yeomanry, the War Office had sent thousands of replacement soldiers to South Africa to maintain formation strength. The pre-war attrition estimates proved off the mark. During the first eleven months of war, the army dispatched 90,039 replacements to South Africa: 18,414 Army Reservists, 13,014 Militia Reservists, and 58,611 recent recruits who had qualified for active service (age 19 with sufficient training).[78] The original estimate for replacements, based on a planned-for expeditionary force of 80,000 regulars, was about 28,000, one-third the

---

[72] Composition of Units Sent to South Africa, *Appendices, South Africa Report*, No. 8.

[73] Regular Troops for South Africa, *Appendices, South Africa Report*, No. 9.

[74] *Appendices, South Africa Report*, Nos. 24-26.

[75] Pakenham, *The Boer War*, 263-64.

[76] Over the course of the war, approximately 71,000 civilians enlisted to fight in South Africa: 35,000 joined the Imperial Yeomanry and 25,000 joined the Regular Army. Over the 32 months of war, standard enlistments for the Regular Army were 11,000 more than the usual peacetime intake. There was no increase in standard militia enlistments. Recruitment, *Appendices, South Africa Report*, No. 13.

[77] *Report of the Royal Commission on the War in South Africa*, 1903, [Cd. 1789], at 40-41.

[78] Ibid.

actual number sent during the first 11-months of war.[79] The primary cause of British Army wastage was disease; especially typhoid fever and other enteric diseases.

Many of the reading public now asked themselves the same question posed by Dillon a year earlier in Commons. If armed amateurs, drawn from a population equivalent to that of Bradford, England, could exhaust British war capacity, what would, or could, the government do if the United Kingdom were to be threatened by a major, European power?[80] This strategic question was not then addressed by government and army leadership. Conquest and pacification of the Boer states was at the time, the only military matter of concern.

In January 1900, draft animals began to arrive in South Africa from the U.K. and North America, along with further troop reinforcements. British troops were arriving from home, the colonies, and India at the rate of 17,000 per month.[81] By February 1900, the Kimberley relief force had grown from one infantry division to four infantry divisions and a cavalry division.[82] In Natal, the British now had two infantry divisions for the relief of Ladysmith instead of one.[83] The British Army in South Africa, 124,396 men organized into seven infantry and two cavalry divisions, and with sufficient animal transport, could now outflank and then overwhelm the Boers in their defensive positions.[84]

It took nine months of conventional warfare for the reinforced British to dislodge and scatter the assembled Boer *commandos*. Following is a timeline of the British Army's advance:

February 15	Relief of Kimberley.
February 28	Relief of Ladysmith.
March 13	Occupation of Bloemfontein, the Free State capital.
May 16	Relief of Mafeking.
May 31	Occupation of Johannesburg.
June 5	Occupation of Pretoria, the Transvaal capital.
September 24	Control of the Pretoria - Portuguese East Africa rail line.

In October 1900, British intelligence estimated that 30,000 Boer troops remained at large. Roberts began a campaign to starve-out these recalcitrant combatants by burning Boer farms and seizing livestock and crops. His implementation of a scorched earth policy was simply an expansion of the on-going civilian reprisal campaign to discourage Boer attacks on British lines-of-communication.[85]

---

[79] Planned for attrition, all causes, of 10% per quarter, or 34% per year, compounded.

[80] The northern industrial city of Bradford, Yorkshire, had a population of 279,767. *Census of England and Wales*, 1901.

[81] *Report of the Royal Commission on the War in South Africa*, 1903, [Cd. 1789], at 36.

[82] Pakenham, *The Boer War*, 326-27.

[83] Haythornthwaite, *The Colonial Wars Source Book*, 199.

[84] Abstract - All Ranks, 1st February 1900, *Appendices, South Africa Report,* No. 10.

[85] Pakenham, *The Boer War*, 465-67, 486.

Buller left South Africa in November 1900 after Roberts dissolved the Natal Field Force command. The War Office returned him to his previous post, commander of the Aldershot Military District. The government appointed Roberts the British Army's Commander-in-Chief and he left South Africa at the end of 1900. The government replaced Roberts with his chief-of-staff, Major-General Herbert Kitchener (Baron of Khartoum and Aspall). The War Office promoted Kitchener to full general and charged him with ending the guerilla war now raging in the former Boer republics.

Guerilla Warfare: 17 Months Until Peace [86]

Throughout 1901, the *commandos* attacked British supply convoys, small encampments, columns of under 1,000 men, and conducted raids into the Cape. The Boers also attacked the railway and telegraph lines and threatened the Rand gold fields. With both the railways and mines insecure, gold output in 1901 averaged only 10% of pre-war levels.[87] Accordingly, the newly acquired British possessions of the Transvaal Colony and the Orange River Colony were unable to support themselves. The expense of colonial administration and the high cost of the war caused increased government pressure on Kitchener to quickly pacify the new colonies.[88] In response, Kitchener modified the search-and-destroy strategy he had employed against the Boer guerillas. That strategy was not working as the British had only about 50,000 men available to seek out and engage the *commandos*. The bulk of Kitchener's 250,000 men consisted of garrison troops in the Cape, Natal, and occupied Boer towns, plus lines-of-communication security forces and the many men in hospital (typhoid fever). To expedite defeat of the Boer guerillas, whom British intelligence now estimated at about 25,000 combatants, Kitchener implemented two new programs: Internment of Boer civilians and maintenance of fortified lines in sparsely populated areas.

To deprive the guerillas of economic support, Kitchener ordered the removal of Boer women and children from their farms and ranches. These civilians, and their non-white domestic servants, were interned in hastily constructed camps where they were housed, clothed, and fed by the British Army. When Kitchener learned that the vacated Boer properties were being worked by African employees on behalf of the guerillas, the "black Boers" were sent to all-black camps. By mid-August 1901, the "refugee concentration camps" held 93,940 whites and 24,957 non-whites.[89] In addition to inmates forcibly interned by the British, the camps held some Boer civilians who entered the camps voluntarily as destitute, war refugees.

Living conditions in the camps varied greatly and accordingly, so did disease mortality rates. At first, public health and medical measures were inadequate and overall mortality rates rose dramatically. They peaked in October 1901, when the annualized death rate was 34%

---

[86] Source: Pakenham, *The Boer War*, 487-606.

[87] Ibid., 588.

[88] To pay for the war, Parliament raised the income tax rate from 3.33% to 5.00% and authorized unlimited government borrowing.

[89] Pakenham, *The Boer War*, 540.

for the 111,619 white internees and 21% for the 43,780 non-white internees. The government then ordered immediate changes to the operation of the camps. The new War Minister, William St John Brodrick, compelled the army to implement the recommendations of a private investigatory committee headed by Millicent Fawcett. By February 1902, the overall, annualized mortality rate for the camps had fallen to 7%. By the summer it was down to 2%, which was lower than the mortality rate in Glasgow.[90] The annual mortality rate for Boers held by the British as prisoners-of-war fluctuated between 1.5% and 2.0%.[91]

To limit *commando* movements, Kitchener ordered that the Boers' area of operations be laced with mutually supporting defensive positions arranged in lines. These defensive positions were small, circular blockhouses, spaced at about 1,000 meters, between which ran barbed wire fencing. Each blockhouse was manned by six or seven soldiers and one or two native scouts. In this manner great swathes of countryside were crossed with fortified, British lines. The open areas between these lines were patrolled by British columns of cavalry and mounted infantry. Blockhouse lines were also constructed to protect railway lines.[92]

In September 1901, the Boers assembled a 900-man *commando* to enter Natal and raid the British Army's eastern base area. After successes against small British columns in the Transvaal, the *commando* was blocked by the British at the Natal border on September 26th. With 15,000 British troops now approaching them, the Boers abandoned their plan to operate in Natal.

By early 1902, it became clear to the Boer leadership that they had lost the guerilla war. The *commandos* were losing 2,000 men a month; killed, wounded, and taken prisoner.[93] These losses could not be made up for as there were no more recruits to be had from the former republics and Cape Afrikaners had lost enthusiasm for the Boer cause. Kitchener; however, did not view the Boers as having been beaten and also feared for the future as the government threatened to withdraw 110,000 troops to reduce war expense. Exacerbating Kitchener's disquiet was the destruction of two British columns, one on February 4th and the other on March 7th. When two weeks later Kitchener was informed that Boer representatives wished to meet the British in Pretoria, he was eager to parlay with them.

Negotiations began on April 11th between a delegation headed by Kitchener and ten leaders of the former republics. High Commissioner Milner joined the British delegation three days later. Milner, unlike Kitchener, believed Boer resistance was about to collapse and did not want a negotiated settlement. Milner's position was bolstered by two major, British military successes in late April. Milner; however, could not prevail upon Kitchener as the Cabinet wanted to bring home the troops as soon as possible. Hostilities ceased on May 31, 1902 when a Boer representative assembly ratified the agreement reached by the delegates in Pretoria. The vote was 54 in favor, 6 opposed.

---

[90] Ibid, 548-49.

[91] Maurice, *History of the War in South Africa*, vol. 4, appx. 20.

[92] Ibid., appx. 2.

[93] Ibid., appx 20.

The Boer War's Legacy

British intelligence estimated that 89,000 men fought for the Boer republics. Transvaal provided about 43,000 combatants and the Orange Free State 29,000. Others that fought with the Boers were 13,000 Afrikaners from the Cape and Natal and 3,400 foreigners. Among the foreigners were up to 1,200 Irish. Most Boer combatants served intermittently and at no time did the Boer Republics field more than 35,000 troops.[94] To defeat the Boer army of amateurs, the British Empire employed almost 450,000 soldiers.

Table 8.
Source of British Imperial Troops in South Africa
(with Extent of Contribution)

Nation	Troops	Percent of Total	Population	Troops per 10,000
United Kingdom	364,674	81.3	32,527,813	112
South Africa (whites)	52,414	11.7	676,850	774
Australia	16,632	3.7	3,774,565	44
Canada	8,372	1.9	5,378,800	16
New Zealand	6,343	1.4	772,719	78
Total	448,435	100.0	43,179,963	104

Sources: Maurice, *History of the War in South Africa,* Vol. 4, appx. 13; Canadian, Australian, and New Zealand Contingents, *Appendices, South Africa Report,* nos. 24-26.

The casualty rates for the British Army were much lower in the Boer War than in earlier, major conflicts. In the Napoleonic Wars the overall British mortality rate was 33.7% and in the Crimean War 21.5%. For the Boer War, 4.6% of all British soldiers deployed in South Africa died, another 4.6% were wounded.[95] As in previous wars, deaths from disease far outnumbered those from combat.

Table 9.
Mortality Rates for Imperial Forces in South Africa
(October, 1899 - May, 1902)

Force	Total Troops	Combat Deaths	Disease Deaths	Total Deaths	Mortality Rate
British Army	364,674	5,679	11,095	16,774	4.6%
South African	52,414	1,473	1,607	3,080	5.9%
Other Colonial	31,347	430	437	867	2.8%
All Forces	448,435	7,582	13,139	20,721	4.6%

Sources: Maurice, *History of the War in South Africa,* vol. 4, appx. 17; Canadian, Australian, and New Zealand Contingents, *Appendices, South Africa Report,* nos. 24-26.

---

[94] *Report of the Royal Commission on the War in South Africa,* 1903, [Cd. 1789], at 157-58.

[95] Maurice, *History of the War in South Africa,* Vol. 4, appx. 17.

The high number of disease deaths relative to combat deaths for the British troops is because of disproportionate British participation in the campaigns to relieve Kimberley and Mafeking and then capture Bloemfontein and Pretoria. Those campaigns experienced severe epidemics of typhoid fever. Note that the disease to combat death ratio for the British Army is about 2:1, while for dominion forces it is about 1:1. In addition to combat and disease deaths, 1,221 deaths from accidental and other causes for all forces is indicated by War Office returns.[96] Including such deaths, the overall, imperial force mortality rate was 4.9%.

The low mortality rate in the Boer War was due primarily to the nature of the conflict. After the initial battles of late 1899 and early 1900, there were few conventional engagements and enemy forces were much smaller than those faced in the Napoleonic and Crimean Wars. Other factors also contributed to the decline in mortality rates. Troops were better clothed, quartered, and fed as the army's logistical organization had expanded and became more effective. Also, the government provided more trained medical personnel than in the past. At the outbreak of hostilities, the Royal Army Medical Corps had 297 personnel in South Africa, 1.4% of total troop strength. By July 16, 1900, when total troop strength was 233,644, medical staff totaled 9,904, 4.2% of total strength. Even with these improvements, sanitary conditions and medical care within the army were substandard through mid-1900. The British government's commission to investigate the care of the sick and wounded drew the following conclusion:[97]

> "The military and medical authorities certainly never anticipated when this war became probable that it would be of the magnitude it has since attained. The Royal Army Medical Corps was wholly insufficient in staff and equipment for such a war, and it was not so constituted as to have means provided by which its staff could be very materially enlarged, or its deficiencies promptly made good. These deficiencies were felt throughout the South African campaign …"

Typhoid fever was the primary killer of British troops. Poor field sanitation during 1899 and early 1900 caused severe fever epidemics within the Kimberley relief force, the newly established British base at captured Bloemfontein, and the columns that headed north to Mafeking and Pretoria.[98] At first, conditions in army hospitals were egregiously poor and numerous patients died from inadequate nursing care and unsanitary conditions. Additionally, most of the medical officers had no experience with typhoid fever as the disease

---

[96] A War Office return dated October 2, 1902, shows total deaths of 21,942 which is 1,221 more than deaths given in General Maurice's official history at Appendix No. 5.

[97] *Report of the Royal Commission appointed to consider and report upon the Care and Treatment of the Sick and Wounded during the South African Campaign*, 1901, [Cd.453], at 4. Hereafter cited as *Report on Care and Treatment of Casualties in South Africa*.

[98] Pakenham, *The Boer War*, 402; *Report of the Royal Commission on the War in South Africa*, 1903, [Cd. 1789], at 106.

had been eradicated in the United Kingdom.[99] Overly bureaucratic RAMC hospital practices also contributed to poor patient outcomes.[100] Lastly, there was the problem of insufficient hospital supply. The advancing British forces were supplied by trains running on congested, single-track railway lines. Movement of troops and war-like stores had transport priority over medical supplies and hospital provisions.[101] Medical care, public health, and field sanitation deficiencies were primarily in the Cape-based forces which were directly under Roberts' command. Buller's force in Natal did not experience egregious health-related deficiencies.[102]

Table 10.
Disease Incidence for Imperial Forces in South Africa

	British Army	Dominion Forces	All Forces
Climatic Disease Cases	138,591	21,223	159,814
Climatic Disease Deaths	8,674	1,043	9,717
Mortality Rate, Climatic	6.3%	4.9%	6.1%
Other Disease Cases	193,326	32,982	226,308
Other Disease Deaths	1,237	256	1,493
Mortality Rate, Other	0.6%	0.8%	0.7%

Source: Simpson, "Medical History of the South African War."

The British disease mortality rate for the Boer War was relatively low: 2.8% compared to 16.5% for the Crimean War. The prevalence of disease among troops in South Africa; however, was extraordinarily high. During the war, diagnosed cases equated to 87.8% of imperial troops that served in the conflict. Those hospital cases included the 36.4% of troops with deadly "climatic" diseases: typhoid fever, malaria, and dysentery.

The total monetary cost of the Boer War to the United Kingdom, including expenses for disabled veterans, widows, and orphans, plus expenses incurred by departments other than the War Office, was probably £200 million.[103]

Though the British public viewed the South African hostilities as a "white man's war" the Cape Coloured and Africans played an important military role. At first there was tacit agreement between the Boers and British that they would not arm the non-white population. Non-whites; however, were employed by both sides as scouts, messengers, and drivers. For

---

[99] *Report of the Royal Commission on the War in South Africa*, 1903, [Cd. 1789], at 105. Note that of the 968 physicians and surgeons then with the army, 481 were British civilians engaged by either the RAMC or Voluntary Hospitals. Strength of the Medical Corps, *Appendices, South Africa Report*, no. 39.

[100] *Report on Care and Treatment of Casualties in South Africa*, at 17-18.

[101] Pakenham, *The Boer War*, 402-04.

[102] Ibid., at 64-66.

[103] Pakenham, *The Boer War*, xix; Expenditure Incurred on Army Votes in consequence of the War in South Africa, *Appendices, South Africa Report*, No. 50.

example, the British Army Service Corps employed 6,771 native wagon drivers.[104] The most important role for African natives was trench diggers and many such laborers were pressed into service by both sides. Probably 100,000 non-whites in total served with the Boers and the British.[105] The British did not adhere fully to the "white combatants only" constraint. Colonel Baden-Powell armed 300 Africans in the defense of besieged Mafeking and Kitchener armed all 10,000 native scouts in 1901.[106] Boers often executed captured, non-white British support personnel, and near always killed such prisoners if they were armed.[107]

British intelligence estimated that 89,000 men served with the Boer forces at some time during the war. Of these, 13,000 were Afrikaners of the Cape and Natal. Deaths from combat and disease through December 31, 1901, were estimated at 4,800 which gives a mortality rate of 5.4%; a rate 17% higher than that for imperial forces. At the end of 1901, the British held in captivity 25,472 Boer prisoners-of-war and 7,587 Afrikaner rebels.[108] In 1902, the Boers incurred about another 4,300 casualties of all types (killed, wounded, captured).

Boer civilian deaths far outnumbered *commando* deaths. In all, about 100,000 to 125,000 civilians of the Boer republics died on account of the war.[109] Of those deaths, 31,000 to 45,000 occurred in concentration camps. This was due to poor living conditions in the camps through which passed at least 250,000 people.[110] From 130,000 to 140,000 Boers were in concentration camps of which 20,000 to 28,000 died (mortality rate of 18-20%). Another 120,000 to 130,000 "Black Boers" were in camps of which 13,000 to 20,000 died (mortality rate of 11-15%).[111] Most of the dead were children. At the end of hostilities, 20,604 men and 117,373 women and children were still interned in concentration camps.[112]

The Irish and the Boer War

The great majority of Irish nationalists opposed British military action against the Boers.[113] Among the anti-war Irish was Joyce's father, John Stanislaus.[114] Early in the war, one prominent Irish MP stated in Commons that "the Irish Nationalist representatives would not dare to face their constituents at the next General Election unless they opposed this

---

[104] Report on Field Transport, *Appendices, South Africa Report*, No. 33B.

[105] Pakenham, *The Boer War*, xxi.

[106] Ibid., 425, 580-81.

[107] Ibid., xxi, 608.

[108] Estimate of Strength of the Boer Forces on 31st May 1902, *Appendices, South Africa Report*, No. 60.

[109] Spies, *Methods of Barbarism*, 148.

[110] Ibid., 215, 262, 265-66.

[111] Ibid.; Pakenham, *The Boer War*, 607-08.

[112] Grant, *History of the War in South Africa*, Vol. 4, appx. 12.

[113] McCracken, *Forgotten Protest*; Diver, "Ireland's South African War."

[114] During a 1900 trip to London by Joyce and his father, John Stanislaus Joyce made clear his pro-Boer sympathies to several jingoist Englishmen. Bowker, *James Joyce*, 74.

infamous war by every means which the forms of the House permitted."[115] Four months after the start of hostilities, John Redmond, leader of the Irish Parliamentary Party, spoke out against the war in the reconvened Commons. "The sympathy of Ireland is with the two South African Republics. We abhor this war; we call for its stoppage…"[116] Separatists who sought an Irish republic were unanimous in opposition to the war. Among home-rulers; however, there was some support for British imperial ambitions in South Africa.[117] The pro-war nationalists, though a minority, were vocal in their opposition to Irish pro-Boer sentiment. Irish unionists, nearly all Protestants, supported the British war effort, though at times such support was lukewarm.

Possibly as many as 1,200 Irish risked prosecution for treason and fought with the Boers. About 400 served in two Irish "brigades" and up to another 800 in Boer *commandos*. In one engagement, an Irish brigade fought Irish regiments of the British Army, including the Royal Dublin Fusiliers.[118] About 30,000 Irish served in South Africa with the British Army and despite nationalist anti-war sentiment their exploits were generally admired at home.[119] Those soldiers included about 2,700 men of five militia infantry battalions and two militia garrison artillery companies that volunteered for service in South Africa.[120]

Most nationalists believed that the British high command in South Africa treated Irish soldiers as cannon fodder.[121] This belief most likely arose from reports of British losses early in the war. What registered strongly in Irish minds were the Boer capture of six companies of Royal Irish Fusiliers in October 1899, the high casualties incurred by the Royal Dublin Fusiliers in the 1899/1900 campaign to relieve Ladysmith, and the Boer capture of the 13th (Irish) Yeomanry Battalion in May 1900. In actuality, the combat casualty rates of Irish regiments in South Africa were about the same as those of English, Scottish, and Welsh regiments. Of the 80 Regular Army battalions that served in South Africa, 10.0% were Irish.[122] As noted previously, the British Army suffered 5,679 combat deaths during the Boer War. Irish regiments incurred 579 of those deaths, 10.2% of the army's total.[123]

Throughout the war, Irish republicans conducted an anti-recruitment campaign to keep Irishmen from fighting the Boers. Tactics included rallies, defacement of recruiting posters, distribution of pro-Boer literature, a press campaign, threats of social reprisal, and physical

---

[115] Michael Davitt, MP Mayo, South, debating in opposition to the government's proposal to mobilize the Militia Reserve. 77 Parl. Deb. (4th ser.) (1899) 392.

[116] 78 *Parl. Deb.* (4th ser.) (1900) 831.

[117] McCracken, *Forgotten Protest*, 100-110.

[118] Ibid., 127-31.

[119] Diver, "Ireland's South African War."

[120] *Appendices, South Africa Report*, No. 14.

[121] McCracken, *Forgotten Protest*, 52.

[122] *Appendices, South Africa Report*, Nos. 7-9. Of the 60 militia battalions that served in South Africa 6 were Irish (10.0%). Ibid., No. 16.

[123] McCracken, *Forgotten Protest*, 134.

assaults on soldiers.[124] The results of this effort were mixed. From 1889 through 1901, the Regular Army enlistment rate in Ireland fell while in the rest of the United Kingdom enlistment rates rose.[125]

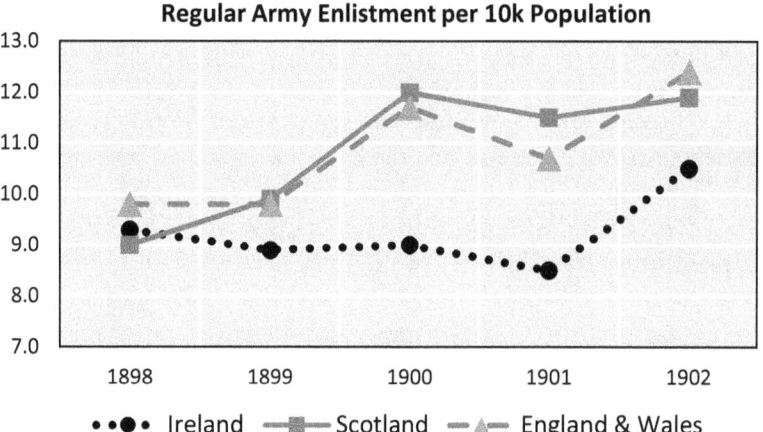

The rate of Imperial Yeomanry enlistment in Ireland corresponded to that for Great Britain.[126] Militia enlistments in Ireland decreased markedly during the war while in Great Britain they held steady. Overall, Irish young men were far less enthusiastic for the war than their British counterparts; however, a great number of Irishmen, mostly nationalists, joined the army to fight the Boers.[127]

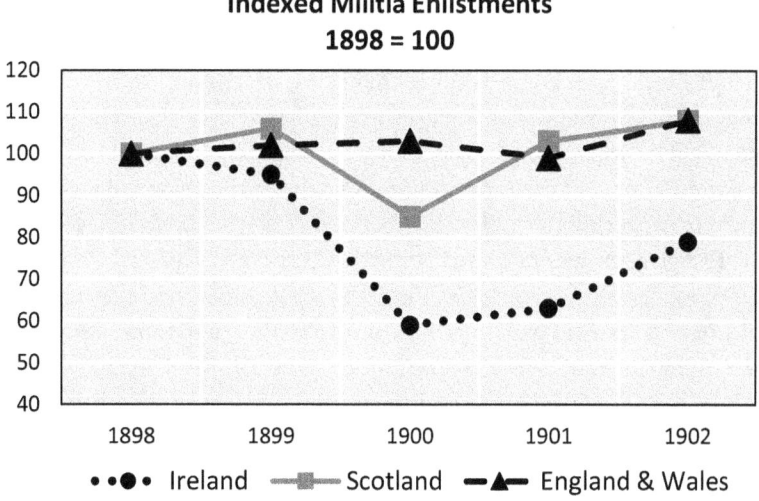

---

[124] Ibid., 45, 50-52, 107-08; Denman, "The Red Livery of Shame."

[125] War Office: *General Annual Return for the British Army for the Year 1898*, 1899, [C. 9426]; *Annual Report of the Inspector-General of Recruiting for the First Nine Months of 1903*, 1903, [Cd. 1778].

[126] Ireland, with 10.8% of the UK population, provided about 10% of the territorial Yeomanry companies (7 of 69). *Monthly Army List, May 1901*.

[127] McCracken, *Forgotten Protest*, 107-08.

Late Victorian Military Campaigns and Army Reform

The Cardwell, Childers, and Haldane Reforms

British Army Service Prior to 1870 [128]

After Napoleon's exile to St. Helena in 1815, the British Army plodded along in its old ways with major interruption by the Crimean War and the Indian Mutiny. The army at home was essentially a factory where working-class recruits were the raw material input, and soldiers ready for dispatch abroad, the finished product. The "goods" were transported to imperial customers (colonial governors and the British East India Company) by army troop ships and Britain's mercantile fleet, protected by the Royal Navy. Acquisition of raw material was the principal concern of the War Office; product delivery was an Admiralty matter. The army's middle-management (staff officers and regimental commanders) dealt with the manufacturing process, termed "musketry instruction" and "drill." Like business executives who seek profit maximization through cost minimization, government ministers strove to minimize army cost to maximize the discretionary income of their constituents, the British establishment. The less the upper and middle classes paid in taxes, the more money they had for themselves. As a result, in the early nineteenth century a soldier's life was dreadful. Accordingly, except in wartime when large bonuses were paid to war-service recruits, the number of enlistees was insufficient to meet strategic demands.

In the early nineteenth century, a British soldier's life was comparable to that of workhouse and prison inmates. Lodgings were uncomfortable and unhealthy; the food was both disgusting and inadequate nutritionally. Accordingly, the mortality rate for soldiers in domestic barracks was about twice that of the comparable civilian population. Discipline was strict and punishment, which included flogging, was brutal. A soldier's daily routine consisted of fatigue assignments (labor details), unnecessary guard duty, close-order drill, some physical conditioning, and needless cleaning and polishing of barracks and uniform accoutrements. Sunday mornings soldiers attended mandatory church services of their chosen denomination ("church parade"). The army provided no sports or other recreational facilities and a soldier spent his spare time, if he had the money, drinking beer in the regimental canteen.

Though promised a shilling a day by recruiters, new enlistees never received such remuneration. The government expected soldiers to support themselves on their pay. After deductions for food, clothing, laundry, uniform upkeep, and barracks damages, privates were left with about two to five shillings a month in discretionary income. Opportunities for a soldier to increase his income through promotion were few. Regiments had a wide span of supervision, there being only 8 NCOs in a company of 100 men, and half of those NCOs were corporals. There were few staff positions for enlisted personnel and commissions from the ranks were rare. In any event, very few soldiers had the literacy and arithmetic skills to qualify as an officer. Financially, a soldier's life was better overseas, especially in India. Abroad, soldiers received subsidized rations and the cost of food and entertainment (alcoholic beverages and prostitutes) was much cheaper than at home. The higher standard

---

[128] Skelley, *The Victorian Army at Home*, 21-300; Spiers, *The Army and Society*, 35-176; Farwell, *For Queen and Country*, 79-104, 177-223.

of living abroad came at a price; the mortality rate overseas, especially in the tropics, was notably higher than at home. Exceptions to the health peril abroad were North America, most of Australia, parts of South Africa, and New Zealand. During campaigns, soldiers had the opportunity to receive legal booty and engage in unlawful looting. This benefit came with a price; risk of combat death or disability.

In the post-Napoleonic Wars army, recruits enlisted for life or for 7-year, renewable terms. After 21-years' service in the infantry, 24-years in the other arms, soldiers were eligible for a pension. In 1829, the government ended the 7-year engagement as it had few takers as life-enlistments came with a substantial cash bonus. In 1847, the government terminated open-ended enlistment and introduced 10-year terms for the infantry and 12-year terms for the other arms. After completion of their initial engagement, soldiers could re-enlist, at the army's discretion, to attain pension eligibility. Half or more of a soldier's service was spent abroad where the risk of disability and death from disease was high. Wounded soldiers unfit to serve received permanent pensions. Soldiers invalided out of the army because of disease received only temporary pensions which lasted at most three years. Disability pensions, like longevity pensions, were meager. Veterans who could not obtain employment, or family support, often begged to survive.

Nearly all army recruits were desperate, uneducated, and unemployed young men. After acceptance for service, recruits for the infantry and cavalry went to regimental depots where they received about three months of indoctrination and rudimentary training. If a regiment was at home, it received recruits upon their completion of this basic training. If a regiment was abroad, it received recruits after they were seasoned at the depot for an additional nine months. Recruits for the engineers and artillery went to specialized training depots. There they received technical training for about a year before dispatch to an engineer company or artillery battery, either at home or abroad.

The life of a British soldier improved somewhat during the mid-nineteenth century. Barracks became healthier, net pay increased slightly, and the quality of food was better than in 1815. The government built separate family quarters for the few enlisted men that had received permission to marry and established army schools for both soldiers and their children. Medical care also improved and remained free of charge for soldiers and their on-the-strength dependents.[129]

The Era of British Army Reform, 1870 - 1912

Unlike the Continental nations, which by the end of the nineteenth century had all adopted compulsory short-term, full-time military service followed by a long-term reserve obligation, the United Kingdom maintained a volunteer army. Among the overwhelming majority of politicians, compulsory military service in any form was unacceptable. Proposals by advocates of the Prussian-style, compulsory reserve system were dead-on-arrival at

---

[129] Wives and children of soldiers that had married with their commanding officer's permission were part of the regimental establishment. Such dependents received free housing, very low-cost schooling, and subsidized food. On-the-strength wives had the opportunity to hold regimental jobs such as laundress, teaching assistant, seamstress, servant for an officer's family, and if educated, governess of an officer's children. Trustram, *Women of the Regiment*, 30-42, 75-79, 105-35.

Westminster. Also rejected by all governments were calls by several senior army officers for reinstatement of the militia ballot (compulsory service by lot). To increase the British army's war-fighting capacity, successive governments introduced slowly, beginning around 1870, a series of changes to the recruitment and management of the army. Historians refer to these reforms by the names of the implementing, Liberal government War Ministers: Edward Cardwell (1868-1874), Hugh Childers (1880-1882), and Richard Haldane (1906-1912). Brian Tweedy's army service encompassed the administrations of Cardwell and Childers.

Both the War Office and Parliament made changes to army life to make a military career more attractive to potential recruits. Some of these Cardwell-Childers reforms that affected the enlisted ranks are as follows:[130]

> Increased net, basic pay for all enlisted ranks.
>
> Instituted longevity and good conduct pay.
>
> Increased the number of NCO positions and thereby improved promotion prospects for enlisted men.
>
> Improved food quality and increased the number of daily meals from two to three.
>
> Converted the senior NCO ranks of each arm of service into "warrant officer" appointments and gave holders of such ranks greater privileges and status than they had as senior NCOs.
>
> Abolished flogging.
>
> Continued barracks modernization and construction of separate family quarters.
>
> Instituted payment of family separation allowances.
>
> Provided enlisted men with libraries, game rooms, and sports facilities.

Cardwell and Childers, and their respective governments, through changes in army administration, tried to increase the army's combat capability. The most notable of these "reforms" were as follows:[131]

> Ended the purchase of infantry and cavalry commissions and promotions.
>
> Granted pension rights to officers and established maximum ages for serving officers based on rank and duties.
>
> Instituted promotion of officers by examination.

---

[130] Skelley, *The Victorian Army at Home*; Farwell, *For Queen and Country*, 87-90, 96-109, 227-32; Trustram, *Women of the Regiment*, 90-91.

[131] Spiers, *The Army and Society*, 177-205; Skelley, *The Victorian Army at Home*, 301-05; Farwell, *For Queen and Country*, 153-64.

Increased the capacity of the army staff college to train more junior officers for senior staff appointments.

Instituted "short-service" to create an army reserve. Under this scheme, men engaged for twelve years of total service split between the Regular Army and the Army Reserve. The War Office set a different length of full-time service for each branch of the army. During the late nineteenth century, minimum service with the Regular Army ranged from one to eight years. During a soldier's initial engagement, he could request to both extend his short period of regular service and re-engage for an additional nine years. This would allow the soldier to serve for 21 years and attain pension eligibility. Extension or re-engagement required command and War Office approval.

Implemented territorialization of infantry regiments. All line infantry battalions were linked with militia and volunteer force battalions as a single regiment with a common depot. Each regiment had an assigned recruiting territory known as a territorial district. The depot staff provided basic training for regulars and militiamen, administered reservists that resided within the district, and cared for stores of uniforms, accoutrements, and weapons. Typically, each regiment had two regular battalions that shared a unitary promotion list for officers, and three to five auxiliary battalions (militia and volunteer force) based in the district. The partial exception to this scheme were the guards units, the King's Royal Rifles, and the Rifle Brigade. These elite regiments had no specified territories and were linked with auxiliary battalions located throughout the United Kingdom. Cardwell believed that through territorialization regiments would form "ties of kindred and of locality" and reverse the decline of enlistments from rural areas.[132] Though regiments now had recruiting districts, young men remained free to choose the regiment in which to enlist. Note that territorialization applied only to the infantry.

Provided more technical training for both officers and other ranks.

Increased somewhat, the relative number of support and service troops.

It was Cardwell who pushed the government to reorganize the War Office. The War Office Act of 1870, as implemented by Orders in Council, rationalized somewhat the existing administrative structure.[133] For example, all arms and departments, other than the ordnance (procurement, manufacturing, and construction), were all made subordinate directly to the Commander-in-Chief. The Permanent Under-Secretary, as in all government departments, was the office's senior civil servant. All department heads subordinate to the Commander-

---

[132] Spiers, *The Army and Society*, 196.

[133] 33 & 34 Vict., c. 17.

Late Victorian Military Campaigns and Army Reform

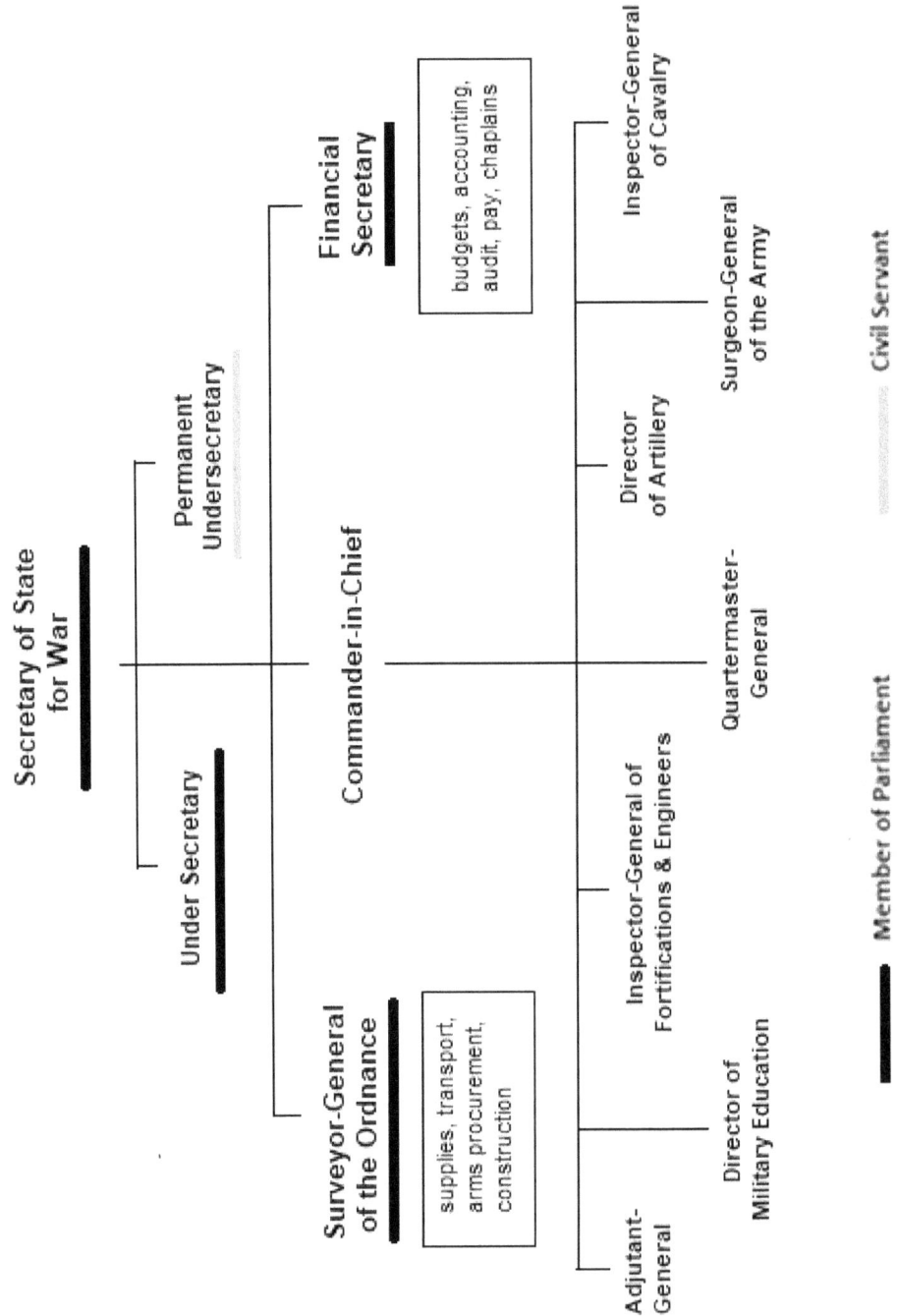

in-Chief were general officers. Military intelligence, recruitment, and supervision of the infantry and auxiliary forces were in the purview of the Adjutant-General, the army's chief administrative officer. This structure was changed in the late 1880s by abolition of the Surveyor-General's position and allocation of his responsibilities between the Commander-in-Chief and the Financial Secretary.[134]

The Boer War showed the shortcomings of the War Office's organization. Accordingly, the post-war Conservative government implemented a few reforms. The subsequent Liberal government made further changes. On February 6, 1904, by Letters Patent, it abolished the position of Commander-in-Chief and created the Army General Staff and the Army Council.[135]

The general staff was responsible for training, officer education, collection of intelligence, and formulation of war plans. In time of war, the Chief of the General Staff would be the immediate superior of any expeditionary force commander.[136] Among the General Staff's duties was cooperation with the Royal Navy in all matters of imperial defense. Note that in 1890 a royal commission concluded that

> "... little or no attempt has ever been made to establish settled and regular inter-communication or relations between them, or to secure that the establishments of one service should be determined with any reference to the requirements of the other."[137]

Within Childers' War Office, on-going, inter-service cooperation was limited to the Director of Military Intelligence's authority "to correspond semi-officially with other departments of state."[138] The Army Council, modeled on the Admiralty Board, was the army's governing body. The Secretary of State for War formulated policy in consultation with the other council members and all army orders and regulations were promulgated in the name of the Army Council.[139]

The capstone of reform that would give Britain its army of the First World War resulted from the Norfolk Commission's investigation into the state of Britain's auxiliary forces and the Esher Commission's review of the army's prosecution of the Boer War.[140] Issues

---

[134] Orders in Council, December 27, 1887, February 21, 1888.

[135] Wheeler, *The War Office Past and Present*, 270-71.

[136] Ibid., 298.

[137] *Preliminary and Further Reports of the Royal Commissioners appointed to enquire into the Civil and Professional Administration of the Naval and Military Departments*, 1890, [C. 5979], at vi.

[138] Ibid., at 95.

[139] *Report of the War Office (Reconstituted) Committee, Part II*, 1904, [Cd. 1968], at 5-7.

[140] *Report of the Royal Commission on the Militia and Volunteers*, 1904, [Cd. 2061]; *Report of His Majesty's Commissioners appointed to inquire into the military preparations and other matters connected with the War In South Africa*, 1903, [Cd. 1789].

addressed by the commissions were resolved during Haldane's tenure as War Minister. This last series of structural reforms began in 1906.[141]

In August 1914, at the outbreak of war, Britain placed in France a 120,000-man expeditionary force organized in seven divisions (six infantry and one cavalry). This force, despite 90,000 casualties, grew to about 220,000 men by the end of the year (plus about 25,000 men remaining in the Indian Army Corp brought to France in October). Britain's volunteer army managed, despite numerous casualties, to grow to 2.0 million men by December 1915. At that point, the well of potential volunteers ran dry. On January 1, 1916 Parliament passed the Military Service Act which, with a few exceptions, imposed war-time, compulsory military service on the young men of England, Wales, and Scotland. The Irish were expressly exempt from compulsory service, both uniformed and industrial.[142]

---

[141] Spiers, The *Army and Society*, 271-72.

[142] Haythornthwaite, *The World War One Source Book*, 211-17, 245-46.

Chapter Bibliography

Bowker, Gordon. *James Joyce A New Biography*. New York: Farrar, Straus & Giroux, 2011.

Davenport, Rodney and Christopher Saunders. *South Africa a Modern History*, 5th Ed. New York: MacMillan, 2000.

David, Julian Saul Markham. "The Bengal army and the outbreak of the Indian mutiny." PhD thesis, University of Glasgow, 2001.

Denman, Terence. " 'The Red Livery of Shame': The Campaign against Army Recruitment in Ireland, 1899-1914," *Irish Historical Studies* 29, no. 114 (November 1994): 208-33.

Diver, Luke. "Ireland's South African War 1899-1902," *Scientia Militaria, South African Journal of Military Studies* 42 (2014):1-17.

Douglas, George and George Dalhousie Ramsay, eds. *The Panmure Papers*. Vol. 2, London: Hodder & Stoughton, 1908.

Farwell, Byron. *For Queen and Country*, London: Penguin, 1981.

Maurice, Frederick and Maurice Harold Grant. *History of the War in South Africa 1899-1902*. Vol. 4, London: Hurst & Blackett, 1910.

Haythornthwaite, Philip J. *The Colonial Wars Source Book*. London: Arms & Armour, 1995.

────── *The World War One Source Book*. London: Arms & Armour, 1992.

Heathcote, T.A. *The Indian Army*. Vancouver: David & Charles, 1974.

Holland, Trevenen and Henry Hozier. *Record of the Expedition to Abyssinia*. London: HMSO, 1870.

Maurice, Frederick and Maurice Harold Grant. *History of the War in South Africa 1899-1902*. Vol. 2, London: Hurst & Blackett, 1907.

────── *History of the War in South Africa 1899-1902*. Vol. 4, London: Hurst & Blackett, 1910.

McCracken, Donal P. *Forgotten Protest*. Belfast: Ulster Historical Foundation, 2003

Meredith, Martin. *Diamonds, Gold, and Wars*. New York: Simon & Schuster, 2007.

Pakenham, Thomas. *The Boer War*. New York: Random House, 1979.

Pati, Biswamoy, ed. *The 1857 Rebellion*. Oxford: Oxford Univ. Press, 2007.

Reid, C. Lestock. *Commerce and Conquest*. London: Kennikat, 1947.

Robins, Nick. *The Corporation that Changed the World*, 2nd Ed. London: Pluto, 2012.

Simpson, R.J.S. "Medical History of the South African War." *Journal of the Royal Army Medical Corps* 16-1 (January 1911): 20-40.

Skelley, Alan Ramsay. *The Victorian Army at Home*. London: Croom, Helm, 1977.

Spiers, Edward M. *The Army and Society 1815-1914*. London: Longman, 1980.

Spies, S.B., *Methods of Barbarism?* Cape Town: Human & Rousseau, 1977.

Thompson, Leonard. *A History of South Africa.* New Haven: Yale Univ. Press, 1990.

Trustram, Myna. *Women of the Regiment.* Cambridge: Cambridge Univ. Press, 1984.

Wagner, Kim A. *The Great Fear of 1857.* Witney, UK: Peter Lang, 2010.

Wheeler, Owen. *The War Office Past and Present.* London: Methuen, 1914.

Chapter 5
The Armies of the British East India Company

Of the eight Irish regiments in the British Army of 1904, four had at least one battalion with a British East India Company antecedent.[1] For two of those regiments, the Royal Munster Fusiliers and the Royal Dublin Fusiliers, both of their regular battalions were former Company units. In 1858, the European regiments of the EIC were transferred temporarily, by statute, to the new Indian Army.[2] Three years later, those regiments were incorporated into the British Army.[3] In *Ulysses*, Joyce unequivocally describes Molly Bloom's father as having served in the Royal Dublin Fusiliers and makes no mention of any other British Army regiment. Accordingly, Brian Tweedy began his military career in service to "John Company" and not the Crown.

History and Organization of the EIC and its Armies [4]

In 1600, Queen Elizabeth I authorized a group of London merchants to organize trading ventures to the East Indies. The merchants' royal charter granted a monopoly on the import of all goods from east of the Cape of Good Hope to the Straits of Magellan. This gave the new "Company of Merchants of London trading into the East Indies" dominion over all English trade with the Indian subcontinent, Southeast Asia, China, and Japan. The initial franchise was for 15 years but it was renewed by subsequent sovereigns. The first trading venture promoted by the Company encompassed four, small vessels that sailed from London on February 13, 1601.

For 50 years the Company of Merchants promoted trading ventures to the east, each venture capitalized separately by outside investors. Though the Company had a royal monopoly, other merchants received similar "monopolies" from Stuart kings and Cromwell's Commonwealth. In 1657, with approval of the Commonwealth government, the Company of Merchants merged with its competitors and reorganized as a permanent stock company. The new United British East India Company raised capital of £369,000 through sale of 3,690 shares at £100 each. The shares represented a continuous unlimited investment

---

[1] Six of the sixteen regular battalions of the Irish regiments were former European regiments of the EIC. The Leinster Regiment and the Royal Inniskilling Fusiliers each had a regular battalion that was a former EIC regiment. *Hart's Annual Army List, 1905.*

[2] Government of India Act, 1858. 21 & 22 Vict., c. 106.

[3] General Order No. 332 of 1861 by the Governor General of India in Council, *The Calcutta Gazette*, Extraordinary, April 22, 1861.

[4] Sources: MacMunn, *The Armies of India*, 1-36; Reid, *Commerce and Conquest*, 1-201; James, *Raj*, 1-206; Robins, *The Corporation that Changed the World*.

without reference to individual voyages. Only shareholders with five more shares could vote for directors and each such shareholder had one vote regardless of holdings. Shareholders with at least 20 shares could serve as directors. Control of the Company was effectively in the hands of a small group of affluent merchants.

In 1611, the Company of Merchants established its first Indian trading station at Machilipatnam on the Bay of Bengal. By 1650, the Company had four permanent facilities, including one at Madras and one at Calcutta. The Company, like all traders in the East, engaged armed guards to protect its warehouses and offices. In 1627, the Company began to mount cannon at its stations. Fortification was a measure to deter attacks by rival Dutch, French, and Portuguese traders.

Seventeenth century international traders sought exorbitant profits through monopoly pricing, both in selling and buying. At home, they sought from their governments national monopolies on imports either of specified goods, or all goods from a delineated overseas territory. If that company was the only importer at home, it could price its goods at whatever the market could bear. Abroad, traders sought from local rulers a monopoly on foreign exports. If a trading company was the only authorized foreign purchaser, it would not have competitors that could bid up the cost of goods. The quest for exclusive export rights abroad let to armed engagements among the British, French, Portuguese, and Dutch. Dutch arms prevailed in what is now Indonesia and the Company abandoned trade with the Spice Islands and their environs to concentrate on the Indian subcontinent.

The armies of the EIC grew out of the corps of guards for its trading stations. The earliest military force consisted of three widely separated detachments. One, near Calcutta, was a force of 31 infantry and a single canon and its crew. Another was the British garrison at Bombay, received by England from Portugal in 1662 and transferred subsequently by the Crown to the EIC.[5] Finally, there were several infantry companies at Madras, which was threatened by nearby French trading stations. The early EIC units consisted of both Indian and European mercenaries. In 1748 wholly European companies "were formed from detachments sent from England, from runaway sailors, men of disbanded French corps, from Swiss and Hanoverians, from prisoners of war, and any white material in search of a livelihood."[6] At first, the Company's mercenaries fought soldiers of other European trading companies, especially the French. Sometimes the EIC defended its stations and trading territories against attacks by rivals; other times it attacked its rivals' stations. The nature of Company warfare changed dramatically in 1757.

In 1756, the new Nawab of Bengal, 21-year old Siraj-ud Daula, a vassal of the Mughal Emperor, broke with the EIC. He terminated the Company's trading concession, seized its Calcutta base, Fort William, and captured the Company's 400-man garrison. London took notice and the Company's share price dropped 30%. In response to Siraj's actions, the EIC planned his overthrow and replacement with a compliant ruler.

---

[5] Tangiers and Bombay were part of the dowry of Catherine of Braganza for her marriage to King Charles II. On March 27, 1868 Charles transferred Bombay, and its garrison, to the EIC.

[6] MacMunn, *The Armies of India*, 4-5.

The plan was formulated and executed by Robert Clive, commander of the EIC's forces in Madras. Clive, who began his Company service as a gentleman clerk, had previously commanded EIC forces in war with the French. At Madras, the Company mobilized a 1,500-man ground force and five armed ships. It conspired with Calcutta merchants and bankers, eager to dethrone the Nawab, to destabilize the regime. Clive soon found a nobleman to replace Siraj; Mir Jafar, a senior general in the Army of Bengal. Within six months the EIC had increased its Madras force to 5,000 men and was ready to march on Calcutta. Clive's plan was to first skirmish with the Army of Bengal at which point Mir Jafar and his troops would defect and attack the Nawab's loyalist soldiers. The decisive battle took place on June 23, 1757 at Plassey. The Company and its Indian allies defeated the Army of Bengal and Mir Jahar's retainers captured and then killed Siraj-ud Daula.

As part of his bargain with the EIC, Mir Jafar granted the Company taxation and administrative rights over part of Bengal and gave Clive and his colleagues personal fortunes. Additionally, the new Nawab excused the Company's trade from excise taxes and duties and ended state regulation of EIC business affairs. Taxation proved to be a lucrative sideline for the EIC and that, in conjunction with the other state concessions, increased Company revenues greatly.[7] Acquisition of territory and favorable trading concessions, by force and threat of force, soon became one of the Company's "trade" practices. Use of force for economic gain (in effect piracy); however, required significant expansion of the Company's armed forces.

During the 100 years that followed Plassey, the EIC's armed forces and territory it controlled, through direct governance and influence over allied native rulers, expanded enormously. Such growth was more haphazard than planned.[8]

> "There was never a masterplan for the conquest of India. No minister in London or governor-general in Calcutta consciously decided that the ultimate goal of British policy was paramountcy throughout the sub-continent. Instead, there was a sequence of tactical decisions made in response to local and sometimes unexpected crises. A backsliding raja who evaded his treaty obligations, a client state in peril from its neighbours, encroachments on British territory, or an independent frontier state making aggressive noises were sufficient justifications for war. When the fighting was over, the Company found itself with additional land, responsibilities, and revenues. The upshot was that by the middle of the century it had acquired a monopoly of power in India."

For administrative purposes, the EIC divided its Indian empire into three "presidencies" each with its own army. Presidency structure resembled national administrations and each presidency army resembled a state army. Army combat units consisted of native regiments with European officers and a few all-European regiments. In 1824, the European units of

---

[7] In the Mughal Empire, taxes took 40 to 50% of a farmer's sales revenue. James, *Raj*, 192.

[8] Ibid., 63.

the EIC armies had about 10,000 men organized in 5 infantry regiments and the equivalent of 10 artillery battalions. By 1857, the Company's European component fielded about 15,000 men organized in 9 infantry regiments and 17 artillery battalions. The European units were an exceedingly small part of the EIC's armed forces. In 1857, native units totaled 205 infantry regiments, 54 cavalry regiments, and 6 artillery battalions. The native units were led by about 3,900 European officers and senior NCOs.[9] Another 2,500 European officers and NCOs held staff appointments.

The three EIC presidencies were Bengal (headquarters in Calcutta), Bombay, and Madras. The Governor of Bengal served as Governor-General India; the Commander-in-Chief Bengal Army served as Commander-in-Chief for India. Following is an organization chart for the Bengal Presidency.[10]

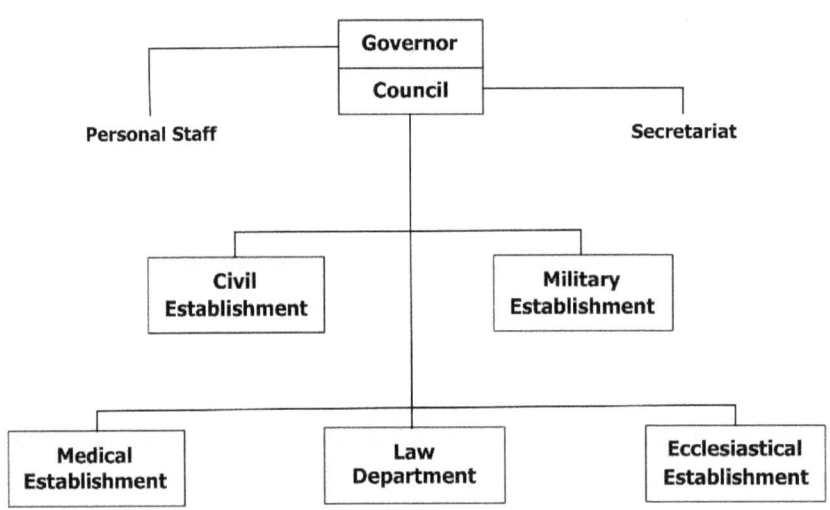

Presidency Government Structure

The Council consisted of the commander-in-chief of the presidency army and four civilian members. Laws were proclamations of the "Governor taken in Council." The Medical Establishment provided civil public health services and health care support to the army. Two or three medical officers were attached to each regular regiment and there were also army general hospitals. The Law Department included the courts and police. The Ecclesiastical Department provided Anglican and Presbyterian houses of worship and chaplains for the military and civilians. In 1856, the EIC employed 140 clergymen and ecclesiastical expenditure totaled £140,000, including £8,500 in allowances to Catholic

---

[9] EIC, *Returns Relating to the Armies in India*, 1857-58, H.C. Accounts & Papers, No. 201.

[10] *East India Register and Army List, 1857*.

priests.[11] Indians served in the lower ranks of the civil service though they were not barred officially from senior positions. Many native civil servants were Christians and of mixed European-Indian ancestry. In 1857, of the 5,928 professional-level civil servants, 48% were native Indians and 52% were Europeans or Indo-Britains.[12] At the time, 343 EIC military officers were on secondment from their regiments to the civil establishment.[13] Such officers accounted for about 5% of the professional staff.

Each presidency's Military Establishment was a complete army with staff offices and support departments. Following is a presidency army organization chart.[14]

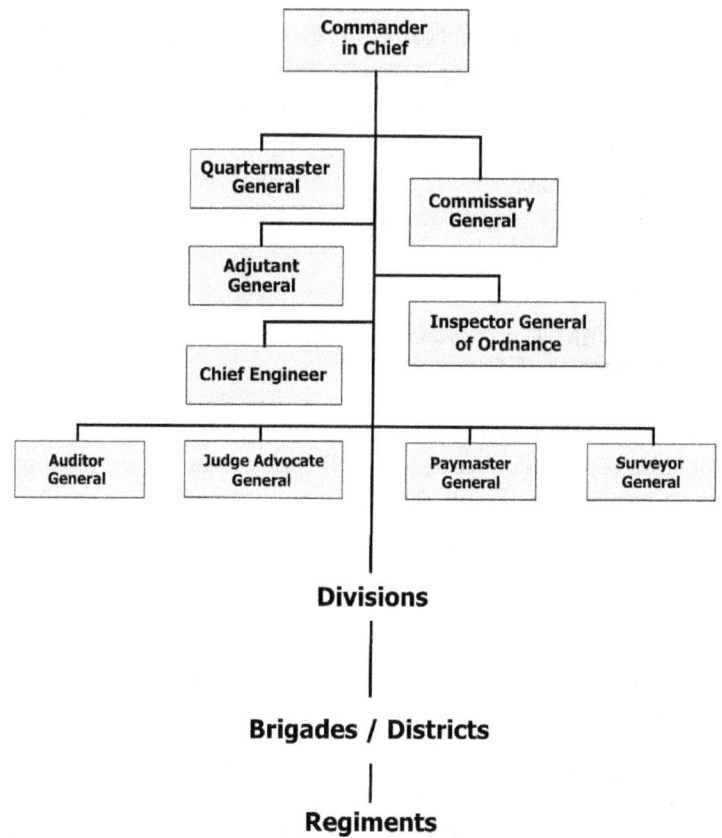

---

[11] EIC, *Return of Annual Expenditure of Ecclesiastical Objects in India*, 1857-58, H.C. Accounts & Papers, No. 33.

[12] EIC, *Statement showing the Number of Natives of India employed in the Civil Administration*, 1857-58, H.C. Accounts & Papers, No. 201-VI.

[13] EIC, *Returns Relating to the Armies in India*, 1857-58, H.C. Accounts & Papers, No. 201.

[14] *East India Register and Army List, 1857*.

Staff and departmental positions were usually filled with EIC officers; however, some British Army officers received staff appointments.[15] In 1857, regimental EIC officers served in 894 staff and departmental positions.[16]

The Military Establishment also provided civil services through its public works and survey departments. Combat units were organized to British Army standards, but EIC regiments employed many more civilians to perform menial tasks than did Crown regiments. The rank structure for native soldiers paralleled that of the British Army; however, ranks for native soldiers had Hindi names. Infantry privates were labeled *sepoys* while cavalry privates were *sowars*. The highest native ranks were *subedar-major* in the infantry and *risaldar-major* in the cavalry. Table 11 shows the typical strength of native regiments.

Table 11.
Composition of EIC Native Regiments in India, 1857 [17]

Unit Type	European Officers	European NCOs	Native, All Ranks
Infantry	17	2	935
Irregular Infantry	3	1	782
Cavalry	13	3	454
Irregular Cavalry	3	0	634
Artillery	23	6	784

Source: EIC, *Returns Relating to the Armies in India,* 1857-58, H.C. Accounts & Papers, No. 201.

At any given time, only about 70% of line officers were present with their native regiments. Furloughs, staff assignments, and civil appointments accounted for the shortfall in officers on duty with their regiments. The officer establishment for regular, native cavalry regiments was 17 line officers, 1 quartermaster, 2 surgeons, and 1 veterinarian. The establishment for regular native infantry regiments was 24 line officers, 1 quartermaster, and 2 surgeons.[18]

From 1773 through 1853, Parliament enacted statutes that progressively limited the Company's trading privileges and its authority to govern in India. Following is a brief description of the most important of those statutes.

> The Regulating Act, 1773 (13 Geo. 3, c. 63). The first statute that interfered with Company governance of India. The act gave the British government sole authority to set political policy in India. The government would control political matters through orders to the Governor-General of India.

---

[15] Only a few British Army officers were assigned to staff and departmental positions. Testimony of MG Thomas Harte Franks, *Report of the Commissioners appointed to inquire into the Organization of the Indian Army*, 1859 Sess. 1, [2515], q. 4244.

[16] EIC, *Returns Relating to the Armies in India,* 1857-58, H.C. Accounts & Papers, No. 201.

[17] Artillery units are foot battalions.

[18] *East India Register and Army List, 1857.*

East India Act, 1784 (24 Geo. 3 Sess. 2, c. 25). Created the Board of Commissioners for the Affairs of India ("Control Board"). It was composed of government ministers. The Control Board set state policy for India and exercised oversight of the Company at home.

East India Company Recruitment Act, 1799 (39 Geo. 3, c. 109). Regulated military recruitment and provided for British Army supervision of enlistments.

East India (Charter) Act, 1813 (53 Geo. 3, c. 155). Ended the Company's monopolies except for tea of any place of origin, and all goods from China. The act also gave the British government greater authority over the EIC's administration of India.

Government of India Act, 1833 (3 & 4 Will. 4, c. 77). Ended all the Company's commercial trading. Henceforth, the EIC would be agent of the Crown for the tea and opium trade and the governance of India.

Government of India (Charter) Act, 1853 (16 & 17 Vict., c. 95). Affirmed that the Company was simply an agent of the Crown and that Parliament could terminate that agency at will.

EIC Indian Service in the 1850s

Brian Tweedy retired from the British Army in 1885 or 1886, after a long period of service. This indicates he enlisted in the Company's armed forces in the 1850s. Following is a description of the officers and enlisted men of the EIC armies and the terms and conditions of military service during the final years of Company administration of British India.

Officers

EIC military officers were recipients of direct commissions or graduates of the Company's military college at Addiscombe, located about 15 kilometers south of central London. Prior to 1853, college admissions were patronage perquisites of Company directors. Beginning that year, admission to Addiscombe was by competitive examination.[19] Like British Army officers, EIC officers were of the upper and middle classes. Unlike Crown cavalry and infantry officers, Company officers of such arms did not purchase their commissions. The main difference between officers of the British Army and the EIC was that Company officers generally had little, or no, personal income. Accordingly, most Company officers had to live off their pay; a practice that was impossible in nearly all cavalry and infantry regiments of the British Army. In that the EIC paid army officers substantially more than did the British government, competition for Company commissions was keen

---

[19] Government of India (Charter) Act, 1853. 16 & 17 Vict., c. 95.

among young gentlemen without private incomes who desired military careers. Table 12 shows pay by rank for EIC and British Army infantry officers.

Table 12.
Annual Pay of Infantry Officers: EIC & British Army, 1857
(1 Rupee = 2 Shillings)

	East India Company	British Army at Home
Colonel	£ 1,518	£ 600
Lieutenant-Colonel	956	310
Major	732	292
Captain	449	211
1st Lieutenant	270	119
2nd Lieutenant	218	96

Sources: *East India Register and Army List 1857*; *Hart's New Annual Army List 1858*.

British Army officers serving in India were paid by the EIC, not the War Office, and they received the same pay and allowances that Company officers received. Accordingly, the financial advantage of an EIC commission was that its holder would spend his entire military career in India and always receive the higher pay. In addition to having a higher income than at home, an officer had a lower cost of living a "gentlemanly" life in India than in the United Kingdom. That cost of living was so much lower that junior officers could live in high style on their army pay. For example, young, bachelor officers at home would live in private rooms in barracks or the officers' mess building. Each officer was attended to, part-time, by a servant soldier. In India, such officers lived in spacious cottages and each was waited on by a staff of servants, who also maintained the premises.[20]

Retirement was another advantage of the Company's commission over the sovereign's commission. EIC officers were entitled to retire at full-pay after 22-years' service, while at the time Crown officers had no longevity pension rights.[21] British Army infantry and cavalry officers could sell their commissions, and a limited number of engineers and artillery officers could receive pensions at the discretion of the War Office. All officers had the option to place themselves on the half-pay, reserve list, which for officers without valuable saleable commission, was the usual method of voluntary retirement. Additionally, unlike Crown officers, Company officers could draw a disability pension on account of disease. The only basis for a British Army disability pension was wounds. Table 13, on the following page, summarizes the pension benefits of EIC and British Army infantry officers, by rank.

---

[20] James, *Raj*, 160-61.

[21] Officers received pension rights in 1871 when commission purchase was abolished by the government as part of a series of measures later termed "the Cardwell Reforms." Edward Cardwell served as war minister 1868-1874.

Table 13.
Annual Pensions of Infantry Officers: EIC & British Army, 1857

	East India Company		British Army	
	Longevity	Disability	Half-Pay	Wounds
Colonel	£ 1,278	£ 200	£ 219	£ 300
Lieutenant-Colonel	656	200	155	300
Major	492	173	137	200
Captain	299	128	91	100

Sources: *East India Register and Army List 1857*; *Hart's New Annual Army List, 1858*; de Fonblanque, *Treatise on the Administration and Organization of the British Army*, 311-12.

The Company also offered its officers supplemental disability plans that included widows and orphans benefits. Annual premiums for Bengal ranged from £12 for a 2nd lieutenant to £75 for a colonel, about 5% of pay.[22] For EIC officers, there was also mandatory membership in the Military Orphan Society, a survivors' benefit plan. Each orphan received £30 to £45 per year until age 18 for girls, and age 19 for boys. Annual subscription fees for Bengal Army officers ranged from £6 for a 2nd lieutenant to £30 for a colonel, plus £3.6 for each girl and £1.2 for each boy, regardless of rank; in all about 2-3% of pay.[23]

When Crown officers sold their commissions to retire, they often used the proceeds to purchase an annuity. Such officers effectively converted their commissions into pensions. The market price of British Army commissions was always higher than the official price and the spread varied over the years. Table 14 shows that even if a retiring, 50-year old officer sold his commission at twice the regulation price, the annual annuity benefit during his remaining life would be less than an EIC pension amount.[24]

Table 14.
Value of Line Infantry Commissions in Terms of Annuity Payments
(3.0% Interest) [25]

	Commission, Official Price	20-Year Annuity			EIC Pension
		1.0 X	1.5 X	2.0 X	
Lt.-Colonel	£ 4,500	£ 300	£ 450	£ 600	£ 656
Major	3,200	215	323	430	492
Captain	1,800	120	180	240	299

Source: *Hart's New Annual Army List, 1858*.

---

[22] Bengal Presidency Military Funds. *East India Register and Army List, 1857*.

[23] Bengal Presidency Military Orphan Society. *East India Register and Army List, 1857*.

[24] Mean life expectancy for 50-year old males was 19.87 years. Registrar General, *English Life Table No. 2*, 12th Annual Report, 1853. Office of National Statistics, File Ref. 005562.

[25] Typical nineteenth century interest rate on perpetual, callable, consolidated British government securities, known as "consols".

In 1876, War Office actuaries calculated that at 4.0% interest (instead of 3%) a life-annuity beginning age 50 would pay a fixed annual amount equal to 8.0% of initial capital. At age 45, such annuity would pay annually 7.2% of initial capital. Accordingly, a 50-year old retired officer could purchase a fixed, lifetime annuity of £360 per year for £4,500. For a 45-year old that same annuity would pay £324 per year for life.[26] As shown in Table 14, EIC officers, depending on rank, received annual pensions of £299 to £656 regardless of age.

Though Company officers were of the same social classes as Crown officers, they generally had a somewhat lower social standing and a much lower financial standing. Also, many EIC officers fell short when it came to "gentlemanly" attributes such as birth, landed status, and education at an elite "public" school. Company officers were typically sons of minor manufacturers and merchants, clerics, and medical doctors. EIC officers of landed families generally did not have private incomes and were not heirs apparent to the estate. In social and command precedence, Company officers were below Crown officers of the same rank. Most British Army officers viewed EIC officers as their social and professional inferiors.[27]

Other Ranks

The EIC filled the enlisted ranks of its European units through recruiting staffs that were in effect, licensed by the War Department.[28] There were seven recruitment districts in the United Kingdom; three each in Ireland and England, one in Scotland. During the ten years preceding 1857, 35% of the recruits enlisted in Ireland. Half of those recruits were obtained in the Cork district. Enlistment standards were the same as for the British Army except that EIC recruits usually had to be at least 20 years of age. The enlistment term was ten years, the same as for the British Army.[29] Men who presented themselves for enlistment were examined by British Army medical officers and enlistment bonuses were the same as for the British Army.[30] Unlike British Army recruits, EIC recruits could not specify a regiment in which to enlist; company recruits could only choose between artillery and infantry service. Additionally, each year the EIC offered engineering training to about 35 recruits.[31]

As with the British Army, the artillery had higher enlistment standards than the infantry and gunners were paid more than foot soldiers. For the four years ended 1856, the artillery units received 25% of EIC recruits.[32] The Company trained all European recruits for four

---

[26] *Report of the Royal Commission on Army Promotion and Retirement*, 1876, [C. 1569], appx. M.

[27] Stanley, *White Mutiny*, 22-31.

[28] EIC, *Returns Relating to the Armies in India*, 1857-58, H.C. Accounts & Papers, No. 201-F.

[29] The EIC term of service could be extended to twelve years if deemed necessary by the presidency army. In the British Army, artillery service was twelve years.

[30] EIC, *Returns Relating to the Armies in India*, 1857-58, H.C. Accounts & Papers, No. 201-F.

[31] Ibid.

[32] Ibid.

to five months at its Warley depot in Essex.[33] Prior to 1843, the training depot was at Chatham in Kent. The few recruits selected for engineering service received specialty training at the British Army's Military School of Engineering at Chatham. Recruits trained as combat engineers served with native companies of "sappers and miners." All such engineers were NCOs.

In nominal terms, European EIC soldiers were paid less than British Army soldiers at home. The disparity varied by rank with EIC quartermaster-sergeants paid 6% less and EIC sergeants paid 25% less than their British Army counterparts.

Table 15.
Annual Pay of Infantry Soldiers: EIC and British Army, 1857
(1 Rupee = 2 Shillings)

	East India Company	British Army at Home
Sergeant-Major	£ 53	£ 58
Quartermaster-Sgt.	46	49
Colour-Sergeant	36	46
Sergeant	27	36
Corporal	19	24
Private	14	18

Sources: *Report of the Commissioners appointed to inquire into the Organization of the Indian Army*, 1859 Sess.1, [2515], appx. 51; de Fonblanque, Edward Barrington, *Treatise on the Administration and Organization of the British Army* (London: Longman, 1858), 265

The principal financial advantage had by the ranker in India over his colleague at home was a much lower cost of living. For example, in the 1850s the army charged soldiers at home for the full cost of their meals. Abroad, meals were subsidized by the army, both British and EIC. Additionally, as wages were much lower in India than in the United Kingdom, soldiers' expenses in India for tailoring, boot repair, haircuts, and laundry were much less than at home. Deductions from pay, termed "stoppages," for most soldiers at home totaled £10 to £11 per year; in India, they were less than half that amount. After taking into account stoppages, the nominal net pay of privates in India was about the same as that of privates at home; however, in India that pay went much further than at home. Alcohol, quality prostitutes, and small luxury items were much cheaper in India than in the United Kingdom.[34]

For privates, life in the EIC's armies was easier than it was in the British Army. The Company made more native servants available to its soldiers than it did to Crown enlisted men. Each infantry regiment had 60 menials paid for by the EIC; porters, water carriers,

---

[33] Stanley, *White Mutiny*, 120.

[34] James, *Raj*, 210-11.

sweepers, and latrine cleaners.³⁵ Additional servants were engaged communally by soldiers at negligible expense. EIC privates had significantly fewer non-military duties than British Army soldiers. For example, artillery batteries had native grooms and stable attendants for the horses. Unlike in the British Army, officers did not assign pointless "busy-work" simply to fill the regimental daily schedule. Discipline was not as stringent in the EIC armies as in the British Army and Company officers were more forgiving of minor transgressions than Crown officers.³⁶

In the 1850s, promotion prospects for British Army privates were slim. Promotional opportunity in the infantry, the largest arm, was more limited than in other corps. An infantry regiment had only 57 sergeant positions out of an authorized strength of 1,057 enlisted men.³⁷ Extra-regimental sergeant positions were few, and generally were at basic training depots, recruiting districts, and the few military specialty training schools. For the relative handful of educated soldiers, there were some academic teaching positions with the Corps of Army Schoolmasters.³⁸

Company soldiers had far more advancement opportunities than did Crown soldiers. First, EIC European regiments had a higher proportion of sergeants than did British Army regiments. In an EIC European infantry regiment, sergeants of all ranks accounted for 6.5% of enlisted strength compared to 5.4% for the British Army.³⁹ Sergeants accounted for 9.2% of the European artillery's authorized 4,000 men and 28.6% of approximately 265 European engineers.⁴⁰ In 1857, the 965 regimental sergeants' positions accounted for 6.5% of total European enlisted strength.⁴¹ Second, and more importantly, the EIC armies had numerous extra-regimental sergeant positions which in 1857 totaled 1,407.⁴² There were 403 senior NCOs (sergeant-majors and quartermaster-sergeants) with native regiments; two per infantry regiment, one per irregular infantry regiment, and three per cavalry regiment. There were also 279 sergeants in the Public Works Departments and 261 in the Commissariat and Ordnance Departments. Overall, sergeants accounted for 16.0% of European enlisted strength. Also, sergeants in the Commissariat and Ordnance Departments were eligible for

---

³⁵ *Report of the Commissioners appointed to inquire into the Organization of the Indian Army*, 1859 Sess. 1, [2515], appx. 51.

³⁶ Stanley, *White Mutiny*, 70-79

³⁷ *Report of the Commissioners appointed to inquire into the Organization of the Indian Army*, 1859 Sess. 1, [2515], appx. 52.

³⁸ Skelley, *The Victorian Army at Home*, 111-14.

³⁹ Ibid., appx. 51.

⁴⁰ Artillery: *Report of the Commissioners appointed to inquire into the Organization of the Indian Army*, 1859 Sess. 1, [2515], appxs. 56, 58, 62. Engineers: Each native company of sappers and miners had 4 European sergeants and 10 European corporals. Ibid., appx. 45. In 1857 there were 19 companies of "sappers and miners" in the EIC armies. *East India Register and Army List, 1857*.

⁴¹ Total European Enlisted Strength of 14,814. *Report of the Commissioners appointed to inquire into the Organization of the Indian Army*, 1859 Sess. 1, [2515], appx. 16.

⁴² Ibid., appx. 28.

promotion to non-combatant officer ranks such as Conductor and Deputy Assistant Commissary. Annual pay for such ranks ranged from £168 to about £350.[43]

While pensions were the same for Company and Crown soldiers, EIC soldiers could better adjust to retirement in India and therefore could have a much higher standard of living than what they would have had at home. Additionally, literate EIC veterans in their early 40s could readily obtain well-paid employment with British merchants in India.[44]

In general, the EIC received a somewhat higher caliber recruit than did the British Army. This was due to two factors. First, the advantages of Company service over Crown service was known by young men who sought a military career. Second, the EIC, as a small force, could be more selective than the British Army as to which volunteers it accepted. For the four years ended 1856, Company enlistments averaged 2,168 men per year.[45] For four years from 1851 through 1856, excluding the two years of the Crimean War, British Army enlistments averaged 10,501 men per year; nearly five times the EIC intake.[46] While recruits of both forces were overwhelmingly unemployed, and of the lower strata of the working class, the Company enlisted fewer unskilled laborers and more educated young men than did the Crown. Note that army standards of the mid-nineteenth century defined "educated" as having mastered the curriculum of the fourth year of primary schooling.[47] In the 1850s about 33% of Company recruits claimed to have been laborers while nearly 10% claimed to have been clerks.[48] For the British Army of that time, a preponderance of recruits were laborers, both agricultural and industrial. Illiteracy rates were somewhat different for the two forces. About 40% of Company recruits were unable to read and write compared to 60% of Crown recruits.[49]

A high proportion of the enlisted men were Irish, at most times 40% to 50%.[50] This was due to several factors. During the eighteenth century, Westminster barred or restricted British Army service of Catholics. This was not the case for the EIC's armies and accordingly, there was a longer Irish tradition of military service to the Company than to the Crown. Ireland, poorer than England and with a higher percentage of poverty-stricken agricultural laborers, was a fertile recruiting ground for both the British and EIC armies. Finally, in that Company soldiers served in India, they would never be required to enforce laws and suppress rebellion at home.

---

[43] Stanley, *White Mutiny*, 17-18.

[44] Stanley, *White Mutiny*, 13.

[45] EIC, *Returns Relating to the Armies in India*, 1857-58, H.C. Accounts & Papers, No. 201-F.

[46] *Report of the Commissioners appointed to inquire into the Present System of Recruiting in the Army*, 1861, [2762], at 325.

[47] Skelley, *The Victorian Army at Home*, 88-90, 94, 311.

[48] Stanley, *White Mutiny*, 118-19.

[49] Ibid., 43.

[50] Cadell, "Irish Soldiers in India;" Kenny, "The Irish in the Empire.".

Many of the Company's recruits viewed EIC military service as "… not simply a refuge from poverty but a route to prosperity and even respectability."[51] Accordingly, the Company, compared to the British Army, recruited a high proportion of persons wanting to better themselves; people whom the British establishment of the time referred to as "the respectable working class."[52] Joyce's Brian Cooper Tweedy was one of those ambitious, respectable, working class young men.

Health and Mortality

To obtain the advantages of EIC military service, the Company soldier had to pay a price. Frequent wars on the sub-continent meant a higher risk of death or disability from combat for EIC soldiers than for British Army soldiers. More importantly, disease incidence made mortality rates much higher in India than in the United Kingdom. For the years 1859 and 1860, a period without notable warfare in India, the annual mortality rate for enlisted men of the former EIC regiments averaged 3.5%. For enlisted men of the British Army in India during those two years, the annual mortality rate was similar, 3.6%.[53] Brian Tweedy served in the 2nd Battalion, Royal Dublin Fusiliers. Its EIC antecedent was the 1st Bombay European Regiment, which saw no active service during the hostilities of 1857 and 1858. For the years 1858 and 1859, the 1st Bombay Europeans experienced an average, annual mortality rate of 3.2%.[54] An annual mortality rate of 3.0% indicates only 52.8% of soldiers would survive 21 years in India. At a rate of 3.5% only 47.3% would survive.[55] Note that enlisted men of the EIC and British Army had to serve 21 years to obtain a full, longevity pension.

In 1859, the overall mortality rate of the British Army was 2.3%. For soldiers at home, the mortality rate was 1.0% and for those abroad, excluding India, it was 1.4%. The home mortality rate for 1859 was markedly better than the rates for earlier years. From 1839 through 1853 the average, annual mortality rate for the British army at home was 1.8%. For the comparable civilian population, the rate was 0.9%.[56] An annual mortality rate of 1.0% indicates 81.0% of soldiers would survive 21 years. At a rate of 2.3%, 61.4% would survive.

---

[51] Stanley, *White Mutiny*, 13.

[52] Ibid., 16.

[53] India Office, *Return of the Mean Strength and of the Deaths and other Casualties among the Non-Commissioned Officers and Men of the European Forces of the late East India Company*, 1865, H.C. Accounts & Papers, No. 486; War Office, *Return of the Mean Strength and of the Deaths and other Casualties among the Non-Commissioned Officers and Men of Her Majesty's Forces in India*, 1865, H.C. Accounts & Papers, No. 341.

[54] D. Costello, M.D., "Annual Report of the 1st Bombay European Regiment," *Transactions of the Medical and Physical Society of Bombay* (1857 & 1858): 288-313; D. Costello, M.D., "Annual Report of the 1st Bombay European Regiment," *Transactions of the Medical and Physical Society of Bombay* (1859):165-88.

[55] (1 − mortality rate) raised to the 21st power.

[56] *Report of the Commissioners appointed to inquire into the Regulations Affecting the Sanitary Condition of the Army*, 1857-58, [2318], at vi; War Office, *Statistical Report on the Health of the Army in 1959*, 1861, [2583]; War Office, *Return of the Mean Strength and of the Deaths and other Casualties among the Non-Commissioned Officers and Men of Her Majesty's Forces in India*, 1865, H.C. Accounts & Papers, No. 341.

From Company to Crown: The White Mutiny [57]

In 1857, when news of the Indian mutinies reached London, the Company's directors anxiously asked Westminster for authority to raise additional European troops. The request was granted and the EIC began almost indiscriminate recruiting. With government permission, the Company increased enlistment bonuses, reduced recruit height requirements, and lowered its enlistment standards.[58] With reduced selectivity, more unskilled laborers and fewer clerks entered Company service. Unlike in previous years, the new recruits were overwhelmingly urban; coming from London, Manchester, Birmingham, Glasgow, Dublin, and Cork. This was due to high urban unemployment from the recession of 1857-58. As a result, more skilled tradesmen joined the Company's armies than in the past.

The EIC sent 3,917 recruits to India in 1857; 5,149 in 1858.[59] About another 1,000 men were recruited locally in India. In all, the EIC enlisted 10,000 new soldiers in two years. In the past, two-year intakes for the Company averaged about 4,500 men. With the additional soldiers, the EIC formed three new infantry regiments and five cavalry regiments, all for the Bengal Army. The 5th Bengal European Cavalry was composed entirely of Europeans resident in India. The new recruits from home received at most five weeks training before dispatch to India. During normal times, as noted previously, recruit training was four to five months. By September 1858, there were about 20,000 European enlisted men serving with the EIC in India.[60]

One of Westminster's many responses to the mutinies and rebellion in India was to strip the EIC of its role in government, effective November 1, 1858. With the Government of India Act of 1858, Parliament terminated the EIC's agency and took direct control of Indian affairs.[61] The presidency administrations were transformed into the Government of India which reported to the new India Office headed by a Secretary of State. Under Section 56 of the Act the Company's armed forces became the armed forces of India and the European soldiers became soldiers of the Crown in service of the Indian Government. Their terms and conditions of employment remained as if they were still employed by the EIC; however, all European soldiers would be under Crown authority as if they had enlisted in the British

---

[57] Source: Stanley, *White Mutiny*, 116-165.

[58] *Report of the Commissioners appointed to inquire into the Organization of the Indian Army*, 1859 Sess. 1, [2515], at 140.

[59] Ibid.

[60] Testimony of MG Robert J. Hussey Vivian, *Report of the Commissioners appointed to inquire into the Organization of the Indian Army*, 1859 Sess. 1, [2515], q. 3779-3962. At the time, official returns of strength for Madras and Bombay were unavailable. September data for Bengal and March data for Madras and Bombay show 18,085 EIC enlisted men in India. Ibid., appx. 19.

[61] 21 & 22 Vict., c. 106.

The Armies of the British East India Company

Army and not the EIC armies. The legal basis for the transfer was the East India Company Recruitment Act of 1799, which imposed dual enlistment on EIC European recruits; such recruits were soldiers of both the Crown and the Company. This dual enlistment was set forth, though unartfully, in the enlistment oath. The second and final paragraph of the oath was as follows:[62]

> "I do also make oath that I will be faithful and bear true allegiance to Her Majesty, Her heirs and successors, and that I will, as in duty bound, honestly and faithfully defend Her Majesty, Her heirs and successors, in person, crown and dignity, against all enemies, and will observe and obey all orders of Her Majesty, Her heirs and successors, and of the generals and officers set over me,
>
> and
>
> that I will also be true to the said East India Company, and will duly observe and obey all their orders, and the orders of their generals and officers who shall be lawfully set over me."

Note that the recruit first swore allegiance to the Crown and agreed to obey all orders of the British military set over him. The recruit then swore further allegiance to the Company and agreed to obey the orders issued through the Company's military chain of command.

Proclamation of the new regime in India immediately caused distress among the Company's European soldiers. The first unsettling realization for the rankers was that they were no longer part of an institution that had existed for 200 years. They now served a new entity "the Government of India." The second and more important realization was that the distinct culture of the EIC armies would end and Company soldiers would become minions of the British Army. Good promotion opportunities, the relatively easy daily routine, and a military career spent in India would no longer be part of their lives.

Disaffection spread through the European ranks. Most troops claimed they never took an oath to the Queen. Some claimed that their sworn allegiance to the Crown was simply a matter of form; *i.e.* they would be loyal British subjects. About half the troops asked for their discharge, while the ones willing to serve in the Indian Army asked for enlistment bonuses. Governor-General Charles Canning was disturbed. The native mutinies had only recently been suppressed and pockets of rebellion existed in northern India. A mutinous situation among European troops in India could prove disastrous. As of February 1, 1859 there were 19,224 former EIC enlisted men in India, 18,580 in European regiments, 644 in native regiments.[63] Canning wrote to Westminster for clarification of Section 56 of the new law. He hoped the government would reconsider its position and meet the demands of the former EIC enlisted men.

---

[62] India Office, *Papers connected with the late Discontent among the Local European Troops in India*, 1860, H.C. Accounts & Papers, No. 169, at 384.

[63] India Office, *Return of the actual Strength, both of the Queen's and the East India Company's Forces*, 1859 Sess. 2, H.C. Accounts & Papers, No. 64.

"UPON the recent transfer of the forces of the late East India Company to the immediate service of Her Majesty, under the provisions of the Act of the 21 & 22 Vict. c. 106, certain European soldiers of the East India Company's forces having claimed their discharge, or their enlistment anew into the Queen's service with fresh bounty, the subject was brought under the consideration of Her Majesty's Government, and referred to the law officers of the Crown. His Excellency the Viceroy and Governor General of India in Council has now to announce to the European soldiers of Her Majesty's Indian forces in the three Presidencies, who were formerly in the service of the East India Company, that Her Majesty's Government have finally decided that the claim made to discharge, or re-enlistment with bounty, is inadmissible."

The response of the former Company troops to this brusque proclamation was termed by the contemporary press "the White Mutiny." The government's position dashed the hopes the soldiers formed since November 1858. Until the proclamation "every man expected he would be allowed to take his discharge."[64] With discharges and bounties "inadmissible," protests erupted at station after station of the European regiments in the Bengal Presidency. Men refused to turn out on parade and perform military duties. Military property was vandalized and radical graffiti appeared on cantonment walls. The men insulted and threatened officers and refused to obey orders. Bengal Army authorities considered the 4th Bengal European Infantry in a state of mutiny. Outbreaks in the Madras and Bombay presidencies; however, were minor.

That European troops should so dramatically assert themselves, especially considering the recent native uprising, alarmed the military and civil authorities. There was also fear among Indian government officials that native troops would soon seek the concessions demanded by the Europeans. Opinion of senior Indian Army officers was split as to whether Crown troops were sympathetic towards their former Company colleagues. Accordingly, British Army troops were not moved openly against the disaffected European regiments.

In May, Canning, on his own initiative, authorized discharges for men of the most disaffected units. He also ordered that courts-martial were to try only the ringleaders of the protests. The next month Canning, recently created Earl Canning, caved in completely to the European soldiers' demands. On June 20, 1859, by General Order 883, he announced that all former Company soldiers who so desired would be granted a discharge and free passage home. Those soldiers who wished to serve in the new Indian Army would receive enlistment bonuses.[65] Somewhat over half the former EIC soldiers, 10,116 men, took discharges which were of honorable character. About 2,800 of those men would within two years enlist in the British Army.[66]

---

[64] Stanley, *White Mutiny*, 130.

[65] India Office, *Papers connected with the late Discontent among the Local European Troops in India*, 1860, H.C. Accounts & Papers, No. 169, at 236-37.

[66] *Return of the Number of Men of the European Local Troops in India, who have taken their Discharge, and the Number who were Re-Enlisted*, 1860, H.C. Accounts & Papers, No. 468.

Canning's concession to the former Company soldiers drew criticism from the conservative elements within government and the armed forces. For example, the Chief of Staff of the new Indian Army, wrote to the Under Secretary of State for War as follows:[67]

> "… whether the Government of India was right or wrong in its original policy and arguments, it has been beaten by its own army; it has yielded to intimidation; it has abandoned the ground it assumed by its acts and orders, and the mutineers have achieved a victory."

European Regiments of the Indian Army

The Indian government continued to recruit Europeans for its army. By June 1860, enlisted strength of the European regiments was 13,884. Additionally, there were 1,600 recruits training at Warley and 338 in transit to India. The 1st Bombay Europeans, Brian Tweedy's regiment, had on its roll 49 officers and 600 other ranks.[68] During 1860 and 1861, the Indian Army disbanded the three new Bengal European infantry regiments and two of the five Bengal cavalry regiments. The army transferred those units' soldiers to other European regiments.

In April 1861, the new Governor-General, James Bruce (8th Earl of Elgin), announced the plan to "amalgamate" the surviving European regiments with the regiments of the British Army. The nine infantry regiments would become the 101st through 109th Regiments of Foot and the three cavalry regiments the 19th through 21st Light Dragoons. The artillery would be consolidated into 14 new battalions of Royal Artillery.[69] By the end of 1862, the European regiments had been incorporated into the British Army.[70]

The Indian government closed the former EIC training depot at Warley. The former EIC regiments utilized recruit depots at Chatham, Colchester, Parkhurst, and Pembroke in England; Birr, Cork, and Fermoy in Ireland. The 106th through 109th Regiments of Foot recruited in Ireland.[71] Following is a genealogy of the EIC European regiments as of 1904:[72]

---

[67] Letter, MG William Mansfield, Chief of Staff Indian Army, to the Earl of Ripon, Under Secretary of State for War, September 26, 1859, India Office, *Papers relating to the future Organization of Her Majesty's European Forces serving in India*, 1860, H.C. Accounts & Papers, No. 330, at 76.

[68] India Office, *Returns of the Strength Regimentally, in Officers and Men and Horses, of the Regular Local Army of the several Arms in India*, 1860, H.C. Accounts & Papers, No. 361.

[69] General Order No. 332 of 1861 by the Governor General of India in Council, *The Calcutta Gazette*, Extraordinary, April 22, 1861.

[70] *Hart's New Annual Army List, 1863*.

[71] *Hart's New Annual Army List, 1864*.

[72] *Hart's Annual Army List, 1905*.

EIC and Indian Army, 1858	British Army, 1862	British Army, 1904
1st Bengal Infantry	101st Regiment of Foot	1/Royal Munster Fusiliers**
1st Madras Infantry	102nd Regiment of Foot	1/Royal Dublin Fusiliers**
1st Bombay Infantry	103rd Regiment of Foot	2/Royal Dublin Fusiliers**
2nd Bengal Infantry	104th Regiment of Foot	2/Royal Munster Fusiliers**
2nd Madras Infantry	105th Regiment of Foot	2/South Yorkshire Regiment
2nd Bombay Infantry	106th Regiment of Foot*	2/Durham Light Infantry
3rd Bengal Infantry	107th Regiment of Foot*	2/Royal Sussex Regiment
3rd Madras Infantry	108th Regiment of Foot*	2/Royal Inniskilling Fusiliers**
3rd Bombay Infantry	109th Regiment of Foot*	2/Leinster Regiment**
1st Bengal Cavalry	19th Light Dragoons	19th Hussars
2nd Bengal Cavalry	20th Light Dragoons	20th Hussars
3rd Bengal Cavalry	21st Light Dragoons	21st Lancers

* Depot in Ireland, 1862.   ** Irish territorial regiment, 1881.

Note that the number preceding the 1904 regimental name is the battalion number. Each of the seven listed regiments had two regular army battalions with strengths that varied from 600 to 1,000 men. The higher strength was for battalions stationed abroad.

Chapter Bibliography

Cadell, Patrick. "Irish Soldiers in India." *The Irish Sword* 1, no. 2 (1953): 75-79.

de Fonblanque, Edward Barrington. *Treatise on the Administration and Organization of the British Army*. London: Longman, 1858.

Kenny, Kevin. "The Irish in the Empire." In *Ireland and the British Empire,* edited by Kevin Kenny. Oxford: Oxford Univ. Press, 2004.

James, Lawrence. *Raj, The Making and Unmaking of British India*. New York: St. Martin, 1997.

MacMunn, G. F. *The Armies of India*. London: Black, 1911.

Reid, C. Lestock, *Commerce and Conquest*. London: Temple, 1947.

Robins, Nick. *The Corporation that Changed the World*. London: Pluto, 2012.

Skelley, Alan Ramsay. *The Victorian Army at Home*. London: Croom Helm, 1977.

Stanley, Peter. *White Mutiny, British Military Culture in India*. New York: NYU Press, 1998.

Chapter 6
Army Life and Retirement: Officers

Socially, the officer corps was homogenous as nearly all officers were of the British establishment: the upper and middle classes. Of the 5% of officers who were not "gentlemen" nearly all held non-combatant commissioned rank. Non-combatant regimental officers (quartermasters and ridingmasters) as ex-rankers, were of the working class. In the mid-nineteenth century junior officers of the "civilian" departments (commissariat and ordnance stores) were usually of the lower-middle class. With rare exception, combatant officers were born, bred, and educated as gentlemen.[1] They were of the aristocracy, gentry, and long established middle class families. The senior civilian department officers, and surgeons, were almost entirely of the middle class.

Successive governments and army high commands found it desirable that officers have a nexus with the land.[2] Accordingly, landed families were over-represented in the highest ranks. In 1899, the peerage, baronetage, and gentry accounted for 38% of colonels and 51% of generals.[3] On the eve of the First World War the landed classes accounted for 64% of all general officers. Additionally, at that time hereditary titles were held by one-third of household cavalry officers and one-fifth of foot guards officers.[4]

Prior to 1871, about 80% of commissions in the combat arms could be obtained only by purchase. The balance was for graduates of high academic standing of the Royal Military College at Sandhurst (infantry and cavalry), all graduates of the Royal Military Academy at Woolwich (engineer and artillery), and enlisted men.[5] Each year only a relative handful of rankers were commissioned as combatant officers. Commissions from the ranks were almost always appointments as quartermasters and ridingmasters.[6]

<u>Commissions in the 1850s</u>

When Brian Tweedy began his military career the British Army officer corps totaled somewhat under 7,000 men of which about 1,000 were in the "scientific" arms of the artillery and the engineers. Infantry and cavalry commissions were obtained by direct appointment,

---

[1] Spiers, *The Army and Society*, 1.

[2] Harries-Jenkins, *The Army in Victorian Society*, 25-37.

[3] Spiers, *The Army and Society*, 7-8.

[4] Razzell, "Social Origins of Officers."

[5] *Report of the Commissioners Appointed to Inquire into the System of Purchase and Sale of Commissions in the Army*, 1857 Sess. 2, [2267], at xx. Hereafter cited as *Purchase Commission Report (1857)*.

[6] Spiers, *The Army and Society*, 2-6.

promotion from the ranks, or graduation from Sandhurst. Engineer and artillery commissions went exclusively to graduates of the Royal Military Academy at Woolwich. For direct commissions, passing marks on an officer entrance examination were required.[7] The examination was no real obstacle for a young gentleman desirous of a commission as the army set passing marks at a level suitable for graduates of the older public schools.[8] The following tables show the distribution of officers by corps and sources of commissions, two years before the Crimean War.

Table 16.
British Army Officer Strength
Combat Arms, 1853

Arm	Number	Percent
Cavalry	874	12.8
Infantry	5,061	73.6
Artillery	652	9.5
Engineers	279	4.1
Total	6,866	100.0

Source: War Office, *A Return of the Number of Officers and Men, distinguishing each Arm of the Service*, 1859 Sess. 2, H. C. Accounts & Papers, No. 88.

Table 17.
Source of Regimental Commissions in 1853

Source	Number	Percent
Sandhurst, Mostly Purchase	32	7.1
Direct, Purchase	339	75.3
Direct and Woolwich, Non-Purchase	58	12.9
From the Ranks, Non-Purchase	21	4.7
Total	450	100.0

Source: *Report of the Commissioners Appointed to Inquire into the System of Purchase and Sale of Commissions in the Army*, 1857 Sess. 2, [2267], appx. XII.

Ridingmasters and quartermasters were all commissioned from the ranks. Paymasters could be commissioned directly though most were former combatant officers. Candidates for infantry and cavalry commissions required approval of the Commander-in-Chief of the British Army; artillery and engineer candidates by the Master-General of the Ordnance.

Cavalry and infantry commissions were expensive. In the 1850s a commission in the horse guards and line cavalry cost £1,000 to £1,600, foot guards £900, and line infantry £400 to £450.[9] This was at a time when well-paid construction workers with year-round work in

---

[7] *Purchase Commission Report (1857)*, at xx.

[8] Spiers, *The Army and Society*, 19.

[9] *Purchase Commission Report (1857)*, appx. I.

London earned £112 annually and in Dublin £75.[10] Infantry and cavalry officers usually had to pay for promotion though the cost was offset in part by the proceeds from the sale of their current rank. For example, in the infantry it cost on average £700 to purchase first-lieutenant rank, £1,800 for captain, £3,200 for major, and £4,500 for lieutenant-colonel. Prices in cavalry and guards regiments were much higher.[11] The exception to promotion by purchase occurred when an officer forfeited his commission to the Crown. This enabled the senior officer of the lower rank to be promoted without purchase. Forfeiture resulted from death, promotion to general, special pension voted by Parliament, or cashierment. From 1849 through 1853 about 30% of promotions were without purchase.[12]

There was no purchase system in the British East India Company. Officers of its armies were commissioned directly or graduated from the Company's military "seminary" at Addiscombe. Direct commissions were by director nomination; entrance to the seminary by competitive examination. About one-quarter of the Company's officers were Addiscombe graduates.[13]

Logistic support for the army's combat units, both in garrison and while on active service, was provided by the army's "civil" departments whose officers were civil servants. For example: Transport and provision of food were arranged for by the Commissariat, an office of the Treasury, procurement and warehousing by the Military Stores Department of the Board of Ordnance, and medical services by the Army Medical Department. The War Office assigned combatant "relative" rank to civil department ranks for purposes of precedent, duties, allowances, and privileges.[14]

Commissions in 1904

The post-Boer War army differed significantly from the army of the Crimean War era. The armies of the EIC had been transferred to a new Indian Army officered by Sandhurst and Woolwich graduates; purchase of commissions and promotions had been abolished; the civil departments with military responsibilities had been incorporated into the army. Many aspects of officer life; however, remained unchanged. As in the 1850s, officers were commissioned into their regiments, not the army as a whole. For promotion purposes, the Royal Engineers and Army Service Corps were both single regiments but there were separate promotion lists for the Royal Horse Artillery, Royal Field Artillery, and Royal Garrison Artillery. Many departmental officers, as in the past, did not have combatant rank, *i.e.* captain,

---

[10] Board of Trade, *Returns of Wages Published between 1830 and 1886*, 1887, [C. 5172], at 354, 388.

[11] *Purchase Commission Report (1857)*, appx. I.

[12] Spiers, *The Army and Society*, 19. Calculated from *Report of the Commissioners Appointed to Inquire into the System of Purchase and Sale of Commissions in the Army*, 1857 Sess. 2, [2267], appx. XII.

[13] E.H. Nolan, *The Illustrated History of the British Empire in India and the East*, Vol 1. (London: Virtue, c. 1860), 338-39.

[14] *Royal Warrant for the Pay and Promotion, Non-Effective Pay, and Allowances of Her Majesty's British Forces Serving Elsewhere than in India, 1866*, Arts. 47-53, 102. Hereafter cited as *Royal Warrant for the Pay*.

major, *etc.* Their ranks remained descriptive of their duties, such as chaplain, surveyor, inspector, and commissary.[15]

To maintain combatant officer strength, the British and Indian armies together required 800 to 900 new officers per year. Sandhurst and Woolwich provided about 500 officers, the auxiliary forces somewhat under 300 (nearly all from the militia), and other sources about 50. Commission sources for the combatant corps were as follows.[16]

Infantry and Cavalry:	Sandhurst, auxiliary force infantry and cavalry officers, university candidates, and the enlisted ranks.
Royal Artillery:	Woolwich and auxiliary force artillery officers. District officers were commissioned from the enlisted ranks.
Royal Engineers:	Woolwich; however, coast battalion officers were commissioned from the enlisted ranks.
Army Service Corps:	Officers of other combatant corps with at least one year's service, Sandhurst, auxiliary force officers, and university candidates.

About ten combatant commissions were awarded annually, in total, to officers of the colonial armed forces and graduates of the Canadian Royal Military College at Kingston.[17] Each year about forty university candidates received combatant commissions. Eligible for such direct commissions were graduates of the seven ancient universities, the Royal University (Ireland), plus the Universities of London, Manchester, Durham, Birmingham, Liverpool, and Wales.[18]

Though officer candidates no longer had to purchase infantry and cavalry commissions, the officer selection process and insufficient pay insured that the officer corps remained the domain of the British establishment. Cadetships at Sandhurst and Woolwich went over-

---

[15] *Royal Warrant for the Pay, 1899.*

[16] *Report of the Committee appointed to consider the Education and Training of Officers of the Army*, 1902, [Cd. 982], at 9-17, appx. xxi, hereafter cited as *Akers-Douglas Committee Report;* Testimony of MG A. Turner, I-G Auxiliary Forces, *Royal Commission on the Militia and Volunteers, Minutes of Evidence*, 1904, [Cd. 2062], at q. 3750; *Royal Commission on the Militia and Volunteers, Appendices*, 1904, [Cd. 2064], no. 54.

Note that in 1902 the only auxiliary force officers who could obtain regular commissions were those of the militia. Beginning in 1903, 5 to 10 commissions annually went to yeomanry officers; in 1904 15 to 20 to volunteer force officers.

[17] *Akers-Douglas Committee Report*, at 9-17, appx. xxi.

[18] *Regulations under which Commissions in the Army may be Obtained by University Candidates*, 1904.

whelmingly to public school graduates. In the last five years of the nineteenth century the public schools provided 78.7% and 55.4% of Sandhurst and Woolwich entrants, respectively.[19] These percentages increased during the following years. In the first decade of the twentieth century, for the combined cadet corps, ten major public schools accounted for 69% of entrants, the other public schools 25%, proprietary schools and tutoring 4%, state and free sectarian secondary schools less than 2%.[20] The most prevalent schools attended by cadets were Wellington, Cheltenham, and Eton, followed by Belford, Harrow, Marlborough, and Rugby. Eton was the school of choice for Sandhurst cadets; Cheltenham for Woolwich.[21] There was considerable snobbery over militia commissions and both county lieutenants and regimental colonels insured that all officer candidates possessed a gentleman's education.[22]

Commissions in the Ordnance Department and the Pay Department, with a few exceptions, went to serving officers of the combat arms. For the Ordnance Department, temporary appointments to vacancies were open to combatant officers with at least four years' service.[23] After an appointee had seven years with the Ordnance Department, the War Office could make permanent his departmental assignment.[24] For the Pay Department, probationary appointments to vacancies were open to combatant officers who were under age 35, had at least five years' service, and held substantive rank not higher than captain. After successful completion of one-year of departmental service, the combatant officer could, at his option, make his Pay Department appointment permanent.

Commissions in the Royal Army Medical Corps and Veterinary Department, other than those of quartermaster, were direct to qualified, civilian professionals. Quartermasters were commissioned from the enlisted ranks. Commissions in the Chaplains' Department were direct to state-recognized clergymen. In 1904, there were only Anglican, Roman Catholic, and Presbyterian commissioned chaplains. The army also had 75 acting chaplains among whom were 14 Methodists.

The initial cost to become an officer through Sandhurst was £350 for the infantry and £750 for the cavalry. The entry cost for artillery and engineering officers was £500. Cadets at Sandhurst and Woolwich, with some exceptions, payed tuition and other fees. In 1904, the one-year Sandhurst course cost £150 and the two-year Woolwich course £300. In 1911, John Ward, trade unionist and Liberal MP, remarked on a proposal to lower fees as

---

[19] Harries-Jenkins, *The Army in Victorian Society*, 140-41.

[20] Otley, "The Educational Background of British Army Officers." Major public schools were those with annual fees of at least £140. An example of a Victorian Era, free sectarian secondary school is the Erasmus Smith High School, Dublin which Leopold Bloom attended for a year or two.

[21] Churchill, "The Army as a Profession.".

[22] Bowman and Connelly, *The Edwardian Army*, 114. Candidates for militia com-missions were nominated by county lieutenants, then recommended to the War Office by regimental commanders. Militia Act, 1882. 45 & 46 Vict., c. 49, §6; Testimony, *Minutes of Evidence taken before the Royal Commission on the Militia and Volunteers*, 1904, [Cd. 2062, Cd. 2063].

[23] *Royal Warrant for the Pay 1899*, Art. 393.

[24] Ibid., Art. 396a.

follows: "A mere reduction of the fees which have to be paid from £150 to £80 a scholar is of no consequence at all so far as being an inducement to the children of the working classes and of the poor, and gives them no opportunity to qualify for positions previously closed to them."[25] A War Office committee that investigated expenses incurred by officers found that in 1903, the initial cost to outfit a new officer was £555 for the cavalry and £155 for the other corps. The standard officer mess entrance charge was £10 and the cost to furnish a junior officer's quarters was £35.[26]

To maintain themselves in the style appropriate for their regiment, new infantry officers required an annual private income of £100 to £150, while cavalry officers required £350 to £700 and upwards.[27] Engineer, artillery, and services officers could usually live a gentlemanly life on their pay as such officers led a much less extravagant life than did those of the infantry and cavalry.[28] The service corps and scientific arms were of relatively low prestige and accordingly, always lacked sufficient officers of the aristocracy and gentry to have developed a tradition of high living.[29] Middle class officer aspirants without private incomes, or who could not rely on their families for support, favored careers in the services, scientific arms, and the Indian Army (where the pay was higher and the cost of living lower than at home).[30]

Officer Training

In 1902, a committee on officer education, chaired by the Conservative MP Aretas Akers-Douglas, found that officers were insufficiently educated in both general and military subjects.[31] Witnesses before the committee were "unanimous in stating that the junior officers are lamentably wanting in military knowledge, and what is perhaps even worse, in the desire to acquire knowledge and zeal for the military art." The committee noted it is not uncommon to find officers unable to write a good letter or to draw up an intelligible report, and found officers deficient in knowledge of mathematics, modern languages, Latin, and experimental science.[32]

---

[25] 23 *H.C. Deb.* (5th ser.) (1911) 422.

[26] *Report of the Committee to enquire into the nature of the Expenses Incurred by Officers of the Army*, 1903, [Cd. 1421], at 7-8.

[27] Ibid., at 8-9.

[28] French, *Military Identities*, 51.

[29] Ibid., Chapter 6.

[30] Spiers, *The Army and Society*, 11, 23; French, *Military Identities*, 51-52, 107-08. For infantry officers of the Indian Army pay for lieutenants and captains was 1.6 to 1.7 times British Army pay, for majors 2.0 times British pay, and for lieutenant-colonels 2.5 times British pay. *Hart's Annual Army List 1904*, 1168-69.

[31] *Akers-Douglas Committee Report*, at 2.

[32] Ibid., at 2-4.

Initial officer training was at Sandhurst (infantry, cavalry, services), Woolwich (artillery, engineers), or the regiment, corps, or department into which the officer was commissioned.[33] The one-year program for Sandhurst cadets was not academically rigorous and provided little useful, tactical training. The two-year program for Woolwich cadets was more difficult as it concentrated on university-level mathematics, but it provided little general education and no instruction in tactics and military history.[34] While the Akers-Douglas Committee was satisfied overall with the quality of education at Woolwich, it found the curriculum and standards of Sandhurst severely deficient.[35] Together, the two military schools met about half the army's combatant officer requirement. For the five years 1886 through 1890, Woolwich graduates accounted annually, on average, for 136 new engineers and artillery officers; Sandhurst graduates for 310 new officers for the other combat arms.[36]

For combatant officers with direct commissions, their initial training was "on-the-job" with their regiment. Each year about 225 regular commissions were awarded to officers of the auxiliary forces (overwhelmingly from the militia; very few from the volunteer force and yeomanry), and about 40 to university graduates.[37] In the latter part of the nineteenth century such commission routes provided most of the line infantry's new officers as financially well-off Sandhurst graduates favored the prestigious guards regiments and cavalry, while impecunious cadets sought appointment to the Army Service Corps and Indian Army.[38] All graduates of War Office approved universities were eligible for commissions.[39] Officer candidates from both the auxiliary forces and the universities had to pass written examinations in general and military subjects.[40] Professional opinions varied as to the difficulty of the academic tests.[41] All combatant officers who received direct commissions had to pass a physical fitness test given at the conclusion of a course in "gymnasia."[42]

---

[33] Note that university candidates could be commissioned directly as second-lieutenants of cavalry, infantry, and the Army Service Corps. *Royal Warrant for the Pay, 1899*, Arts. 1A, 3.

[34] Harries-Jenkins, *The Army in Victorian Society*, 113; *Akers-Douglas Committee Report*, at 16-17.

[35] *Akers-Douglas Committee Report*, at 15-17, 19-23.

[36] Ibid., appx. xxi.

[37] Ibid.

[38] Harries-Jenkins, *The Army in Victorian Society*, 153. This changed between 1900 and the First World War. During that period Sandhurst provided about one-half the line infantry's officers, the militia about one-third, other sources, mostly the universities, the balance. Bowman and Connelly, *The Edwardian Army*, 12.

[39] Prior to 1904, university candidates were not required to hold degrees. They only needed to have passed the first-year examinations (Oxford, Cambridge, Dublin, and Durham), intermediate examinations (London, Manchester, Royal University of Ireland) or academic military examination (Scottish universities). *Regulations under which Commissions in the Army may be obtained by University Candidates*, 1899.

[40] Ibid.; *Regulations under which Commissions in the Army may be obtained by Officers of the Militia*, 1899.

[41] Harries-Jenkins, *The Army in Victorian Society*, 154-55.

[42] Goodenough and Dalton, *Army Book for the British Empire*, 425.

Novice officers received technical training both on-the-job and through course attendance at army training establishments. All engineers, upon leaving Woolwich, commenced the full-time, two-year engineering course of the School of Military Engineering at Chatham. All novice artillery officers took the six-week course of the School of Gunnery at Shoeburyness. Infantry and cavalry officers upon reaching three years of service took the four-week riflery course of the School of Musketry at Hythe.[43] All other formal training was optional, and few junior officers applied for the technical courses offered by the army.[44]

Newly commissioned combatant officers had to sit for oral examinations in "drill" and regimental financial management (A and B exams), and written examinations in law, tactics, army organization, topography, engineering, and army equipment (C, D, and G exams). Passing all examinations was required for promotion to first lieutenant.[45] Preparation for these examinations was self-study of military textbooks that were of little practical value, on-the-job training, and coaching by veteran lieutenants.[46] The oral examinations were not rigorous and the written examinations were simply a test of memory.[47] In a survey of commanding officers by the Akers-Douglas Committee, one-third of respondents thought the oral examinations a farce and the written examinations merely tests that were crammed for and the memorized material quickly forgotten.[48] The attitude of novice officers who had passed the mandatory examinations was generally "I have got nothing to do, thank goodness, now for six years" when they would be eligible for the captaincy examination.[49]

After officers passed the lettered examinations, they were eligible for further formal, military training through a limited number of professional training courses. Very few officers availed themselves of such opportunities. Victorian army mores marked an intellectual officer as "an oddity and potential deviant."[50] Officers keen on military studies were held in low regard by their mess-mates and were termed "mugs." Those were the few officers who neither rode nor shot, did not play games, and never over-indulged in wine and whiskey. In their free time, mugs studied algebra, fortification, and French, and then went to bed early.[51]

---

[43] Ibid., 225, 240, 425; French, *Military Identities*, 154; *The King's Regulations and Orders for the Army, 1908*, ¶798-800. Hereafter cited as *King's Regulations* or *Queen's Regulations*.

[44] Testimony of General Evelyn Wood, former Adjutant-General of the Army, *Minutes of Evidence taken before the Committee appointed to consider the Education and Training of Officers of the Army*, 1902, [Cd. 983], at q. 17.

[45] *Akers-Douglas Committee Report*, at 30-32.

[46] The textbooks were "unequivocally condemned" by witnesses before the Akers-Douglas Committee. Ibid., at 33.

[47] Ibid., at 30-32.

[48] Ibid, appx. I.

[49] Testimony of General Evelyn Wood, former Adjutant-General of the Army, *Minutes of Evidence taken before the Committee appointed to consider the Education and Training of Officers of the Army*, 1902, [Cd. 983], at q. 17.

[50] French, *Military Identities*, 145-46; Harries-Jenkins, *The Army in Victorian Society*, 105, 167, 278.

[51] Younghusband, *A Soldier's Memories*, 115-16.

Expert knowledge of the military arts rarely increased an officer's promotion potential as the military establishment valued good character much more than professional merit. What senior officers usually meant by "good character" was birth into a socially prominent family, attendance at the right public school, possession of the social graces, ability to play polo well, and behavior not egregious enough to warrant imprisonment or cashierment. Furthermore, seniority, not merit, was in practice the principal factor in advancement.[52]

The technical courses available to British army officers in the early twentieth century were as follows:[53]

School of Military Engineering, Chatham	Several short courses in specific aspects of military engineering open to all officers.
School of Gymnastics, Aldershot	A short course open to all officers with more than two-years' service.
School of Gunnery, Shoeburyness	Short courses open to all officers and a long course for artillery officers.
Army Service Corps School, Aldershot	A ten-week course for captains and higher who had been approved for entry into the Staff College.
School of Signaling, Aldershot	A three-month course open to all officers.
Artillery College, Woolwich	Short courses open to all officers and a one-year professional course for artillery officers.
Ordnance College, Woolwich	A one-year professional course that accepted 16 officers per year through examination. Enrollees had to have at least five-years' service. A one-year advanced course given annually to the top half of basic course graduates. Officers who completed the professional course had "o" in their army list entries; the advanced course "p.a.c." The college also offered various short courses open to all officers.

Advanced officer training was provided by the Staff College. The course consisted of two-years' instruction at the college's facility in Camberley, then followed by a one-year

---

[52] Ibid., 104; Spiers, *The Army and Society*, 249.

[53] *Queen's Regulations, 1899*, ¶¶771-72, 1155-60, 1177, 1245, 1250, 1260, 1295; *King's Regulations, 1908*, ¶¶765-68; Goodenough and Dalton, *Army Book for the British Empire*, 423-24, 426; Spiers, *The Late Victorian Army*, 101.

internship of two six-month postings to different staff offices.[54] The Staff College accepted 32 enrollees per year; 24 through competitive examination, 8 by appointment.[55] To be eligible for entry a candidate had to hold the rank of captain or higher, have five-years' service, and be recommended by both his regimental commanding officer and general-officer-commanding.[56] Officers who completed the full course had "p.s.c." in their army list entries. The Staff College opened in 1858 and relatively few officers eligible for enrollment applied. Many senior officers viewed the course as of little military value and believed its students attended to have a multi-year holiday from regimental routine. Some colonels viewed a staff college appointment as an opportunity to rid their regiment of dead wood and recommended for enrollment officers they viewed as useless. Other colonels viewed Staff College aspirants as "disloyal" officers eager to absent themselves from the battalion for three years.[57]

By 1890, the attitude of junior officers to continuing professional education had changed. Staff College students were no longer viewed as "mugs" as young officers with the p.s.c. credential often received well-paid and socially prestigious staff assignments. Also, Staff College graduates were in general promoted more rapidly than other officers. By Bloomsday, most captains coveted a place at the Staff College.[58]

In 1904, the adage that an infantry or cavalry officer never opened a book unless it was the *Queen's Regulations* or *Hart's Army List* was no longer near universally true. The Boer War had shocked many officers out of their belief that gentlemanly virtues and expert knowledge of regimental routine, were all an officer needed to fill war-time command and staff positions.[59] Consequently, a large number of young army officers developed a keenness for their profession, something the Akers-Douglas Committee noted as lacking just a few years earlier. Publishers had a thriving business providing army officers with technical books, pamphlets, and periodicals.[60]

Pay, Promotion, and Types of Rank

Officer Pay

Officers received base pay, allowances, and supplemental pay. Supplemental pay depended on an officer's assignment. For example, commanding a battalion or serving as a

---

[54] Spiers, *The Army and Society*, 154.

[55] *Queen's Regulations 1899*, ¶1144.

[56] Ibid., ¶1146.

[57] Younghusband, *A Soldier's Memories*, 115; Tulloch, *Recollection of Forty Years' Service*, 145; French, *Military Identities*, 153.

[58] Younghusband, *A Soldier's Memories*, 115-16; Duncan, *Military Education of Junior Officers*, 179-86.

[59] Farwell, *For Queen and Country*, 152; French, *Military Identities*, 178.

[60] Duncan, *Military Education of Junior Officers*, 130-43.

regimental adjutant, came with supplemental pay. Engineers received an additional 2s. 9d. to 14s. per day for military engineering work. An engineer who worked professionally ten months in the year earned annually an extra £41 if a second lieutenant, £210 if a lieutenant-colonel.[61] The supplemental engineering pay was the main reason Woolwich cadets favored engineer commissions over artillery commissions.

Base pay for officers, like for enlisted men, varied by rank and corps. The Ordnance Department and Royal Army Medical Corps were the highest paying; the line infantry and Army Services Corps the lowest paying. For officers below the rank of lieutenant-colonel, there was usually longevity pay based on years in rank. Base pay of officers was well below that of civil servants in equivalent positions. Combatant lieutenant-colonels received about half the annual pay of their civilian counterparts, Clerks 1st Class (£670 to £800). Combatant captains received about one-third to one-half the annual pay of their civilian counterparts, Clerks 2nd Class (£315 to £500).[62] For engineers, supplemental professional pay narrowed the gap between army and civil service pay.

Table 18 shows typical officer pay for selected corps and departments. To illustrate the financial advantage of service on the sub-continent, the table includes infantry and cavalry pay scales of the Indian Army at the 1904 exchange rate of 1 Rupee = 16d. Infantry and cavalry pay are for officers of line regiments. Pay includes standard *"batta"* which was an allowance all officers received. Also, note that medical officers with the rank of first lieutenant were newly appointed and after completion of a probationary period were promoted to captain.

Table 18.
Typical Annual Base Pay in Pounds Sterling, 1904
British and Indian Armies, Selected Corps & Departments

	British Arms			British Support			Indian Army	
	Infantry & ASC	Cavalry	Artillery & Engineers	Ordnance	Medical	Pay Dept.	Infantry	Cavalry
Colonel	329	392	475	730	730	548	1,033	1,170
Lt.-Col.	329	392	329	547	602	456	823	923
Major	292	310	292	365	365	319	535	741
Captain	211	237	211	273	250	273	331	449
1st Lieut.	137	140	143	NA	146	NA	204	291
2nd Lieut.	96	122	102	NA	NA	NA	161	247

Source: *Hart's Annual Army List, 1904.*

Officers who held certain staff appointments received pay in accordance with the staff pay schedule, not the base pay schedule. Such pay was dependent on staff position, not rank. These positions were with the War Office, army districts, garrisons, brigade and higher

---

[61] *Royal Warrant for the Pay, 1899*, Arts. 210, 213.

[62] Harries-Jenkins, *The Army in Victorian Society*, 87.

formation echelons, *etc*. Following are examples of annual staff pay, including allowances, in pounds:⁶³

Staff Position Title	Annual Pay
Inspector General of Fortifications	2,100
Military Secretary	1,500
Deputy Director-General of Ordnance	1,200
Military Attaché	800
Deputy-Assistant Quartermaster General	650
Brigade Major	600
Staff Captain	500
Staff Lieutenant	400

If staff pay was less than an officer's base pay, then he received "half-pay" plus the staff pay. Half-pay was not 50% of base pay but pay set by the half-pay schedule. Following is an abbreviated version of the 1904, half-pay schedule, in pounds paid annually:⁶⁴

	Cavalry	Infantry	Other
Colonel & Lt.-Col.	228	201	213
Major	182	173	182
Captain	137	128	152

The highest-paid and socially most desirable staff positions were those in London at the War Office's premises. The lesser-paid positions were widespread geographically. For example, military secretaries were assigned to colonial governors as well as army district commanders and departmental heads. All camp and garrison staffs, such as those of the Curragh Camp in Ireland and the Gibraltar Garrison, had a "deputy-assistant adjutant & quartermaster general." Each brigade had an administrative officer termed "brigade major" though that position was usually held by a captain. Staff captains and lieutenants were found on departmental, camp, garrison, and higher echelon staffs (districts, divisions, and field army corps).

There were many monetary allowances to which officers were entitled.⁶⁵ For example, when on active service or otherwise living "under canvas" officers received a daily field allowance. The amount depended on rank, and for regimental officers it was from 2s. for lieutenants to 4s. for colonels; for staff officers 3s. to 6s. There were also allowances for purchase and maintenance of a horse, employment of servants, official travel, and transportation of personal possessions due to a change of station.

---

⁶³ *Royal Warrant for the Pay, 1899*, Arts. 115-17.

⁶⁴ Ibid., Arts. 307, 309.

⁶⁵ *Regulations relating to the Issue of Army Allowances, 1884*.

Types of Rank [66]

Officers of the Edwardian army could hold simultaneously three different ranks: substantive, army (which includes brevet), and temporary (which includes local). It was not unusual for a temporary lieutenant-colonel to be a regimental captain and a brevet major.

There were four types of substantive rank: Regimental (through lieutenant-colonel for the combat arms, colonel for the support departments), colonel commandant designation of infantry and cavalry officers (colonel), special appointment, and general officer ranks above brigadier-general. Note that brigadiers held an appointment, not a rank; a brigadier's regimental rank was his substantive rank. Pension amounts were determined by substantive rank.

Army rank was an officer's rank when posted to an extra-regimental position. For example, a regimental captain could hold the rank of major while in the Adjutant-General's office or on a division's staff. Army ranks were permanent ranks that were not substantive and included regimental brevet ranks. The brevet ranks of major, lieutenant-colonel, and colonel were army ranks obtained as an award for outstanding service. All lieutenant-colonels in command of battalions and regiments became army colonels after four years in command.[67] Pay, with a few exceptions, was not based on army rank. One such exception was that captains with higher rank by brevet received additional pay.[68]

Temporary rank was for officers temporarily in a position that warranted a rank higher than their substantive rank. It was typically awarded by the War Office to officers filling positions where the incumbent was on extended medical leave or had died. Temporary promotion to a higher rank often brought with it higher pay. Local rank was a form of temporary rank assigned to officers detached from their unit for temporary assignment elsewhere. Such officers held the temporary rank, and received its pay, while in their new, time-limited posting.

Promotion, Substantive Rank

Prior to August 1877, promotion in substantive rank was strictly by seniority. There was no "up or out" rule and officers could serve until death or severe infirmity.[69] An 1875 royal commission on army promotion and retirement noted that senior combatant officers were generally middle-aged with many in their late 50s, and most generals were elderly. Average age at time of promotion was as follows:[70]

---

[66] Sources: Farwell, *For Queen and Country*, 52-53; *Royal Warrant for the Pay, 1899*, Arts. 1314-16; Testimony of Various Officers, *Report of the Royal Commission on Army Promotion and Retirement*, 1876, [C. 1569].

[67] *Royal Warrant for the Pay, 1899*, Art. 39.

[68] Ibid, Arts. 199, 201, 218, 224.

[69] Spiers, *The Army and Society*, 195.

[70] *Report of the Royal Commission on Army Promotion and Retirement*, 1876, [C. 1569], appx. E.

Army Life and Retirement: Officers

	Infantry & Cavalry	Artillery	Engineers
Major-General	71	69	62
Colonel	58	54	54
Lt.-Colonel	54	48	47
Major	50	40	40

Company commanders, captains who on average attained that rank in their mid-thirties, could serve as such well into their fifties. Major-General Parr, in his 1917 memoirs, described his first company commander as a fat, old man in his forties "who was of no possible use as an officer, either as instructor or leader of men." The company's lieutenant was not much better. He had been promoted from the ranks for gallantry and Parr found him "extraordinarily stupid, obstinate, pig-headed, and tactless."[71]

By Bloomsday, officer promotion was in part merit-based and there were age limitations for serving officers. Other than screening by written promotional examination and rejection of promotion candidates with egregious faults, advancement in rank was effectively through seniority.[72] Promotion board members rarely had first-hand knowledge of promotion candidates (other than the notorious cases) and could not rely on flattering or irrelevant letters of recommendation.[73] Retirement for combatant officers was compulsory upon reaching the following ages: Captain 45, Major 48, Lieutenant-Colonel 55, Colonel 57, Major-General 62, Lieutenant-Generals and Generals 67.[74]

For quartermasters, ridingmasters, district officers of artillery, and officers of the engineer's coast battalion, all former enlisted men, the mandatory retirement age was 55.[75] Officers of the non-combatant departments faced compulsory retirement between age 55 and 65 depending on department and rank.[76] Rank, for purposes of compulsory retirement, was substantive and brevet rank. This allowed the many captains who were brevet majors to remain in the army for an additional three years while majors who were brevet lieutenant-colonels gained seven years.

Promotion was constrained by the rank distribution set in the annual, army appropriation. Departmental officers had the best promotion prospects; all could expect to retire at age 55 with the equivalent rank of major. Officers of the Army Service Corps and line infantry regiments had the worst promotion prospects, and most would suffer compulsory retirement at age 45. These forced-out captains would receive upon retirement the gentlemanly rank of honorary major; however, their pensions could only provide them an ungentlemanly lower-middle class standard of living.

---

[71] Parr, *Recollections and Correspondence*, 70-71.

[72] Bowman and Connelly, *The Edwardian Army*, 34; French, *Military Identities*, 149-50.

[73] French, *Military Identities*, 152.

[74] *Royal Warrant for the Pay, 1906*, Art. 516.

[75] Ibid., Arts. 530-31.

[76] Ibid., Arts. 550, 558, 566, 575.

Table 19.
Substantive Rank Distribution, 1904
Selected Corps & Departments, Percent

Rank	Line Infantry	Line Cavalry	ASC	Engineers	Field Artillery	Garrison Artillery	Ordnance Dept.	Pay Dept.
Colonel	NA	NA	NA	NA	NA	NA	6.4	7.2
Lt.-Col.	4.4	4.2	5.6	9.0	4.1	6.3	6.4	31.8
Major	13.0	16.7	13.2	18.4	18.4	14.7	27.3	
Captain	30.4	20.8	38.2	25.6	22.4	28.3	40.6	61.0
1st Lieut.	34.8	33.3	32.9	28.1	55.1	30.4	19.3	NA
2nd Lieut.	17.4	25.0	10.1	18.9		20.3	NA	NA

Source: War Office, *Army Estimates for the year 1904-05,* 1904, H.C. Accounts & Papers, No. 73.

Line infantry officers could obtain substantive rank of colonel through command of a territorial district. Any officer could obtain such rank through assignment to a staff position the War Office had designated for substantive colonels.[77] Promotion to substantive colonel was rare. For the 36 months ended June 1905, such promotions, excluding those of medical professionals and chaplains, averaged just 30 per year.[78] Of those promotions, 49% went to line infantry officers, 16% each to engineers and artillery officers, 7% to the cavalry, 3% to the ASC, 4% each to the Foot Guards and Ordnance Department, and 1% to the Pay Department. Infantry promotions to colonel were overwhelmingly from appointment as officer commanding a territorial district. Promotion to substantive colonel was not particularly important to a combatant officer's career as the pay and pension for that rank were generally the same as for lieutenant-colonel. Additionally, full colonels could remain in the army for only two years longer than could lieutenant-colonels.

The crucial substantive promotion in an officer's career was from captain to major. This was because of the significant difference in compulsory retirement age and pension amount for holders of those two ranks.[79] For officers of the support departments, who rarely had private incomes, the difference in pay between captains and majors was also important. Officers with successful careers (substantive colonels) were on average promoted to major at age 37 after 8 years as a captain.[80] Accordingly, of great interest to army officers was the ratio of authorized captains to majors. As shown in Table 20, this ratio varied greatly by corps and department and explains why impecunious infantry and services officers sought transfer to the Ordnance and Pay Departments.

---

[77] *Queens Regulations, 1899*, ¶40.

[78] *Quarterly Army List, June 1905.*

[79] Captains were forced out at age 45 or 48 and received pensions of £200 annually; majors at age 48 or 55 with pensions of £300 annually. *Royal Warrant for the Pay, 1906*, Arts. 516, 524.

[80] Ibid.

Table 20.
Ratio of Substantive Captains to Substantive Majors, 1904
Selected Corps & Departments

Corps/Department	Percent Captains	Percent Majors	Ratio
Royal Field Artillery	22.4	18.4	1.22
Cavalry, Line	20.8	16.7	1.25
Royal Engineers	25.6	18.4	1.39
Ordnance Department	40.6	27.3	1.49
Pay Department	61.0	31.8	1.92
Royal Garrison Artillery	28.3	14.7	1.93
Infantry, Line	30.4	13.0	2.34
Army Service Corps	38.2	13.2	2.89

Source: War Office, *Army Estimates for the year 1904-05*, 1904.

There were some non-professional, departmental commissions not available to serving officers. Inspectors of Army Schools were commissioned from the senior schoolmaster ranks.[81] Superintending Engineers and Inspectors of Mechanical Transport of the Army Service Corps were all former civilians who had received direct commissions through competitive examination. Inspectors of Ordnance Machinery were also commissioned directly.[82] Such officers, like the regimental quartermasters and ridingmasters, were viewed by both civilians and combatant officers, as of much lower social standing than officers commissioned through the usual channels. The following table shows the distribution of rank for these specially commissioned, tradesman-like, officers.

Table 21.
Rank Distribution, Number of Positions
Inspectors and Superintending Engineers, 1904

Relative Rank	Inspectors Army Schools	Inspectors Ordnance Machinery	Inspectors Mechanical Transport	Superintending Engineers
Major	0	5	1	1
Captain	2	15	1	0
1st Lieutenant	25	5	1	1

Sources: War Office, *Army Estimates for the year 1904-05*, 1904, H.C. Accounts & Papers, No. 73; *Royal Warrant for the Pay, 1906*, Art. 318; *Monthly Army List, December 1904*.

---

[81] *Royal Warrant for the Pay, 1899*, Art. 8. Senior schoolmasters were warrant officers. Junior schoolmasters were rank-equivalent to infantry colour-sergeants.

[82] *Royal Warrant for the Pay, 1906*, Art. 409G.

The all-male army had a professional nursing organization of female, civilian employees: Queen Alexandra's Nursing Service. Nurses, like army schoolmistresses, were subject to military discipline but lacked relative rank.

Table 22.
Queen Alexandra's Nursing Service
Staffing and Annual Pay, 1904

Position	Number	Minimum Pay	Maximum Pay	Infantry Rank Pay Equivalent
Matron-in-Chief	1	£ 250	£ 300	Major
Principal Matron	2	150	180	1st Lieutenant
Matron	29	70	120	Sergeant-Major
Sister	102	37	50	Sergeant
Staff Nurse	208	30	35	Corporal

Source: War Office, *Army Estimates for the year 1904-05*, 1904, H.C. Accounts & Papers, No. 73.

As shown in Table 22, to receive a lower-middle class wage a nursing service employee needed promotion to the supervisory, matron positions. Holders of such positions accounted for only 9.4% of the total staff. A hospital's chief nurse received additional pay of up to £30 annually. Note that nursing sisters, all educated professionals, received the same pay as did infantry sergeants who generally had incomplete, primary school educations. Staff nurses received four weeks of paid annual leave; nursing sisters five weeks; matrons of all ranks six weeks.[83]

Housing

All officers received housing at state expense and they were usually quartered in barracks. Such quarters were in a segregated section of a large barrack, a two-storey apartment building, or the officers' mess building. Rank determined the size of an officer's apartment.[84]

	Personal Rooms	Servant Rooms
Colonel	5	1
Lt.-Col., Commanding	4	2
Lt.-Col, Other	2	1
Major	2	1
Captain, Lieutenant	1	½

---

[83] *Royal Warrant for the Pay, 1906*, Art. 682F.

[84] *Regulations relating to the Issue of Army Allowances, 1884*, ¶304.

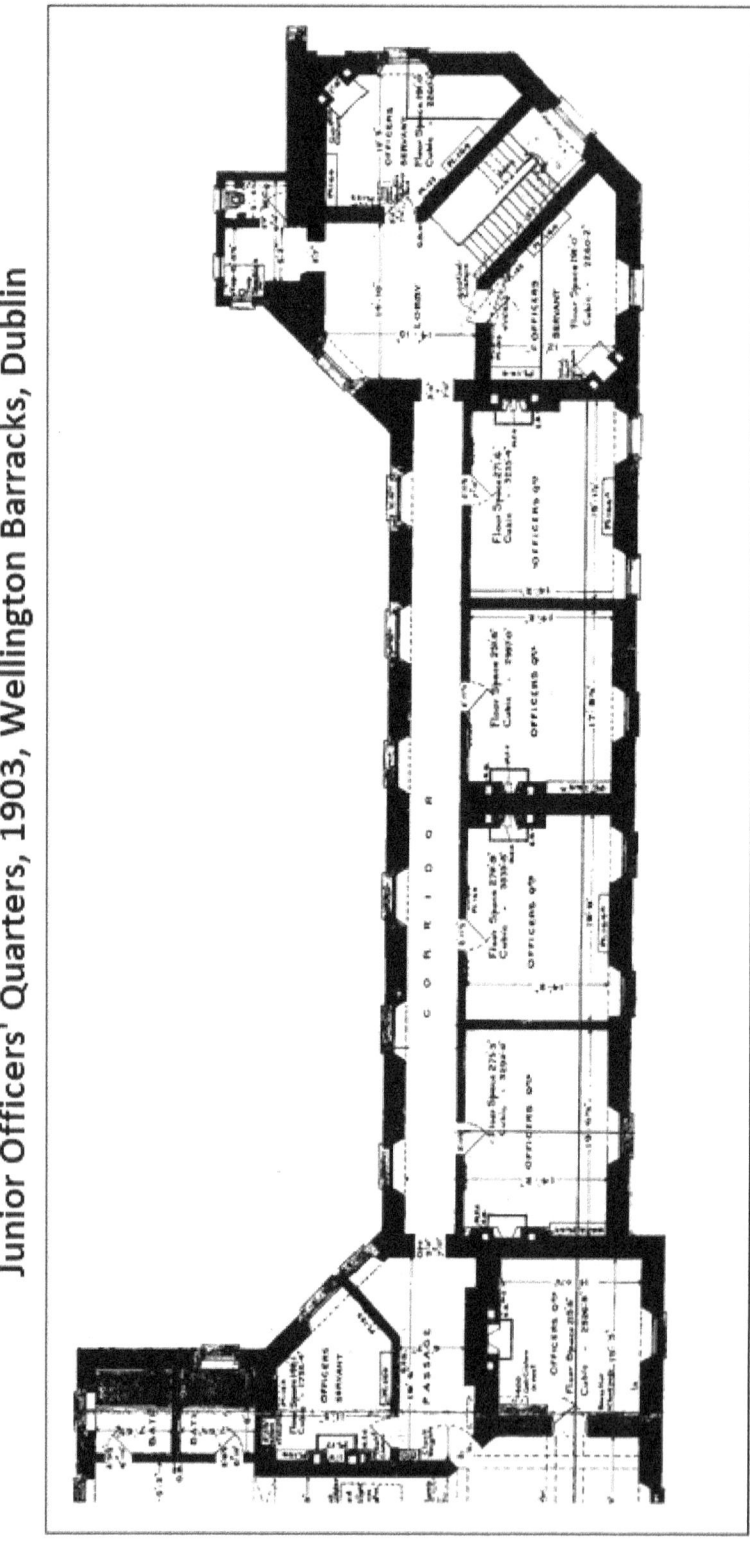

Junior Officers' Quarters, 1903, Wellington Barracks, Dublin

Military Archives, Defence Forces Ireland, IE/MPD/ad134143-010, used with permission.

Servants' rooms adjoined or were near, their officer's room. Servants of junior officers shared a single room while servants of senior officers had private rooms. A colonel that served as "post commandant" usually resided in a house on the barracks or camp grounds.

When army housing was not available, officers received a housing, fuel, and light allowance, the amount determined by rank. Lieutenants received £45 annually, captains £61, majors £81, lieutenant-colonels £90, and colonels £108.[85]

On the previous page is the architectural plan for the upper level, right wing of the officers' mess building at Wellington Barracks, Dublin. Shown is accommodation for five junior officers and six servants. The left wing, not shown, housed one officer and two servants plus contained the mess kitchen, pantries, and a billiard room. The lower level had the mess rooms, 2-room apartments for senior officers, plus accommodation for other junior officers, and servants. Captains and first lieutenants had rooms of 275 square feet; second lieutenants 215 square feet. Two soldier-servants shared a 140 square foot room. All officers were entitled to live off-post. With a few exceptions, officers who lived in civilian housing by choice, including married officers, did not receive housing allowances.[86]

The Garrison Workday

Officers who served with regiments in garrison typically had very abbreviated workdays. Staff officers; however, always had full work schedules as the army was highly bureaucratic. Engineers were usually occupied with professional duties. They surveyed land at home and abroad, made maps, drafted architectural plans for barracks and other improvements on army property, and inspected buildings and fortifications. Artillery officers assigned to batteries instructed gun crews and inspected their weapons and equipment. Army Service Corps officers arranged for and monitored the movement of foodstuffs, equipment, and supplies, both at home and abroad, for units in garrison and on active service. It was the infantry and cavalry officers who while not on active service, led a life of leisure.[87]

Nearly all command work required for an infantry battalion or cavalry regiment was done by the colonel, the adjutant, and the quartermaster. The second-in-command, the senior major, was essentially the commander's understudy. The quartermaster, assisted by the regimental quartermaster-sergeant, was responsible for supply and equipment management, as well as receipt of rations. The adjutant worked with the regimental sergeant-major and orderly room sergeant on all training and administrative matters. The colonel was final arbiter for minor disciplinary matters and reviewed and signed all regimental papers and the many, routine written orders. Company commanders, in the infantry mostly captains and in the cavalry all majors, disciplined the rankers and cursorily inspected their persons and quarters, signed documents prepared by the company sergeant-major and company clerk, and audited

---

[85] Ibid., ¶¶255, 336.

[86] Ibid., ¶346. Exceptions were for chaplains and officers assigned to the permanent staffs of the auxiliary forces. *King's Regulations, 1908*, ¶1050.

[87] Spiers, *The Army and Society*, 22-23.

the company's accounts. Of primary importance to company commanders were financial matters as they were responsible for their units' expenses, down to the penny. Except for the musketry officer, junior officers could accomplish all their routine military tasks in a few hours between breakfast and lunch at the mess.[88] Companies and squadrons of the Edwardian army, like the early Victorian army, were trained and managed by their senior NCOs.

A Royal Irish Fusiliers officer, who was Secretary of the Akers-Douglas Committee, noted in his 1903 book on army life that lieutenants had almost no military duties: After a novice officer passes his post-commissioning examinations "he has a good deal of spare time on his hand" and his few duties pose "no great tax on his intelligence or on his physique."[89] For junior officers, boredom was a major occupational hazard.

Leave and Social Life

Officers were not only subject to military law both on and off duty, their social conduct in the field, in the mess, and away from barracks, was governed by the expectations of the British establishment and the social mores of the mess.

Time Off: Officer Leave

The duration of paid leave for officers was not fixed by regulation. Officers commanding units were free to grant leave to subordinate officers at such times and for such periods as their services could be spared.[90] The only explicit restrictions imposed by regulation were an officer could not be on leave when his company, troop, or battery was in training or on tactical exercise, or for periods that extended past his unit's departure date for overseas deployment.[91] The length of annual leave varied by regiment and was determined by custom and the colonel commanding. Generally, officers received at least two months of annual leave. In guards regiments, four to six months of leave were the norm.[92] As there were two officers per company, cavalry squadron, and artillery battery, a battalion's commanding officer could allow a half-year off to each company officer. Battalion staff officers, such as quartermasters and adjutants, were not easily replaced and as their work was critical to a unit's well-being and efficiency, they received much less annual leave than did company officers.

Officers on leave were expected by colleagues, and civilians of the British establishment, to participate fully in the social activities of their class. There was a popular militarism in

---

[88] Spiers, *The Late Victorian Army*, 107; Bowman and Connelly, *The Edwardian Army*, 39; Parr, *Recollections and Correspondence*, 65-66.

[89] Cairnes, *Social Life in the British Army*, 14, 19.

[90] *Queen's Regulations, 1899*, ¶1913.

[91] Ibid., ¶1914; *King's Regulations, 1908*, ¶1279.

[92] Cairnes, *Social Life in the British Army*, 19-20.

Edwardian and Late Victorian Britain and the civilian elite generally invited naval and military officers to their social events.[93] While away from barracks for extended periods, officers did the rounds of balls, country shooting, hunt meets, and dinners in the town and country houses of the aristocracy and gentry. Some engaged in adventuresome travel which was an acceptable alternative to mingling with society.[94]

The Mess

Officers dined and socialized among themselves at the officers' mess. Each battalion-sized unit had its own mess while garrisons and camps had a mess for departmental and staff officers. For officers of foot guards stationed in London, there was a combined mess known as the Guards Club.[95] By regulation, all officers belonged to a mess and they paid an initiation fee and a monthly subscription. Additionally, the mess billed officers for their drinks from the bar, tended to by a soldier-barman.

The mess was run by a committee of officers and there was a full-time club manager ("mess man"), an NCO receiving extra-duty pay from the mess or a civilian employee.[96] The mess man managed the mess stores, plate and tableware, wine cellar, and a permanent staff that numbered at least seven.[97] The chef was always a civilian and usually French. The barman and three or four waiters were serving soldiers assigned to the mess full-time. There was also a private who polished the silver plate.[98] A mess sergeant supervised the soldier mess attendants and also performed butler duties. At least two soldiers had mess kitchen "fatigue" duties on a rotating basis (cleaning and assisting in the kitchen). For special events where there would be many dinner guests, the regular staff was augmented by the officers' soldier-servants. All mess attendants worked in livery provided by the officers.

Mess rules required officers to conduct themselves as they would in a gentleman's house. Behavior likely to cause dissension, such as gambling at cards and practical jokes, was forbidden. Discussion of politics, named women, and religion, as well as "talking shop," were also prohibited.[99]

The mess was funded by initiation fees, monthly subscriptions, meal and bar charges, and a small War Office allowance. The initiation fee could not exceed thirty-days' pay and the monthly subscription was limited to eight-days' pay. Monthly charges for on-going incidental expenses could not exceed 10s.; special assessments 15s.[100] Meal charges were typically 4s.

---

[93] French, *Military Identities*, 234.

[94] Cairnes, *Social Life in the British Army*, 19-21.

[95] Ibid., 14.

[96] *Queen's Regulations, 1899*, ¶936; *King's Regulations, 1908*, ¶1121.

[97] Cairnes, *Social Life in the British Army*, 47-50.

[98] Ibid., 42.

[99] French, *Military Identities*, 126.

[100] *Queen's Regulations, 1899*, ¶¶948, 955, 965.

Army Life and Retirement: Officers

per day.[101] Accordingly, the maximum, annual basic mess expense for an infantry captain was as follows:

Subscription	£ 55
Incidentals, regular	6
Incidentals, special	9
Meals for 9-months	55
Total Maximum	£125

In 1904, the annual base pay of an infantry captain was £211, so mess expense could be 60% of pay. Messes of corps with low social standing were less extravagant than those of the infantry and accordingly, the cost to its members was much lower than the maximum.

To reduce the financial burden of mess expense for impecunious officers, the War Office provided an "Allowance in Aid of Officers' Mess Expenses."[102] The standard annual allowance was £25 per company or squadron so the mess for an infantry battalion received £200. For the artillery, the allowance was £37.5 per battery, for the engineers £6.25 per member.

Mess facilities were in a free-standing building or the building that housed officers. They consisted, at a minimum, of a dining room, ante room, kitchen, pantry, wine cellar, mess man's quarters, and toilet facilities. Below is the 1913 floor plan of one of two, small, officers' messes at Richmond Barracks in the Kilmainham area of South Dublin.

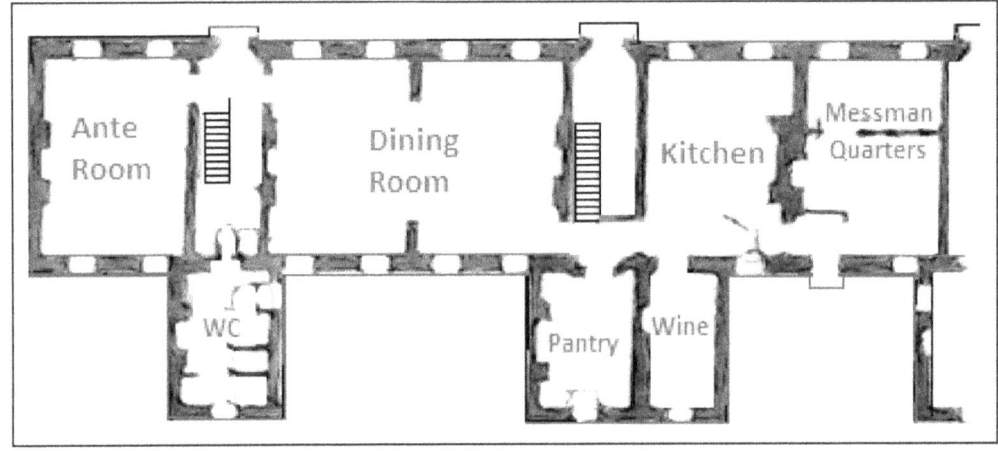

Plan of a Small Officers' Mess
Richmond Barracks, Dublin

UK National Archives, WO 78/3124

---

[101] *Report of the Committee to enquire into the nature of the Expenses Incurred by Officers of the Army*, 1903, [Cd. 1421], at 8.

[102] *Regulations relating to the Issue of Army Allowances, 1884*, ¶¶892-904.

Late Victorian Mess Scenes

Ante Room

Dining Room

R. Caton Woodville from Cairnes, *Social Life in the British Army*.

Social Aspects of Commissioned Life at Home

As noted previously, few officers had full-time military duties and so could engage in gentlemanly pursuits during what for civilians were working hours. The most popular ways in which officers spent their day-time, off-duty hours, were as follows:[103]

> Attending horse race meets and betting on the results.
>
> Gambling at cards.
>
> Horseback riding, fox hunting, and playing polo.
>
> Steeple-chase riding and amateur horse racing.
>
> Playing cricket and tennis.

The extra-military activities in which an officer was expected to participate were determined by a regiment's social status. As officers of the scientific arms, Army Service Corps, and low-status line infantry regiments (such as the Manchester Regiment), rarely had

---

[103] Cairnes, *Social Life in the British Army*, Chapter 2.

large private incomes, those messes did not require participation in the costly, equestrian sports. Note that for the foot guards and all cavalry regiments, custom dictated that each officer maintains two horses.

Evening hours were dominated by mess activities such as celebratory parties, balls, concerts by the regiment's band, and the weekly guest nights. If there was no scheduled mess event, officers remained after dinner to drink, play cards and billiards, and chat. Officers rarely spent evenings at a concert, lecture, the opera, or a theater or dance performance.[104]

Officers of means stationed in London and large cities nearly all belonged to private clubs. Civilian clubs favored by military officers were Bachelor's, Travelers, White's, and the Carlton. Officers, not unexpectedly, also joined the military clubs: Army and Navy, Junior Army and Navy, Naval and Military, United Service, and Junior United Service.[105] For household cavalry officers it was *de rigueur* to join the Cavalry Club. All officers' careers would benefit from United Service membership.[106]

In social situations, an officer had to act in accordance with the norms of gentlemanly behavior and conform to the etiquette, dress, and deportment of persons in "polite" society.[107] To the generals of the nineteenth century army, the social life, customs, and extra-military life-style of the officer corps were considered essential to *esprit de corps*.[108] Accordingly, failure to socialize into the mess brotherhood probably ended prematurely more careers than failure on the lettered, post-commissioning examinations. New officers who fell short of social standards were bullied and harassed by their colleagues into leaving the regiment or quitting the service. Colonels in command usually approved of such conduct as it ridded their regiment of the "wrong sorts" without official action.[109] One general, in his memoirs, characterized misfit officers as "prigs, cads, and bounders" and noted that they quietly disappeared from the service.[110]

Regimental mores addressed all aspects of an officer's personal life, including marriage. Marriage by junior officers was frowned on by all. The occasional married captain was tolerated, but married lieutenants were pressured by their peers to leave the regiment.[111] Also, in that marriage brought with it a financial burden, only the well-off officers could afford to marry while below the rank of major.[112] Should an officer's mess-mates and colonel deem

---

[104] Ibid., 26.

[105] Ibid., 16; Farwell, *For Queen and Country*, 31.

[106] Bellairs, *The Military Guide*, 158.

[107] Spiers, *The Late Victorian Army*, 103.

[108] Spiers, *The Army and Society*, 22.

[109] Ibid., 25.

[110] Paraphrase of Younghusband, *A Soldier's Memories*, 311.

[111] Cairnes, *Social Life in the British Army*, 31.

[112] The minimum annual expense to maintain in suitable style an officer and his family was £600 plus £50 per child. Bellairs, *The Military Guide*, 135-36.

him ready for marriage, he had to take a wife from his social class; a woman who was schooled in etiquette and whose character was unblemished by scandal. Furthermore, the regimental officers had to be assured that should the new wife's husband reach senior rank, she could take on a role that was somewhat between that of a clergyman's wife and lady of the manor. An officer who was adamant about marriage to an "unapproved" woman was expected to leave his regiment.[113]

Retirement and Pensions Before 1871

Prior to 1871, army officers had no statutory pension rights. The British establishment presumed them to be gentlemen of independent means and accordingly, successive governments, always parsimonious with regards to army expenditure, never guaranteed officers a state-paid retirement. The few pensions available were either granted specially by Parliament as a reward for war service (typically to generals) or granted at the discretion of the War Office. The number of discretionary pensions granted was small as they were funded by an extremely limited, annual appropriation.[114] Artillery officers and engineers not receiving pensions could be granted lump-sum, retirement gratuities. Like pensions, they were few in number as gratuities were also paid from a fixed, annual appropriation.

Most retired infantry and cavalry officers received income from annuities purchased with proceeds from the sale of their commissions. In the infantry for example, a 55 year-old lieutenant-colonel who sold his commission for £7,000 could obtain a life annuity of about £600 annually. Most officers would receive much less. A 50 year-old major who sold for £4,700 received about £370 annually and a 45 year-old captain who sold for £2,500 received only £180.[115] Those representative amounts were for officers of line infantry regiments; guards and cavalry officers, especially those of the household cavalry, received much more.

Some infantry and cavalry officers received pensions at the discretion of the War Office, but payment, in total, could not exceed the annual appropriation of £60,000. In 1857, such pensions were paid to 254 retired infantry and cavalry officers and averaged £236 annually.[116] As pensioners died, funds became available for new retirees to be selected by the War Office in rank-descending order.

For artillery officers and engineers, the law provided for some pensions as well as lump sum gratuities. Funding was never enough to provide payment for all officers too old for

---

[113] Venning, *Following the Drum*, 11-12.

[114] Office of the Commander-in-Chief, *Committee on Army Retirement (1867)*.

[115] Average infantry commission prices from the *Report of the Commissioners appointed to Inquire into Over-Regulation Payments on Promotion in the Army*, 1870, [C. 201]. Annual annuity amounts are at 90% of the actuarial table amounts shown in Appendix 9 of the *Report from the Select Committee on Army (System of Retirement)*, 1867, H.C. Accounts & Papers, No. 482, hereafter cited as *Committee Report on Army Retirement (1867)*.

[116] de Fonblanque, *Treatise on the Administration and Organization of the British Army, 303-06*.

active service (over 45 for majors and captains, over 50 or 55 for lieutenant-colonels and colonels). Somewhat under half of artillery and engineering retirees received full-pay pensions. Such pensioners needed 30-years' service and were selected at the discretion of the War Office, in rank-descending order. Accordingly, colonels accounted for most pension recipients. The number of full-pay pensions was limited by the annual appropriation of £32,000 for artillery officers and £16,000 for engineers.

Officers of the scientific arms with fewer than 25-years service were eligible for lump sum gratuities of £4,500 to lieutenant-colonels and £100 for every year of service to captains and majors.[117] Such gratuities were in lieu of a pension and the amounts were much less than that realized from the sale of infantry commissions. Accordingly, annuities purchased with such gratuities provide technical officers with a pension lesser in amount than that received by cavalry and infantry officers who sold their commissions. The number of retirement gratuities was limited by the annual appropriation for the Army Reserve Fund.

All combatant officers with 25-years' service were entitled to go on retired half-pay status. While in such status they performed no military duties, received about half their service pay, and were subject to recall by the War Office. Five years prior to the abolition of commission purchase, about 40% of retired artillery officers and engineers were on the half-pay list. Following is a tabulation of artillery and engineer officers in retirement, as of December 31, 1866, by type of remuneration.[118]

	Pension List			½-Pay List		Gratuities in	Total	Total
	COL	LTC	Other	LTC	Other	Prior 15 Years	Retired	Serving
Artillery	18	36	35	6	92	30, est.	217	835
Engineers	16	10	11	1	12	211	61	393
Totals		126 (45.3%)		111 (39.9%)		241 (14.8%)	278	1,228

Officer Pensions After Abolition of Commission Purchase

As part of the series of Cardwell army reforms begun in 1868, the government gave all officers pension rights in 1871. In general, there were three types of retirement that came with pensions: voluntary, compulsory due to age, and compulsory due to disability. Rank was the principal determinant of pension amount and only lieutenant-colonels and higher received pensions that could provide a middle class standard of living.

Generally, nearly all officers of the combat arms could receive retirement pensions after 15-years' service. Few did so for as shown in Table 23, pensions for majors and captains were insufficient for a gentlemanly life.

---

[117] At the time, the artillery and engineers had no rank denominated "major." The rank "First Captain" was equivalent to the modern era rank of major; the rank of "Second Captain" to captain.

[118] Office of the Commander-in-Chief, *Committee Report on Army Retirement (1867)*, appxs. 1, 2.

Officers of the Combat Arms Holding Combatant Rank

Table 23.
Annual Pensions of Combatant Officers, 1904
Voluntary Retirement

Rank	Requirements	Minimum	Maximum
Colonel	3 years in rank, 52 years of age	£ 450	£ 490
Lt.-Colonel	5 years in rank, 15 years' service	420	450
Lt.-Colonel	3 years in rank, 15 years' service	250	365
Major	3 years in rank, 15 years' service	120	200
Captain	15 years' service	120	120

Source: *Royal Warrant for the Pay, 1906*, Art. 515.

Officers forced out for age received pensions in amounts determined by substantive rank and corps. Though pension amount was based on substantive rank, the minimum retirement age was based on brevet rank. In Table 24, where an age cell contains two ages, the older is for brevet rank one grade higher.

Table 24.
Annual Pensions of Combatant Officers, 1904
Compulsory Retirement, Age

Rank	Age	Corps	Amount
Colonel	57	Eng., Atty., ASC	£ 450
Colonel	57	Cavalry, Infantry	420
Lt.-Colonel	55-57	All	365
Major	48-55	All	300
Captain	45-48	All	200

Source: *Royal Warrant for the Pay, 1906*, Arts. 516, 524.

Disability pensions were classified according to the cause of the infirmity. Combat wounds gave rise to life-time pensions while line-of-duty injuries and diseases a small lump-sum payment or life-time pension. Infirmities that were not service-connected permitted a temporary pension for up to five years.

Officers with non-combat, service-connected disabilities could be given, in lieu of a pension, a lump sum "gratuity" of from three to twelve months pay. Officers below the rank of lieutenant-colonel who were invalided out for other than service-connected reasons, needed twelve-years' service to qualify for a temporary pension. If they had fewer than twelve-years' service, they received nothing. Because of the severely limited disability benefit, it behooved officers to obtain commercial disability insurance that covered causes other than combat wounds.

Army Life and Retirement: Officers

Table 25.
Annual Pensions of Combatant Officers, 1904
Compulsory Retirement, Disability

Rank	Combat Min.	Combat Max.	Line-of-Duty Min.	Line-of-Duty Max	Other Min.	Other Max.
Col., Lt.-Col.	£ 250	£ 365	Lump-Sum	£ 250	£ 250	£ 365
Major	150	200	Lump-Sum	150	173	182
Captain	150	200	Lump-Sum	75	128	152
1st Lieutenant	150	200	Lump-Sum	50	82	85
Term	Lifetime		NA	Lifetime	5 Years Max.	

Source: *Royal Warrant for the Pay, 1899*, Arts. 526, 624, 625.

Departmental Officers

Officers of the support departments received pensions based on their rank and years of service. Table 26 shows the minimum and maximum voluntary pension amounts for officers of each support department.

Table 26.
Annual Pensions of Departmental Officers, 1904
Voluntary Retirement

Department	Minimum	Maximum
Medical	£ 365	£ 639
Veterinary	200	420
Pay	120	420
Ordnance	120	365
Chaplains	228	319
Army Schools	150	200

Source: *Royal Warrant for the Pay, 1899*, Arts. 557, 585, 574, 566, 549, 547.

Quartermasters in all support departments received the same pension as inspectors of army schools (commissioned officers of the school system). Those amounts were the same as received by retired non-combatant rank-holders of the combat corps. Compulsory age and disability retirements came with pensions of reduced amount as set forth in separate schedules.

Chapter Bibliography

Bellairs, William. *The Military Career: A Guide to Young Officers, Army Candidate and Parents.* London: Allen, 1889.

Bowman, Timothy and Mark Connelly. *The Edwardian Army: Recruiting, Training, and Deploying the British Army, 1902-1914.* Oxford: Oxford Univ. Press, 2012.

[Cairnes, William E.] *Social Life in the British Army.* New York: Harper, 1899.

Churchill, A.B.N. "The Army as a Profession," *Journal of the Royal United Service Institution* 54, (January to June, 1910): 166-198.

de Fonblanque, Edward Barrington. *Treatise on the Administration and Organization of the British Army.* London: Longman, 1858.

Duncan, Andrew George. "The Military Education of Junior Officers in the Edwardian Army." PhD Thesis, University of Birmingham, 2016.

Farwell, Byron. *For Queen and Country.* London: Lane, 1981.

French, David. *Military Identities: The Regimental System, the British Army, and the British People c. 1870-2000.* New York: Oxford Univ. Press, 2005.

Goodenough, W.H. and J.C. Dalton. *The Army Book for the British Empire.* London: HMSO, 1893.

Harries-Jenkins, Gwyn. *The Army in Victorian Society.* Toronto: Univ. of Toronto Press, 1977.

Nolan, E.H. *The Illustrated History of the British Empire in India and the East.* Vol 1, London: Virtue, c. 1860.

Otley, C.B., "The Educational Background of British Army Officers," *Sociology* 7, no. 2 (May 1973): 191-209.

Parr, Henry Hallam, *Recollections and Correspondence with a Short Account of His Two Sons.* Edited by Charles Fortescue-Brickdale. London: Unwin, 1917.

Razzell, P.E. "Social Origins of Officers in the Indian and British Home Army: 1758-1962." *The British Journal of Sociology* 14, no. 3 (Sep. 1963): 248-60.

Spiers, Edward M. *The Army and Society 1815-1914.* New York: Longman, 1980.

——— *The Late Victorian Army, 1868-1902.* Manchester: Univ. of Manch. Press, 1992.

Tulloch, Alexander Bruce. *Recollections of Forty Years' Service.* London: Blackwood, 1903.

Venning, Annabel. *Following the Drum.* London: Headline, 2005.

Younghusband, George. *A Soldier's Memories in Peace and War.* London: Jenkins, 1917.

Chapter 7
Army Life and Retirement: Other Ranks

In the early nineteenth century, British Army soldiers led a dreadful life which resembled that of workhouse and prison inmates. By the 1850s, when Brian Tweedy enlisted, the common soldier's conditions of service were much improved, but he remained poorly paid, badly fed, inadequately housed, and in receipt of sub-standard medical care. By Bloomsday, army life had changed greatly, and soldiers no longer lived a penurious and degrading life. The enlisted man's social status; however, had not changed. All classes of society held the army ranker in low regard and accordingly, as in earlier years, the great majority of young men who took the King's shilling were poorly educated, unskilled, unemployed, and desperate.

Enlistment Terms

Generally, prior to 1847 recruits enlisted for life, though they could retire with pensions after twenty-one years' service. In time of war; however, Parliament authorized short-term enlistments for the duration of hostilities. As pensions were meager and employment prospects for unskilled ex-soldiers were poor, most veteran soldiers remained with the army until death or infirmity. Disability pensions were at the War Office's discretion and were offered only to invalided soldiers with at least fourteen-years' service. Disability pension amounts were based on rank and severity of the infirmity and ranged from £12 to £55 per year.[1] As late as the 1890s, one-third of soldiers discharged for medical reasons received no compensation.

With the Army Service Act of 1847, initial enlistment terms were fixed at ten years for the infantry and twelve years for the cavalry and artillery.[2] At expiration of his initial term, with War Office approval, an infantryman could re-engage for eleven years, and a cavalry trooper or artillery gunner for twelve years. Note that infantrymen became eligible for a pension after twenty-one years' service; other soldiers after twenty-four years' service. Like officers, who held commissions in specified regiment, rankers had engagements with either the infantry, cavalry, or artillery and the War Office could not transfer them among corps involuntarily. In 1867, Parliament fixed all initial engagements at twelve-years, subsequent engagements at nine-years, and granted pension rights to all soldiers with twenty-one years' qualifiable service.[3]

---

[1] Skelley, *The Victorian Army at Home*, 205-06.

[2] 18 & 19 Vict., c. 4.

[3] Army Enlistment Act, 1867. 30 & 31 Vict., c.34. Note that service while under eighteen years of age did not reckon towards retirement eligibility.

To both spur enlistment and expand the reserve, Parliament in 1870, greatly revised the terms of army service and imposed a reserve obligation on many new recruits. The new Army Enlistment Act, part of the Cardwell Reforms, authorized short-term enlistments and made assignment to the Army Reserve mandatory for the short-service soldiers.[4] Under the act, the War Office could tailor service terms by corps and could limit the number of both short-service and full-service engagements. The army implemented the act in 1874 and established the following options for initial twelve-year engagements:

> Cavalry, Artillery, and Engineers - eight years with the Regular Army and four years in the Army Reserve;
>
> Infantry and Army Service Corps - six years with the Regular Army and six years in the Army Reserve.

Only the artillery and foot guards could offer an unlimited number of short-service engagements.[5]

On Bloomsday, the minimum terms of regular service were nine years for Household Cavalry, two years for drivers of the Army Service Corps, and three years for all other short-service recruits.[6] In 1904, only 4% of recruits entered the army on a full-service engagement (12 years with the regulars), down from 6% in 1898, the year prior to the Boer War.[7]

In early 1904, the War Office determined that the Army Reserve was manned at an acceptable level but that the Regular Army would soon be critically short of experienced soldiers. Later that year the War Office began to lengthen the full-time service period for new, short-service recruits. In May 1905, infantry and garrison artillery short-service required nine years in the Regular Army; cavalry short-service eight years.[8]

Training

Initial Training

Infantry and Artillery Recruits

Infantry recruits received three months of basic training at their regiment's depot; artillery recruits two months at one of fifteen artillery depots (six garrison artillery, seven field artillery, two horse artillery).[9] At the depot, recruits received training for four hours each morning, Monday through

---

[4] 33 & 34 Vict., c. 67.

[5] Army Orders, 1874, Nos. 34, 51, 53; Army Circulars 1874, No. 67.

[6] General Annual Report on the British Army for the Year Ending 30th September, 1906, 1907, [Cd. 3365]. Hereafter cited as *Annual Army Report*.

[7] *Annual Army Report, 1904*, 1905, [Cd. 2268].

[8] *Annual Army Report, 1906*, 1907, [Cd. 3365].

[9] *Monthly Army List, July 1904*.

Friday. In the nineteenth century British Army, barrack cleaning, unskilled food service work, grass cutting, *etc.* were performed by soldiers and such work was termed "fatigue" duty. Recruits spent their afternoons on fatigues or in cleaning their uniforms and polishing equipment and accoutrements. On Saturday mornings, the recruits cleaned the barracks which were then inspected by an officer. Afterwards, they had the afternoon free to themselves. Sunday was a day off except for mandatory church attendance.

Basic training consisted of physical conditioning, drill, basic musketry (maintenance and used of a rifle) and classroom instruction (army organization, regulations and discipline, customs and courtesies, *etc.*). Physical conditioning accounted for one-third of the training hours. Strength and endurance standards were not very high as recruits upon passing out were expected only to do fourteen chin-ups on a horizontal bar, jump over a three-foot obstacle, and run a mile in nine minutes.[10]

After basic training, new infantry and artillery soldiers went in batches to their assigned units for further training. Infantry battalions put their novice soldiers through three months of advanced training which consisted of drill, physical conditioning, marching, musketry, and bayonet exercises. Such training occupied four to five hours prior to the 1:00 pm meal. Artillery brigades also had training programs for soldiers newly arrived from depot. Gunners' training within their brigade lasted seven or eight months.[11]

Cavalry Recruits

Cavalry recruits were trained by their regiment's reserve squadron. Cavalry regiments had no fixed depots and when posted abroad, their reserve squadrons went to the cavalry facility at Canterbury, Kent where they served as the regimental depots. Cavalry initial training was for eight months of which two months were devoted to mounted training and included ten to thirteen hours riding instruction weekly. Cavalry troopers received training in the care of horses plus all training received by new infantry soldiers.[12]

Engineer, Service Corps, and Departmental Recruits

These recruits were sent to their corps' specialized training establishment. Sapper (engineer) aspirants received initial training at the School of Military Engineering, Chatham, Kent. Their training lasted from four and one-half to eleven months, depending on specialty.[13] Note that the engineers accepted

---

[10] War Office, *Infantry Training Manual* (London: HMSO 1900, §177-78.

[11] Spiers, *The Late Victorian Army*, 260-61; Cairnes, *The Army from Within*, 64-70.

[12] Ibid.

[13] Ward, *The School of Military Engineering*, 44-49.

only recruits who were already proficient in a trade.[14] Recruits for the Army Service Corps, the transport and supply organization, trained for three months at its School of Instruction, Aldershot, Hampshire.[15] Recruits for the Ordnance Department, Royal Army Medical Corps, Veterinary Corps, and Pay Department were trained at departmental establishments.

After recruits completed initial training, the War Office considered them qualified soldiers, though in reality, they were simply apprentices. To reach the journeyman level of expertise, infantrymen usually needed three-years' service, cavalry troopers five-years, and artillery gunners at least five-years.[16] Table 27 shows the training periods and locations for recruits of the five combatant arms: Infantry, Cavalry, Royal Artillery, Royal Engineers, and the Army Service Corps.

Table 27.
Training of Recruits, Combatant Corps, 1904
Months and Location

	Basic Months	Location	Advanced Months	Location	Total Months
Infantry	3	Depot	3	Battalion	6
Cavalry	3	Regiment	5	Regiment	8
Artillery, Garrison	2	Depot	7	Company	9
Artillery, Other	2	Depot	8	Brigade	10
Engineers, Fortress	2	S.M.E.	2.5	S.M.E.	4.5
Engineers, Field	2	S.M.E.	3	S.M.E.	5
Engineers, Electrical	2	S.M.E.	6	S.M.E.	8
Engineers, Mining	2	S.M.E.	9	S.M.E.	11
Army Service Corps	1	ASC	2	ASC	3

Sources: Spiers, *The Late Victorian Army*, 260-61; Cairnes, *The Army from Within*, 64-70; Ward, *The School of Military Engineering*, 44-49; Grierson, *The British Army*, 186.

Annual Training

For soldiers who had completed their initial training (basic and advanced), there was an annual training routine whose nature depended on the amount of open ground at barracks. Routine training for soldiers quartered in cities was severely limited and of little military value, as the only open space readily available was the parade ground. The War Office divided the

---

[14] New recruits were tested for trades proficiency in the S.M.E.'s workshops. Ibid.

[15] Grierson, *The British Army*, 186.

[16] Vivian, *The Army from Within*, 11.

year for soldiers in garrison into two seasons; the furlough season of four cold weather months and the training season of the remaining eight months.[17]

During furlough season, soldiers trained for two hours daily. Such training consisted of classroom instruction, company and smaller formation drill, formation marching, and practice of weapon care and equipment use, such as setting up then breaking down multi-man tents. During the furlough season soldiers received about 170 hours of military training (10 hours per week), excluding physical conditioning. In the training season infantry soldiers, for seven weeks, took part in weekly route marches in company and battalion formation, underwent a twelve-day gymnasium course, and spent ten days in musketry training where they fired 200 rounds at targets. There was an eighteen to twenty-four day field exercise which included skirmishing and defensive practice over broken ground and rifle firing of 400 rounds.[18] During those exercises, soldiers spent a minimum of four nights "under canvas" (*i.e.*, in tents). In training season, infantrymen received a minimum of 270 hours of specialized military training. On days when annual training was not scheduled, soldiers underwent the ordinary two hours of military training.

Cavalry troopers, in addition to the training received by infantrymen, took a three-week course twice each year at the Cavalry School, Canterbury (Kent).[19] Artillery gunners underwent a prescribed annual course within their batteries, and in addition to the field exercise, had an annual fourteen-day firing practice during which they fired from 35 to 65 rounds.[20] Engineers underwent an annual field works course.[21]

For soldiers stationed at the army camps of Aldershot (Hampshire), the Curragh (Co. Kildare, Ireland), Colchester (Essex), and Shorncliffe (Kent), there was an annual brigade-formation drill.[22] Such drill was limited to those locations as only they had sufficient room in which a brigade could deploy in both column-march and line-engagement formation. Of units stationed at home in 1904, those four large camps housed 38% of the infantry battalions, 50% of the cavalry regiments, and 26% of the field and horse artillery batteries.[23]

Maneuvers

Having accepted the necessity of large-formation maneuvers, the War Office in 1897, paid £396,576 to purchase 32,055 acres of flat land of the Salisbury Plain in Dorset and

---

[17] War Office, *Infantry Training Manual* (London: HMSO 1900), §177.

[18] Spiers, *The Late Victorian Army*, 142, 260-61; Cairnes, *The Army from Within*, 82; War Office, *Infantry Training Manual* (London: HMSO, 1905); *Queen's Regulations, 1899*, ¶1064-65.

[19] Spiers, *The Late Victorian Army*, 260-61.

[20] Five weeks for garrison artillery, twelve weeks for horse and field. *Queen's Regulations, 1899*, ¶1066; Cairnes, *The Army from Within*, 108.

[21] *Queen's Regulations, 1899*, ¶1067.

[22] Grierson, *The British Army*, 239; Spiers, *The Late Victorian Army*, 262.

[23] *Hart's Annual Army List, 1904*; *Annual Army Report, 1904*, 1905, [Cd. 2268].

Army Life and Retirement: Other Ranks

Wiltshire, for such purpose.[24] The first large exercise on Salisbury Plain was the "Military Manoeuvres of 1898."[25] The maneuvers employed 55,000 soldiers, including militiamen, divided between an attacking "Blue" force and a defending "Red" force. The exercise took place during the first six days of September after several weeks of preparatory drill.[26]

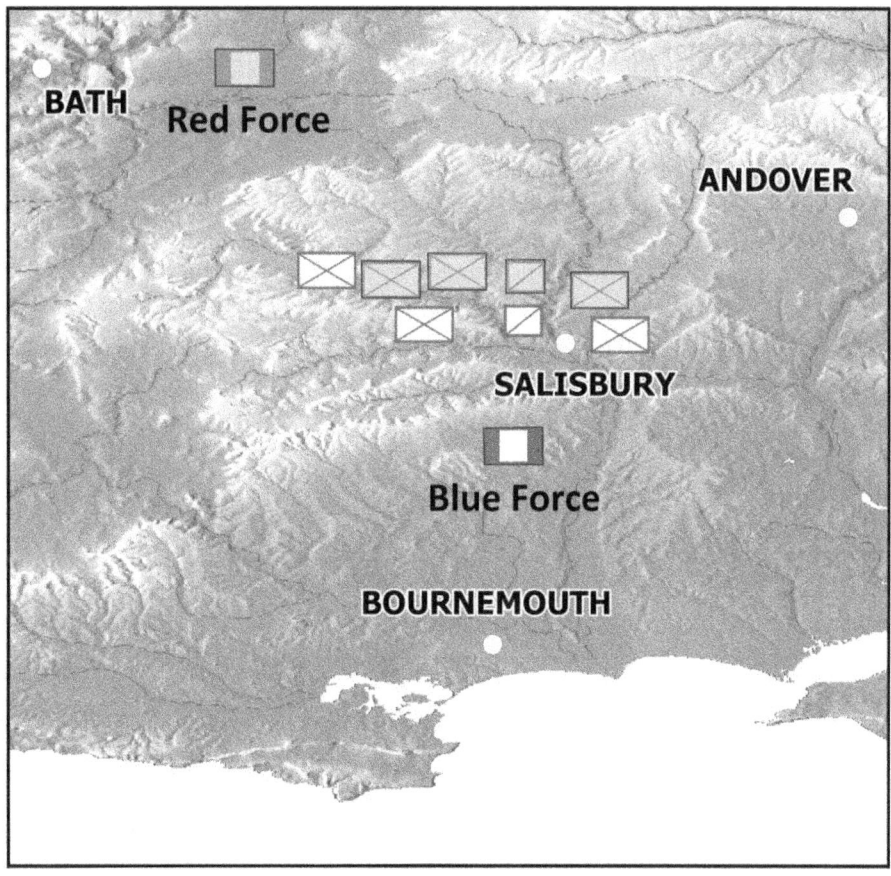

1898 Maneuvers, Positions on September 5th

War Office Sources: *Report on the Manoeuvres Held in the Neighbourhood of Salisbury in August and September, 1898,* 1899, H.C. Accounts & Papers, No. 551; *Maps of 1898 Manoeuvres,* U.K. National Archives, WO 279/4.

---

[24] War Office, *Schedule of Purchases of Property at Salisbury Plain*, 1898, [C.9032].

[25] Leeson, "Playing at War."

[26] *Army and Navy Gazette*, July 16, August 13, 20, and September 17, 1898.

Army Life and Retirement: Other Ranks

Each force was organized as a corps of three under-strength infantry divisions and a small cavalry brigade. These "armies" maneuvered against each other as well as against "invisible soldiers" of hypothetical formations. Engagements were evaluated by 54 umpires who determined the results, *i.e.* who "won" the battle. The "general idea" of the exercise was that "an invading force has landed between Weymouth and St. Alban's Head, to co-operate in an advance on London with another force (imaginary), which is disembarking on the southern shore of the Bristol Channel."[27] The maneuvers were covered intensively by the press which overall viewed the exercise as a farce of little military value and of unnecessary expense. One writer estimated the cost at £180,000, about the same as to maintain five infantry battalions for a year.[28]

The demands of the Boer War precluded large-formation maneuvers for the years 1899 through 1902. Though public opinion judged the 1898 exercise a failure as it seemed not to have prepared the army for the war in South Africa, the War Office staged another large maneuver in September 1903. It involved 45,000 troops organized in opposing army corps each of two infantry divisions and a cavalry brigade.[29] In 1904, there were brigade formation maneuvers in Ireland and on Salisbury Plain, and a corps/division maneuver in Essex. These exercises involved about 45,000 to 50,000 troops (one-third of the army at home).[30]

Army Pay

Soldiers received base pay plus additional pay from a bewildering array of schedules published in the 300 to 400 page *Royal Warrant for the Pay of the Army*. Nevertheless, a soldier's total army pay, plus the value of free lodging, low-cost medical care, and subsidized meals, remained below the remuneration of his employed, civilian counterparts other than Irish agricultural laborers.

Base Pay

Base pay was determined by rank, corps, and occupational classification. The technical corps had a multitude of ranks that reflected a soldier's trade and bore titles such as "Armament Sergeant" and "Lithographer." Even in the combatant corps, there were trades-related ranks such as "Saddler," "Farrier," "Smith," and "Bugler." For rank equivalency and pension purposes, the War Office grouped all ranks into classes (Warrant Officer, Class I, II, II, IV, V). Following is an abbreviated base pay schedule for the combatant corps.

---

[27] Leeson, "Playing at War."

[28] *Army and Navy Gazette*, August 20, 1898. Five battalions represented 3% of total infantry strength. War Office, *Army Estimates for the Year 1904-05*, 1904, H.C. Accounts & Papers, No. 73.

[29] Spiers, *The Army and Society*, 249; *Irish Times*, September 10, 1903; *Army and Navy Gazette*, September 19, 1903.

[30] *Army and Navy Gazette*, August 6, 20, 27, September 24, 1904; *Annual Army Report, 1904*, 1905, [Cd. 2268].

Table 28.
Army Base Pay, 1904
(Day s./d., Annual Pounds)

Class	Infantry Rank	Infantry		Cavalry		ASC		Artillery		Engineers	
		Day	Yr.	Day	Yr.	Day	Yr.	Day	Yr.	Day	Yr.
WO	Sergeant-Major	5/0	92	5/4	97	5/6	100	5/10	106	6/0	109
I	Quartermaster-Sgt.	4/0	73	4/4	79	4/3	78	4/2	76	4/6	82
II	Colour-Sergeant	3/6	64	4/4	79	3/9	68	4/0	73	3/9	68
III	Sergeant	2/4	43	2/8	49	2/7	47	3/2	58	3/3	59
IV	Corporal	1/8	30	2/0	36	2/0	36	2/6	46	2/6	46
V	Lance Corporal	1/3	23	1/6	27	1/5	26	1/7	29	2/2	39
V	Private	1/0	18	1/2	21	1/2	21	1/2½	22	1/1½	21

Source: *Royal Warrant for the Pay, 1906.*

Corps and Engineer Pay

Soldiers of certain corps received in addition to base pay what was termed "corps pay" or in the Royal Engineers, "engineer pay." Amounts were based on a soldier's rank, proficiency in his trade, and years of service.[31] Following are descriptions of corps pay which was available only to soldiers of two combatant corps and two support corps.

Royal Engineers

All soldiers in Pay Classes I - V received 4d. to 2s. daily. Upon successful completion of their year-long, initial training at Chatham, sappers were granted the minimum amount. As soldiers gained experience their engineer pay increased to 1s. or 1s. 4d. for qualified journeymen, and then 1s. 8d. or 2s. for master artisans. Daily engineer pay greater than 1s. 4d. could be granted only by a soldier's general-officer-commanding and only under special circumstances. For journeyman sappers, the maximum annual engineer pay was £19 4s; master artisans £28 16s. Accordingly, skilled corporals received annual pay of £75, about the same as that of infantry quartermaster-sergeants.

Army Service Corps

Soldiers with rank below colour-sergeant received daily an additional 3d. to 1s. 8d. based on proficiency. Maximum annual corps pay was £25 (six days per week for 48 weeks). Accordingly, highly skilled ASC corporals received annual pay of £61, about the same as that of infantry colour-sergeants.

---

[31] *Royal Warrant for the Pay, 1906*, Arts. 915-18 (engineers), 788 (service corps), 799 (medical), 809 (ordnance).

Royal Army Medical Corps

Soldiers of all ranks were eligible for daily corps pay in amounts based on rank and proficiency. Daily base corps pay was 4d. to 6d. for privates, 8d. for corporals, and 10d. to 1s. for sergeants. Maximum annual corps pay was £14 16s. (six days per week for 48 weeks).

Ordnance Corps

Soldiers with rank below colour-sergeant were eligible for daily corps pay in amounts based on proficiency. Daily corps pay was from 3d. to 1s. 2d. Maximum annual corps pay was £16 16s. (six days per week for 48 weeks).

Soldiers received corps and engineer pay only for days on which they performed their regular work. Corps pay was not received during furloughs, while in hospital, and on regular non-duty days such as Sunday. For days where soldiers received extra-duty or working pay they did not receive corps pay.[32]

Service Pay

All warrant officers and senior sergeants received an additional 6d. for each day they performed their duties. After five years in the higher ranks, they received 7d. daily. Allowing for a one-month furlough, their service pay amounted annually to £8 7s. or £9 15s.[33]

Soldiers below the rank of colour-sergeant on long-service enlistments, and not suffering from the effects of venereal disease or alcoholism, received an additional 4d. to 5d. daily once they met one of the following conditions: Two years of satisfactory service if their enlistment term was for at least eight years as a regular; or six-months of satisfactory service if their enlistment term was for at least nine years as a regular and they were age twenty or older. Soldiers in receipt of basic service pay could receive a further 1d. or 2d. daily if their commanding officers evaluated them as highly proficient. Accordingly, for lower ranking soldiers, service pay amounted annually from £5 7s. to £9 15s.

Good Conduct Pay [34]

Privates and lance-corporals who had served before April 1, 1902 were eligible for good conduct pay. Such pay was based on the number of good conduct badges they held. Badges were received for specified intervals where the soldier had no entries in the Regimental Defaulters' Book, the logbook for infractions punished by fines, confinement to barracks for more than four days, and detention.[35] A truncated good conduct pay schedule follows:

---

[32] *Royal Warrant for the Pay, 1906*, Arts. 817-19.

[33] Ibid., Arts. 1084, 1084A-L.

[34] Ibid., Arts. 1085-1108B.

[35] Cairnes, *The Army From Within*, 112.

Years Service	No. of Badges	Award Interval	Annual Good Conduct Pay
2	1	2 Years	£1 10s. 5d.
6	2	4 Years	£3 0s. 10d.
12	3	6 Years	£4 11s. 3d.
16-18	4	4-6 Years	£6 1s. 8d.

Soldiers could receive a maximum of eight good conduct badges which brought annually an additional £12 3s. 4d. It was rare however for soldiers to have more than five such badges as the fifth badge required 21 years of service, the eighth 33 years. Receipt of good conduct pay ended upon promotion to corporal, but the new NCOs retained their badges. Displayed on the left sleeve of a soldier's tunic was an upward-pointing chevron for each badge received.

Soldiers could lose their good conduct badges for serious disciplinary infractions or conviction by court-martial. Good conduct pay ceased once a soldier was granted service pay. Pay for new badges ended April 1, 1902 when service pay took effect. Even when newly awarded good conduct badges brought no additional pay, they still had value. For example, certain retirements came with pensions based in part on the retiree's good conduct as indicated by badges.[36] Possession of two badges was a requirement for permission to marry. Also, most regiments granted badge holders "standing passes" to leave the barracks when off-duty, a privilege normally restricted to senior sergeants.[37]

Privates Compton and Carr of *Ulysses* were most likely infantry soldiers and as such were at the bottom of both the army's social and pay hierarchies.[38] Of the approximately 4,600 Regular Army soldiers stationed in the city during Joyce's last year in Ireland, about 70% were infantrymen. If Compton and Carr received either service or good conduct pay, their total annual remuneration would have been from £20 to £28. As the army provided soldiers with food for 3d. per day, free shelter, and low-cost medical care, compared to many Dubliners, Compton and Carr were financially well-off. Accordingly, the city's prostitutes would have sought their trade, as is portrayed in the "Circe" episode of *Ulysses*.

Extra-Duties Pay

In the Edwardian British Army, there were many military duties that were not performed by officially-sanctioned, specialized soldiers. For example, there was no military occupational specialty such as clerk or cook. Also, regulations allowed rank and file soldiers to earn fees for personal services provided to both other soldiers and officers. Men who performed these extra-establishment duties worked full-time as tradesmen, domestic servants, clerks, and grooms. For most such soldiers, their only military training was the annual weapon firing and

---

[36] Pensions for 2nd Class reservists.

[37] *Queen's Regulations, 1899*, ¶725; Cairnes, *The Army from Within*, 114.

[38] If Joyce had placed them in the artillery or engineers their rank title would have been "gunner" or "sapper."

field exercise. Cavalry troopers with extra-military duties also took the semi-annual riding course.[39] Soldiers who worked outside of their remit as set forth in the tables of organization, received what the War Office termed "extra-duties pay." Such pay was in amounts prescribed by both regulation and custom. Examples of some duties and extra-establishment positions that brought extra pay follow:

Extra Military Duties - Regulation Payments [40]

NCOs
  Cookhouse Supervision: 6d.-9d. per day.
  Financial Account-Keeping: 6d. per day.
  Recruiting: 2s. per day

NCOs and Privates
  Specified Duties of Medical Corps Personnel: 4d.-1s.6d. per day.
  Clerical Work: 3d. to 1s. per day.
  Regimental Policing: 6d.-1s.2d. per day.

Privates
  Horse-shoeing, Saddle Work, Wheel Repair, *etc.*: 4d.-1s. per day.
  Savings Bank Record-Keeping: 2s.6d.-10s. per month.
  Telephone and Telegraph Operation: 3d.-9d. per day.
  Library Management: 3d.-4d. per day.
  Assist at Regimental Schools: 9d.-1s. per day.

As shown above, privates could receive from £3 12s. to £21 12s annually for extra military duties; NCOs £7 2s. to £28 16s. For infantry privates with an annual base pay of about £18, extra duties could significantly improve the quality of their lives.

Domestic Services - Regulation Payments [41]

In Service to a Commissioned Officer: 1s.6d.-2s.6d. per week.
Horse Care for Warrant Officers and NCOs: 1s.6d. per week.

Officers usually paid their servants more than the regulation amount, especially in the cavalry and guards regiments. In practice, personal servants received an extra £6 to £7 annually plus gifts.[42]

---

[39] Cairnes, *The Army from Within*, 86-88.

[40] *Royal Warrant for the Pay, 1906*, Arts. 835-68.

[41] *King's Regulations, 1906*, ¶1343.

[42] *Report of the Committee to enquire into the nature of the Expenses Incurred by Officers of the Army*, 1903, [Cd. 1421], at 8-9.

Domestic, Food Service, and Trades Positions
Customary Payments [43]

Personal Servant of a Sergeant: 1d.-2d. per day.
Canteen Steward: £40 per year.
Canteen Assistant: 1s. per day plus beer ration.
Mess Attendant: 6d. to 1s. per day.

Hairdresser: £10 - £15 per year.
Shoemaker: £15 - £30 per year.
Tailor: £20 - £40 per year.

Note that soldiers were compelled by regulation to use soldier-tradesmen to mend and alter their clothing and boots. As such tradesmen had a monopoly, the appropriate commanding officer (station or regiment), set prices. Payments for haircuts, shoe repair, and tailoring were deducted from soldiers' pay as "stoppages" and remitted to tradesmen by the regimental paymaster-sergeant.

Working Pay

Soldiers could be assigned work of a civil nature not customarily performed as part of their military duties. For such work they received additional pay known as "working pay." The work typically was on military infrastructure projects or involved the movement of heavy items, such as multi-ton, fixed position, artillery guns.

Work that gave rise to working pay was usually performed by civilian employees of firms with War Office contracts. Accordingly, working pay opportunities were rare, especially for soldiers at home. For the twelve months ending March 31, 1905, the War Office requested a Parliamentary appropriation of £20,266 for working pay with 55% allocated to work abroad.[44] This equates to 2s. 3d. per soldier for the year; somewhat more for those overseas, somewhat less for those at home. Following is the working pay schedule:[45]

Type of Work	Hourly Rate
Skills Required of an Artificer	2.00d.
Skilled or Physically Demanding	1.50d.
Semi-Skilled	1.00d.
General	0.75d.
Unskilled and Light	0.50d.

---

[43] Sergeants engaged privates to clean personal equipment, polish boots, and perform other domestic services. Skelley, *The Victorian Army at Home*, 197. Earnings of hairdressers, shoemakers, and tailors derived from War Office, *Return on the Total Amount of Stoppages from the Pay of Privates*, 1890-91, H.C. Accounts & Papers, No. 209. Mess attendant earnings from Cairnes, *The Army from Within*, 112. Canteen and mess employment: Wyndham, *The Queen's Service*, 98-102, 203-04.

[44] War Office, *Army Estimates for the Year 1904-05*, 1904, H.C. Accounts & Papers, No. 73, appx. 3.

[45] *Royal Warrant for the Pay, 1906*, Arts. 871-79.

NCOs who supervised soldiers earning working pay received supervisory pay of from 8d. to 1s. 4d. per day. The amount was based on the nature of the work and the number of men supervised.

Soldiers probably volunteered for paid work assignments not so much for the extra money but to escape the drudgery of garrison routine.

Staff Pay

On Bloomsday, about 1 in 200 enlisted men (1 in 1000 privates) held a staff position, either temporary or permanent. There were about 1,150 staff positions available to rankers. Positions open to both NCOs and privates were 750 "staff clerkships;" open to NCOs only were 400 miscellaneous "staff appointments" (training schools, garrisons, command districts, *etc.*).[46] Staff appointments were permanent while staff clerkships were temporary. Following is the enlisted staff pay schedule (annual) for NCOs in staff positions.

Engineers		Others		All Corps
Sergeant	£ 85	Quartermaster-Sgt.	£ 82	£91 after 3-years on staff.
Corporal	£ 73	Colour-Sergeant	£ 73	No longevity increase.
2nd Corporal	£ 66	Sergeant	£ 64	£73 after 3-years on staff.

Corporals (other than engineers) and privates were eligible only for temporary staff clerkships. While in such positions they received their regular pay plus staff pay of £18 annually.[47] If a clerk was a qualified stenographer, he received an additional 6d. for each day working as such; about £7 per year.[48]

Stoppages

Contrary to public belief, the state did not provide soldiers an expense-free life. Though housing and pension plan came without charge, with two exceptions, soldiers bore the cost of all other living necessities. Food was subsidized in that the army provided a free ration of meat and bread, and by 1898, provided nearly all soldiers an allowance for all other food items, termed "groceries." Generally, soldiers paid 3d. daily for groceries, but those age nineteen and over received an offsetting 3d. messing allowance.[49] Army meals; however, were of insufficient portions and nearly all privates and corporals purchased additional food

---

[46] War Office Pamphlet, *The Advantages of the Army (1896)*, 1898, H.C. Accounts & Papers, No. 81; *Royal Warrant for the Pay, 1906*, Arts. 742-53.

[47] *Royal Warrant for the Pay, 1906*, Art. 851.

[48] Ibid., Art. 865.

[49] Cairnes, *The Army from Within*, 26; Grierson, *The British Army*, 235. Soldiers at least 19 years of age were eligible for service abroad other than in India (minimum age of 20).

(sergeants, like officers, ate at their own mess). Medical care was also subsidized as soldiers were charged only 7d. to 9d. daily for hospital and infirmary stays.[50]

Soldiers paid their non-discretionary living expenses through pay deductions termed "stoppages." To some extent, the War Office utilized stoppages to minimize army expenditure. For example, uniforms were state property and soldiers were charged for excessive wear. The Ordnance Stores Department set useful lives for uniform items artificially long, and soldiers paid part, or all, the cost of items replaced "prematurely."[51] Examples of the disparity between actual and official lives of clothing follow.

	Regulation Life	Actual Life
Fatigue Shirt	12 months	3 months
Fatigue Trousers	12 months	3 months
Parade Trousers	2 years	1 year
Great Coat	5 years	3 years

Other stoppages that soldiers resented were those for hospitalization and damage to barracks and other army property. Until October 1917, soldiers in hospital paid a stoppage of 7d. daily and forfeited their 3d. daily messing allowance.[52] Accordingly, most hospitalized privates were effectively on half-pay. The hospital stoppage was to discourage malingering, though War Office officials claimed it was to offset the additional cost of special medical diets.[53] For soldiers injured in the line-of-duty, the brigade commander, after a Court of Inquiry, could waive the 7d. hospital stoppage but could not restore the 3d. messing allowance.[54] As for property damage, quartermasters rarely attributed defects in furnishings, fixtures, and equipment to ordinary wear and tear, so there were frequent, small stoppages that generally were not itemized.[55]

In 1890, the War Office surveyed stoppages for infantry privates and found they averaged annually £7 3s. 3d., including £4 3s 9d. for groceries.[56] Following is a summary of the 1890 findings on stoppages for infantry privates.

---

[50] Cairnes, *The Army from Within*, 114-15;

[51] Wyndham, *The Queen's Service*, 25-27; Skelley, *The Victorian Army at Home*, 184; Spiers, *The Late Victorian Army*, 138; Cairnes, *The Army From Within*, 15-16.

[52] *Regulations relating to the Issue of Army Allowances, 1884*, Sec. I, ¶¶29, 69.

[53] Cairnes, *The Army from Within*, 115.

[54] *King's Regulations, 1908*, ¶674.

[55] Skelley, *The Victorian Army at Home*, 183; Wyndham, *The Queen's Service*, 258.

[56] War Office, *Return of the average Number, during 1890, of Private Soldiers in each of the No. 1 Companies of Line Battalions now at Aldershot, and details of Stoppages from Pay*, 1890-91, H.C. Accounts & Papers, No. 209.

	Annual Average	
Excessive Wear of Clothing	2s.	8d.
Repairs to Arms and Accoutrements		2d.
Barrack Damages	4s.	6d.
Tailor's and Shoemaker's Bills	9s	11d.
Regimental Necessaries	14s.	
Laundry	11s.	11d.
Haircutting		11d.
Library	1s.	8d.
Fines		5d.
Other	13s.	9d.
Total, Excluding Messing	£3	4d.

At the time, infantry privates received £27 5s. annually including 6d. daily in lieu of rations while on furlough. Accordingly, stoppages accounted for 26.3% of base pay. Regulations limited the weekly charge for messing and laundry to 3s. 2½d.[57] Note that because of stoppages for damage to barracks, uniforms, and equipment, plus disciplinary fines (primarily for drunkenness), soldiers could become indebted to the War Office as by law they had to receive in cash at least 7d. weekly.[58]

Promotion Opportunities

Promotion was constrained by Parliamentary appropriations which fixed the number of men at each rank by corps and department. Service corps and medical corps soldiers had the best chance for advancement; infantry soldiers the worst. Table 29 shows the distribution of ranks in 1904 for the five combatant corps, the medical corps, and the entire army.

Table 29.
Percent Distribution of Ranks for Enlisted Men, 1904

	Infantry	Cavalry	Artillery	Engineers	ASC	RAMC	Army
Quartermasters*	0.2	0.2	0.2	0.6	1.1	1.3	0.4
Warrant Officers	0.2	0.3	0.4	1.6	3.9	2.1	0.8
Sergeants	7.2	8.3	8.1	14.2	14.9	16.1	9.9
Corporals	4.2	3.5	5.5	6.2	8.6	4.4	5.9
Privates	88.2	87.7	85.8	77.4	71.5	76.1	83.0

* Plus other commissioned ranks to which only enlisted men were appointed.

War Office Sources: *Annual Army Report, 1904*, 1905, [Cd. 2268]; *Army Estimates for the Year 1904-05*, 1904, H.C. Accounts & Papers, No. 73; *Monthly Army List, September 1904*.

---

[57] *Royal Warrant for the Pay, 1899*, Art. 957.

[58] *Royal Warrant for the Pay, 1884*, Arts. 606, 755; Army Act, 1881, 44 & 45 Vict., c. 58.

Included in Table 29 are the permanent staff members of the auxiliary forces (all regulars); excluded are bandmasters and schoolmasters. Also included in the table are commissioned officer ranks to which only enlisted men were appointed (quartermaster and equivalent).[59] All army commissions were at the discretion of the War Office's Department of the Military Secretary.[60] Following were the requirements to be eligible for those special commissions:[61]

1. Have a 1st Class Army Certificate of Education (met the educational standard for twelve-year olds with about six to seven years of primary schooling).

2. Be under 40 years of age; 45 for certain enlisted men to be commissioned as quartermasters.

3. Have an exceptional conduct record.

4. Be nominated by his commanding officer.

5. Be approved by a brigade commander after an intensive personal interview.

Though the table of organization indicates otherwise, it was relatively easy for a somewhat educated infantry or cavalry private stationed at home to be promoted to sergeant after only five years of total service.[62] To be eligible for such promotion a soldier had to possess a 2nd Class Army Certificate of Education, which indicated he met the educational standard for 10-year olds. In 1904, such certificate and the 1st Class Certificate, were held by only 21% of infantry privates and corporals, 24% of those rank-holders in the cavalry.[63] Furthermore, many soldiers, especially those with short-service engagements, did not seek promotion as they didn't want the responsibilities that came with NCO rank.[64]

An infantry private with a 3rd Class Certificate of Education (met the education standard for an eight- or nine-year old), of good character, and who sought promotion could expect corporal's stripes after two years of service.[65] As prior to 1905 over 90% of soldiers entered the army on short-service engagements and would transfer to the reserve after three years of full-time service, there were few corporals competing for promotion to sergeant. Accordingly, an infantry corporal of good character, after serving three years in rank, and attainment of a 2nd Class Army Certificate of Education, would be promoted into one of his

---

[59] Equivalents were Ridingmaster, District Officer of Artillery, Officer of the Coast Battalion of Engineers, Indian Staff Corps Commissary.

[60] *King's Regulations, 1908*, ¶215; Wheeler, *The War Office Past and Present*, 296.

[61] *King's Regulations, 1908*, ¶213; *Royal Warrant for the Pay, 1906*, Arts. 6, 8.

[62] Wyndham, *The Queen's Service*, 263, 296.

[63] *Annual Army Report, 1904*, 1905, [Cd. 2268].

[64] Skelley, *The Victorian Army at Home*, 196.

[65] War Office Pamphlet, *The Advantages of the Army (1896)*, 1898, H.C. Accounts & Papers, No. 81.

regiment's 100 sergeant positions.⁶⁶ Note that under the territorial system each regiment consisted of two regular battalions (27 sergeants each), a depot (2 sergeants), one to four militia battalions (on average 9 sergeants each), and usually three volunteer force battalions (about 8 sergeants each). As there were no Regular Army corporals assigned to the auxiliary force's infantry battalions, there was one regular army sergeant for every corporal (40 corporals in each regular battalion, 17 in the depot).⁶⁷ Promotion to the higher NCO ranks was difficult as nearly all sergeants were career soldiers who would remain in the army for twenty-one years. For the next rank up, titled colour-sergeant in the infantry, there was about one position for every three sergeants.⁶⁸

Living Conditions and the Garrison Workday

Housing

Unmarried soldiers were housed in spartan rooms in foreboding looking barrack buildings of two or three storeys. Standard accommodation for privates and corporals was a large barrack-room shared by 12 to 24 soldiers with 50 to 60 square feet per man. Sergeants had private rooms of 130 to 150 square feet.⁶⁹ Severe overcrowding was no longer the major problem it had been in the 1850s when only 2.6% of the rank-and-file at home had the recommended 600 cubic feet of space for each man; 45.4% had fewer than 400.⁷⁰ The 1901 census shows the five largest barracks in Dublin occupied as follows:⁷¹

	Standard	Actual	Utilization
Royal	950	1,089	115%
Wellington	510	550	108%
Marlborough	640	567	89%
Richmond	1,108	617	56%
Portobello	1,551	613	40%

Note that Richmond and Portobello Barracks were each designed for two infantry battalions but on Census Day 1901, each housed only one battalion.

---

⁶⁶ Cairnes, *The Army from Within*, 119.

⁶⁷ War Office, *Army Estimates for the Year 1904-05*, 1904, H.C. Accounts & Papers, No. 73, Vote 15.

⁶⁸ Ibid.

⁶⁹ War Office, *Record Plans, West Block, Wellington Barracks*, 1902.

⁷⁰ *General Report of the Commission appointed for Improving the Sanitary Condition of Barracks and Hospitals*, 1861, [2839], at 32.

⁷¹ War Office, *Tables of Accommodation, Record Plans of Barracks*.

Army Life and Retirement: Other Ranks

Enlisted Men's Quarters, 1899, West Block Wellington Barracks, Dublin

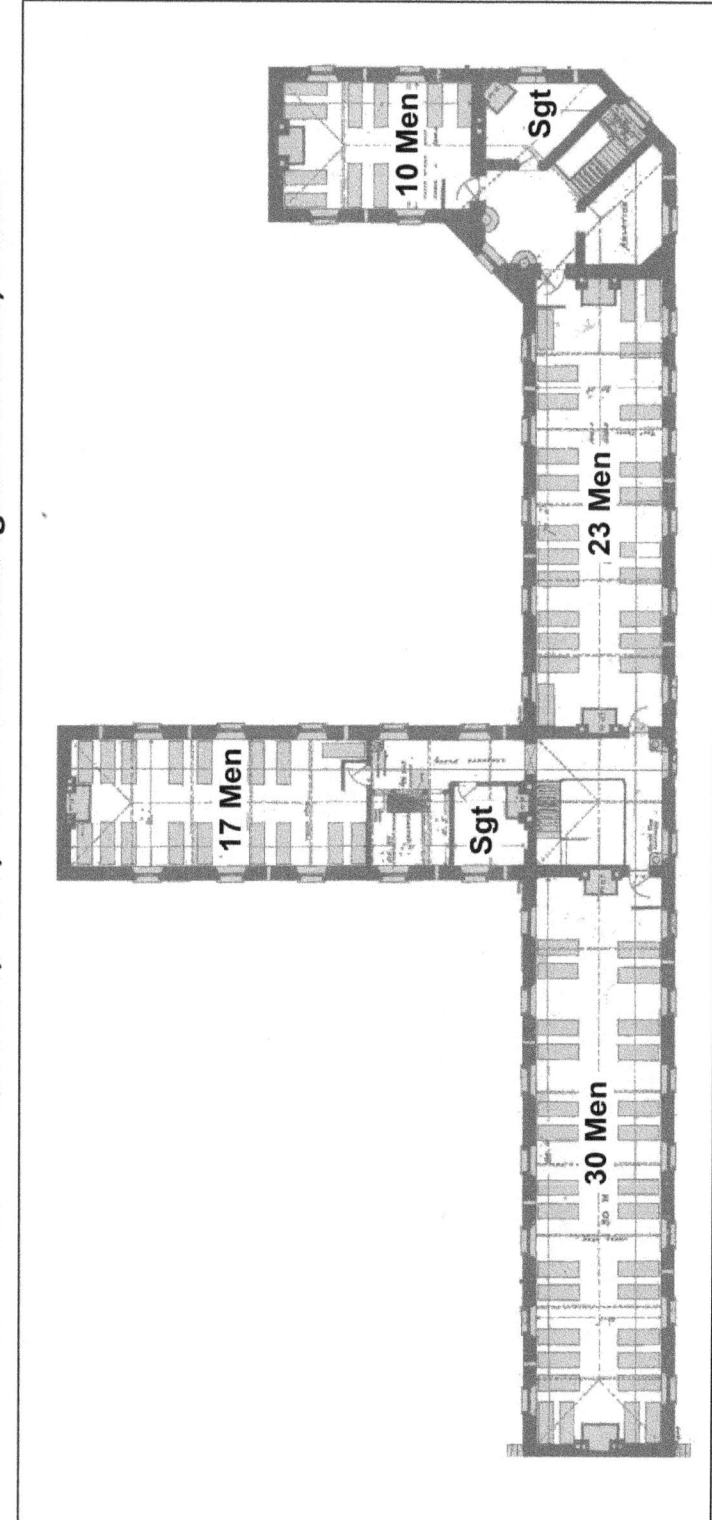

Military Archives, Defense Forces Ireland, IE/MA/MPD/AD119438-001, used with permission.

This floor could accommodate a home-strength infantry company of eighty privates and corporals plus two sergeants. The company's other two sergeants, and two of the corporals, would be married and housed in family quarters. Privates employed as officers' servants lived in servants' rooms located in the officers' quarters.

Army Life and Retirement: Other Ranks

Though barracks were no longer unventilated, foul-smelling, and pestilent, few had the amenities necessary to attract other than the most desperate recruits.[72] There were no indoor toilets and extremely limited bathing facilities. Barracks had on average one slow-filling, bathtub for every 50 to 100 men which if all soldiers used regularly, would allow for only a monthly bath.[73] Barracks did have an adequate number of washrooms, termed "ablution rooms," which were equipped with sinks. Hot running water was a rarity in civilian homes, so such amenity was not found in barracks. Few barracks had dining rooms so nearly all soldiers dined in their sleeping quarters. Each morning soldiers closed their foldable beds and moved them to the barrack-room's perimeter. They then moved large tables and benches to the room's center which were used for eating meals, mending uniforms, writing letters, *etc*. Meals were brought from the cookhouse to the barrack-room in pots carried by "orderly-men" who were soldiers assigned to that duty on a rotating basis.[74]

Orderly-Men at Cookhouse Door

R. Caton Woodville from Cairnes, *Social Life in the British Army.*

Idealized Depiction of Barrackroom Dining

"The Domestic Life of Tommy Atkins," *The Navy & Army Illustrated,* June 4, 1898.

---

[72] Arnold-Foster, *The Army in 1906*, 275-75.

[73] Cairnes, *The Army from Within*, 34-35.

[74] Ibid., 28-31.

Army Life and Retirement: Other Ranks

As was the cookhouse, recreational facilities (regimental library, game room, lounge, and canteen) were typically located in out-buildings. While a few barracks were spacious, comfortable, and with convenient amenities, more were cramped, unpleasant and some cases insanitary. The worst accommodation was in castles, forts, and barracks in the center of cities.[75] H.O. Arnold Forster, Secretary of State for War 1903-1905, wrote that overall, conditions in the very old barracks were awfully bad and in some, a national disgrace. He also found that the newer barracks were "devoid of comfort or convenience, and even the most modern structures leave much to be desired."[76]

When soldiers were stationed where there were no available quarters, the War Office provided them room and board in cheap hotels and boarding houses, or a housing and subsistence allowance of from £16.7 annually for unmarried privates, corporals, and the lowest-ranked sergeants, to £44.7 for married warrant officers.[77] In *Ulysses*, Bloom's rent at 7 Eccles Street was about £35 yearly, excluding fuel and lighting expense (coal and electricity).

Army Food

Army meals were plain and made with low-quality ingredients, and the menu changed infrequently; however, the food was rarely insanitary.[78] The army had no specialist cooks other than the bread-bakers and butchers of the Army Service Corps. Meals were prepared by enlisted men assigned permanently to the cookhouse, one per company.[79] Each battalion-sized unit had a sergeant-cook who supervised the cookhouse, trained the amateur soldier-cooks, and kept the cookhouse records.[80] Beginning in 1885, all sergeant-cooks had to complete the three-month course of the School of Cookery at Aldershot. Corporal-cooks, deputies to the sergeant-cooks and supervisors of small cookhouses, also had to complete the course.[81]

In 1850, soldiers paid nearly the entire cost of their subsistence. At home, the War Office charged enlisted men 6d. daily for the meat and bread ration and regimental cookhouses charged up to 3.5d. daily for all other food.[82] As the army provided only breakfast and lunch

---

[75] Spiers, *The Late Victorian Army*, 138-39.

[76] Arnold-Foster, *The Army in 1906*, 273-74.

[77] Subsistence: 9d. daily. Housing: £3 10d. for unmarried privates, corporals, and sergeants, £7 if married; £13 13s. 9d. for colour-sergeants; £24 6s. 9d. for quartermaster-sergeants; £27 7s. 6d. for warrant officers. Fuel & Light: £3 13s. 6d. for colour-sergeants and higher. *Regulations relating to the Issue of Army Allowances, 1884, ¶¶255, 336.*

[78] Spiers, *The Late Victorian Army*, 142; French, *Military Identities*, 122; Skelley, *The Victorian Army at Home*, 67-68.

[79] *King's Regulations, 1908,* ¶686.

[80] *Queen' Regulations, Part II 1889,* ¶184.

[81] General Order No. 44, March 1884; *King's Regulations, 1908,* ¶763.

[82] *Report of a Committee on Ration Stoppages,* 1867, H.C. Accounts & Papers, No. 130.

(termed "dinner") soldiers purchased a tea or supper, usually at the regimental refreshment room. For privates, who received only 13d. per day (a shilling in pay plus a 1d. beer allowance), 80% of their income went towards subsistence. Throughout the next fifty years, governments progressively decreased the amount soldiers paid for their meals. Beginning in 1873, the meat and bread ration came free of charge, and charges for other food items were limited to 5d. per day, though in practice they were 3d. to 4d.[83] In 1898, the government granted enlisted men age nineteen and over (those eligible for deployment abroad) a daily 3d. meal allowance. As nearly all cookhouses charged each man 3d. daily for groceries, the allowance effectively ended messing charges.[84] Army-provided meals were breakfast (tea and buttered bread, with some meat or an egg three or four times a week), a large mid-day dinner (pea soup with meat, meat pies and potatoes or cabbage, stewed meat and potatoes, and sometimes vegetables), and a late afternoon tea (tea and buttered bread).[85] Nearly all soldiers found those meals deficient in food quantity and they continued to purchase evening suppers.[86]

Sergeants, like officers, had their own mess; however, the sergeants' mess was provided with meat and bread by the Army Service Corps at no charge as such ration was for all enlisted men. The sergeants' mess was a modest version of the officers' mess. Furnishings and plate were not as elaborate, meals not as refined, and the sergeants drank beer and gin, not wine, port, and whiskey. Like the officers, sergeants in mess were waited on by soldier mess attendants who received extra-duties pay. Unmarried sergeants took all their meals at the mess; married sergeants just the mid-day dinner. The maximum daily food charge was 1s. to which the 3d. messing allowance was applied. As married sergeants usually had only five mess meals weekly, their monthly subscription charge was half that of their unmarried colleagues; 9d. rather than 1s. 6d.[87]

The Garrison Workweek and Sunday Church Parade

Regimental soldiers assigned extra duties (working as tradesmen, clerks, warehousemen, regimental police, *etc.*), had civilian-like jobs with a work-week of 40 to 50 hours. Situated similarly were the few specialists on the infantry battalion and cavalry regiment establishments. They were the bandsmen, drummers and trumpeters, and in the cavalry the equestrian specialists such as farriers and harness-makers. In an infantry regiment, these worker-soldiers totaled 95 to 110 men. Performance of their regular extra-regimental duties was interrupted only by the military training required by regulation: annual target shooting, occasional musketry and drill, and the annual field exercise.[88] Specialists and men with extra

---

[83] *Report of the Committee Appointed to Enquire into the Question of Soldiers' Dietary*, 1889, [C. 5742].

[84] *Queen' Regulations, 1899*, ¶707; Cairnes, *The Army from Within*, 111; Grierson, *The British Army*, 285.

[85] French, *Military Identities*, 123; "Service Canteens and Messes," *Navy & Army Illustrated*, November 7, 1903.

[86] Bowman and Connelly, *The Edwardian Army*, 53; Cairnes, *The Army from Within*, 62-63.

[87] Grierson, *The British Army*, 242.

[88] Cairnes, *The Army From Within*, 87-88; Mansfield, *Soldiers as Workers*, 70-137.

duties were exempt from guard and "fatigue" duties. The other soldiers with "9 to 5" jobs were the staff personnel specified in the unit establishments approved by Parliament through the appropriation process. Those staff men were sergeants who held positions such as orderly room sergeant, quartermaster-sergeant, and sergeant-cook. The enlisted staff of an infantry battalion had two warrant officers (regimental sergeant-major, bandmaster) and seven sergeants; a cavalry regiment two warrant officers and eight sergeants. Together, the tradesmen-solders, specialists, and staff sergeants constituted a unit's cadre, and usually accounted for 15% of authorized, enlisted strength. Additionally, in large camps and garrison towns, some regimental soldiers (as many as 10%), were assigned permanently to garrison duties.[89] As cadre and garrison echelon soldiers had regular duties and working hours, their workday was like that of their civilian counterparts: tradesmen, clerks, domestic servants, and semi-skilled laborers. The remaining 75% to 85% of regimental personnel were the rank-and-file and their daily life was much different from that of the cadre.

Rank and file infantrymen, cavalry troopers, and artillery gunners, have no civilian counterparts: They only perform their actual jobs during armed hostilities. Note that the British term for a soldier's combat service is "active" service; that is actively employed as a soldier. Accordingly, a combat soldier in garrison is without active employment and must be otherwise occupied by his employer. All soldiers participated in their battalion's annual, tactical exercise of eighteen to twenty-four days, and musketry training for ten days. Cavalry troopers and artillery gunners had additional annual mandatory training. On all other weekdays, soldiers trained in barracks. In 1904, routine military training for the British soldier consisted of physical conditioning ("gymnasia"), repetition of small formation movements ("drill"), practice in weapon maintenance and taking firing positions, practice in the use of equipment, and classroom instruction. As the law of diminishing returns applies to all those endeavors, military training accounted for only a few hours of a soldier's day. While the War Office was willing to allow officers to go off duty each day at noon (or earlier) and to grant them multi-month leaves, it would not permit enlisted men the same amount of liberty. The middle and upper class army leadership believed that working class soldiers with too much free time would only cause trouble. Accordingly, most of the military work-day for the rank and file combat soldier, was spent at busy-work or menial tasks.

Each weekday, for about six hours, a good portion, or all, of a company's rank-and-file were assigned fatigues. For soldiers not so assigned, their day's work consisted of up to an hour of gymnasia, about two hours of repetitive military training, and four hours of "busy-work." Work whose only purpose was to pass the time of day included buffing leather straps and belts to a shade of brown uniform throughout the battalion, polishing uniform buttons and belt buckles, cleaning rifles that hadn't been fired, and cleaning kit (canteens, cartridge pouches, *etc.*) that wasn't dirty. This make-work of up to four hours daily was a continued source of irritation to enlisted men and was the primary cause of discontent within the ranks. In some regiments; however, when no training was scheduled, the rank-and-file had the

---

[89] Cairnes, *The Army from Within*, 86-87.

afternoon to themselves, though confined to barracks.[90] The weekday schedule for rank and file soldiers was as follows:[91]

6:00 am   Wake, wash, shave, and dress; fold and move beds to the barrack-room perimeter; move tables to the center. Tidy up living quarters and undergo inspection by an officer.

Morning parade (formation) then the following:

Infantry	Cavalry, Artillery, ASC
Drill or Gymnasia	Clean Stables & Tend to Horses

8:00 am   Breakfast.

9:00 am

4 Hours Fatigues	**or**	1 Hour Military Training followed by Busy-Work in Quarters

1:00 pm   Dinner.

2:00 pm   Afternoon parade then the following:

2 Hours Fatigues	**or**	1 Hour Military Training followed by Busy-Work in Quarters

4:30 pm   Tea followed by release from duty.

4½ Hours Free Time

9:30 pm   Roll call in barrack-room.

10:00 pm   Lights out.

Saturday was a work half-day and soldiers' duties ended at dinner time. Fatigues were rarely assigned on Saturday morning and there was no training. Instead, soldiers cleaned the

---

[90] Wyndham, *The Queen's Service*, 49.

[91] Spiers, *The Late Victorian Army*, 142; Cairnes, *The Army from Within*, 28, 79-80.

barrack hallways, offices, store-rooms, windows, *etc.* They also removed litter from the vicinity of their barrack building and cut the grass, if there was any. After such housekeeping, soldiers donned their parade uniforms and they, and the barrack, were inspected by an officer

Sunday was the soldier's day off, though there was mandatory attendance at divine services of his professed denomination. The few Jews and Seventh Day Adventists had their "church parade" on Saturday. An officer marched Anglican soldiers to a nearby church or army chapel; an officer or sergeant Catholics, and a sergeant or corporal others. As a soldier was "not obliged to attend the service of any other religious body than his own," at times a corporal would accompany a solitary private to a place of worship.[92] The officer or NCO in charge of the church parade had to remain with the men throughout the service and report seditious or otherwise inflammatory sermons to the general-officer-commanding.[93] Sunday's schedule follows:

Saturday Inspection

R. Caton Woodville from Cairnes, *Social Life in the British Army.*

    8:00 am   Wake, wash, shave, and dress; fold and move beds to the barrack-room perimeter; move tables to the center.

    9:00 am   Breakfast

    9:30 am   Church parade (march in formation to church or army chapel).

    11:00 am   Release from duty.

> 10½ Hours Free Time
> Meals available at 12:00 pm and 4:30 pm.

    9:30 pm   Roll call in barrack-room.

    10:00 pm   Lights out.

---

[92] *King's Regulations, 1908,* ¶1321-22.

[93] Ibid., ¶1325.

Army Life and Retirement: Other Ranks

Exempt from the daily routine were the rank-and-file on guard duty. Seven days a week, twenty-four hours each day, guards stood at facility and building entrances, in room doorways, and in corridors. In garrison towns, soldiers were assigned guard duty in facilities other than their barracks. For example, infantry battalions quartered in Dublin provided guards for Dublin Castle, the Vice-Regal Lodge, the Military Hospital, the civil Bank of Ireland building and Mountjoy Convict Prison.[94] These almost entirely needless security measures were taken in twenty-four hour shifts and guards alternated between two hours at post and four hours in a guard room where they relaxed, ate, and slept. A soldier had guard duty from three to eight days each month, depending on the practices of the regiment, garrison, and camp.[95] The officer in charge of the day's guard was attended to by his soldier-servant for twenty-four hours. Officers were assigned such duty about once a month.

Adult Education and Army Schools [96]

British Army schools date back to the late seventeenth century and by the 1850s schools were operating at all permanent stations. The core curriculum for soldiers was reading, writing, arithmetic, English history, geography, and army bookkeeping. The objective of such adult education was to ensure the army had enough soldiers with the literacy and numeracy required for NCO positions.

In 1861, the War Office established an incentive system for educational attainment that utilized "certificates of education." These certificates were academic credentials necessary for promotion. The 3rd Class Certificate (basic literacy and numeracy) was a prerequisite for promotion to corporal. The 2nd Class Certificate (mastery of subjects presented in the fourth year of primary school plus regimental bookkeeping) was required for promotion to sergeant. A 1st Class Certificate was necessary for commissioning from the ranks. Soldiers who held such certificate were able to read and write at the level of a primary school graduate, had full knowledge of regimental accounting and record-keeping, and had a basic knowledge of imperial geography and English history. In 1889, the War Office made possession of a 1st Class Certificate a requirement for promotion to the most senior NCO ranks (quartermaster-sergeant and sergeant-major).[97]

Formal army education was of little importance to most soldiers. Many who attended class did so simply to avoid fatigue and other duties and never attempted to learn the material presented by the schoolmasters. Short-service soldiers, especially those serving only three years, usually were not interested in promotion and accordingly never sought the 3rd Class Certificate. In the decades prior to Bloomsday, fewer than three percent of enlisted men held 1st Class Certificate while fourteen to twenty-one percent held 2nd Class Certificates.

---

[94] Wyndham, *The Queen's Service*, 56.

[95] Cairnes, *The Army from Within*, 75-79.

[96] Skelley, *The Victorian Army at Home*, 91-98.

[97] *Queen's Regulations, 1899*, ¶¶740, 746.

Table 30.
Enlisted Men with Certificates of Education
1884 – 1904

	1st Class Number	Pct.	2nd Class Number	Pct.	3rd Class Number	Pct.	None Number	Pct.	Strength
1904	7,429	2.8	54,419	20.3	50,080	18.7	155,673	58.2	267,601
1903	6,368	2.4	48,705	18.2	36,763	13.8	175,517	65.6	267,353
1902	5,618	1.9	43,257	14.2	30,985	10.2	223,740	73.7	303,600
1894	3,770	1.8	44,944	21.3	30,250	14.3	135,791	62.6	214,755
1884	944	0.5	30,508	16.9	28,708	15.9	120,848	66.7	181,008

War Office Sources: *Annual Army Reports, 1884,* 1885 [C. 4570]; *1894,* 1895 [C. 7885]; *1904,* 1905, [Cd. 2268].

Regimental Institutes, Libraries, and Soldiers' Homes

Over the centuries, several organizations developed to afford soldiers recreational opportunities, spiritual and intellectual enlightenment, and inexpensive supplemental meals. There also were organizations that provided the army with foodstuffs other than the bread and meat ration and operated on-post taverns. Most of these organizations were collectively known as regimental institutes but there were also libraries and the off-post facilities known as "soldiers' homes."

Regimental Institutes

Regimental institutes existed within battalion-sized units, garrisons, and camps. Managed by a board of officers, they helped the army feed its soldiers and their on-the-strength dependents, and endeavored to maintain the enlisted men's morale. The institutes were a malt liquor bar ("wet canteen"), grocery shop ("dry canteen"), dining establishment ("supper bar"), and recreation rooms. Regimental institutes received no state funds and paid the War Office for use of army premises. The institutes' revenue came from soldiers' purchases of meals and beer, dependents' purchases of meat and groceries, and soldiers' monthly subscriptions (or contractors' franchise payments). For each regiment, the annual profits amounted to several hundred pounds. The operating surplus was used for the soldiers' benefit: Christmas dinners, bread and cheese snacks, prizes for athletic competitions, and funding regimental charities.[98]

The regimental canteens and supper bar were operated on either the "Tenant System" or the "Regimental Management System." Under the tenant system, a regiment's commanding officer contracted with a firm to operate the canteens and supper bar. The contract set minimum quality standards for goods offered and maximum prices that could be charged the soldiers and the cookhouse. For this privilege, the contractor paid an annual, per capita franchise fee. The reg-

---

[98] Wyndham, *The Queen's Service*, 103.

imental board used the fee to provide the previously described soldiers' benefits. Under most contracts, the tenant firm was also given the exclusive right to provision the sergeants' mess. Two firms together had a near monopoly of tenant-managed institutes: R. Dickeson & Co. (nearly two-thirds of contracts) and Lipton, Ltd. (nearly one-third).[99] Under the regimental system, the institutes were self-managed and the board hired a manager to operate the canteens and supper bar. This manager was either a civilian ("canteen steward") or serving soldier who received extra-duty pay ("canteen sergeant"). In some military districts, the general-officer-commanding contracted with firms to provision all regimentally-managed institutes within the district; usually, R. Dickeson or Lipton.

Institute boards of tenant system regiments selected a contractor from those on the War Office's approved list. The selected contractor received a monopoly for provisioning the wet and dry canteens, the sergeants' mess, and operation of the supper bar. All contractor employees at canteens and supper bars were civilians; about one-third were ex-servicemen.[100]

Boards of regimentally managed institutes hired a manager known as "Canteen Steward" if a civilian, "Canteen Sergeant" if a serving NCO. Canteen stewards were paid on average 6s. per day, canteen sergeants received extra-duty pay from institute funds that brought their total remuneration close to that of canteen stewards.[101] The canteen steward or sergeant hired serving soldiers and military dependents as canteen assistants. Soldiers employed as such received extra-duty pay from institute funds. The commanding officer selected a canteen vendor who was usually also awarded the sergeants' mess contract. Canteen sergeant was the most desirable enlisted appointment in the army as it came with many opportunities for graft and embezzlement. Senior grade NCOs would surrender stripes to obtain the canteen sergeant appointment as an unscrupulous canteen sergeant could expect illicit annual earnings from £50 to £100, depending on the size of the canteen. Soldiers on extra-duty as canteen assistants also had opportunities "make a bit" through watering beer and short-weighting groceries.[102]

Canteen provisioning was highly profitable as firms usually charged canteens more than they did civilian counterparts and provided foodstuffs of lesser

---

[99] *Minutes of Evidence take before the Committee appointed to consider the Existing Conditions Under Which Canteens and Regimental Institutes are Conducted*, 1903, [Cd. 1494]. Hereafter cited as *Minutes of the Canteen Committee Report* or *Canteen Committee Report*.

[100] Testimony of W.O. Kennett and A.W. Prince, R. Dickeson & Co., *Minutes of the Canteen Committee Report,* qq. 2230-37.

[101] Wyndham, *The Queen's Service*, 102.

[102] Ibid., 102, 111, 203-04; French, *Military Identities*, 112-13; Edmondson, *John Bull's Army from Within*, 144-45.

quality.[103] This situation prompted three army officers to form the Canteen and Mess Co-operative Society in 1894 to provide canteens and sergeants' messes with quality goods at reasonable prices. While not a true consumer cooperative, it limited individual investments to £200 and dividends to 5% of invested capital.[104] By Bloomsday, the Co-operative Society was also in the canteen management business. In 1913, the Co-operative Society, R. Dickeson, and Lipton together provisioned 82% of the 293 army canteens in the U.K.[105]

Regimental Institutes and Provisioning Soldiers and Dependents

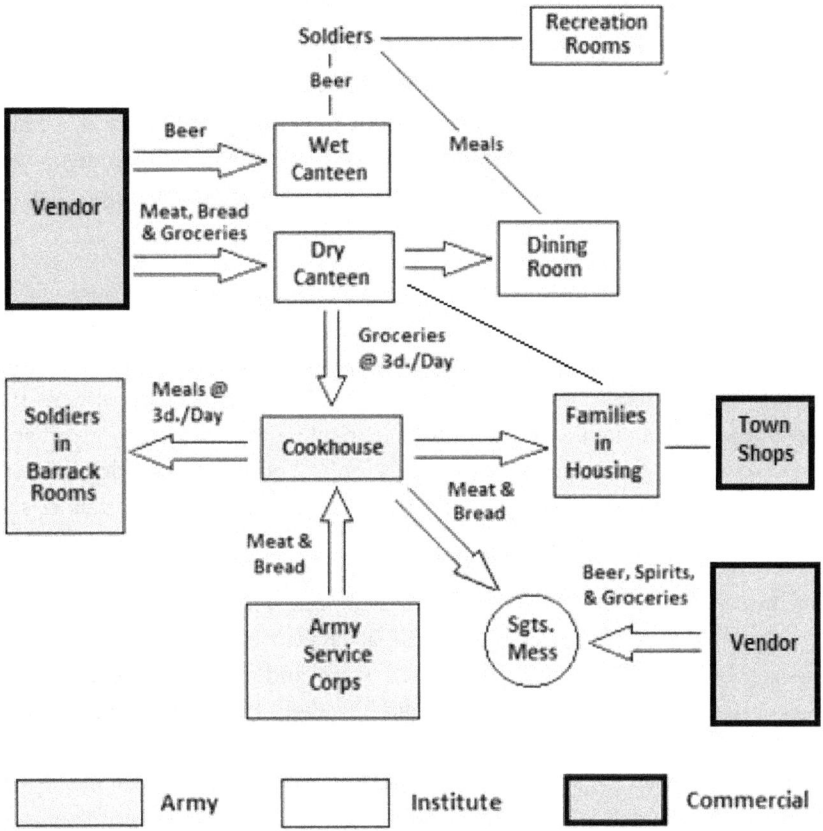

In the Winter of 1913-14 "The Canteen Scandal" broke when Crown prosecutors revealed that Lipton had from 1904 onwards systematically bribed

---

[103] Skelley, *The Victorian Army at Home*, 162; Edmondson, *John Bull's Army from Within*, 141-52; French, *Military Identities*, 112-13.

[104] *Canteen Committee Report*, at xi

[105] Captain Percy Clive, Liberal-Unionist MP, Herefordshire, Army Estimates Debate, 56 H.C. Deb. (5th ser.) (1913) 593. R. Dickeson - 40%, the Co-operative Society - 24%, Lipton - 18%, and about a half-dozen firms the remaining 18%.

civilian and military employees of canteens as well as non-employee NCOs and commissioned officers. "Between 1904 and 1912 their illegal activities touched at least sixty-nine units, and involved nearly 150 military personnel. Bribes were given to Quarter-Masters, Warrant Officers, and NCOs to influence their COs against existing contractors and in favour of Liptons. In at least three instances there were suspicions that unit commanders themselves had received bribes ranging from £100 to £300."[106] In April 1914, eighteen persons (nine military and nine civilians) were committed for trial. A former commanding officer of the 2/KOYLI was found guilty of accepting bribes and imprisoned and several civilians were fined.[107]

Supper Bar: Here soldiers could obtain a light meal of soup and buttered bread for 2d. or one pound of buttered bread for 2d. to 3d. A full meal of meat and potatoes with an eight-ounce glass of ale cost 4.5d; meat and two vegetables 6d. Coffee could be had for ½d. per cup. The premises were usually well appointed, lighted, and heated. In the evenings, the supper bar was always crowded; proof of the insufficiency of army meals.[108]

Dry Canteen: This institute provisioned the regimental cookhouse with all foodstuffs other than meat and bread, sold groceries to "on-the-strength" dependent families, and usually provisioned the sergeants' mess with food and drink. It also was the soldiers' convenience store as here they could buy sundries such as knife polish, hair oil, socks, razors, toothbrushes, foot powder, over-the-counter medications, *etc*.[109] About 75% of a dry canteen's sales were to the regimental cookhouse. For an 800-man infantry battalion, such sales would total £3,650 annually.[110]

Wet Canteen: This was the soldiers' pub and was well utilized as prices were lower than those at civilian establishments. The War Office authorized these canteens to sell wine and malt products, but not whiskey. Beer was the preferred alcoholic beverage. Bitter was 2.5d. per pint, stout 2d., and ale 1.5d.[111] Wet canteens were open daily from 1:30 pm to 9:30 pm, were supervised by an NCO, and staffed by privates.[112]

---

[106] French, *Military Identities*, 112-13.

[107] *The Times*, April 3, and May 28, 1914.

[108] Wyndham, *The Queen's Service*, 101; Cairnes, *The Army from Within*, 63; Testimony, *Minutes of the Canteen Committee Report,* MAJ H.R.J. Willis at qq.1469-71, PVT Fred Cooper at q.1644, Mr. W.T. Lewis at qq.4471-73.

[109] "Service Canteens and Messes – III," *Navy & Army Illustrated*, August 15, 1903.

[110] Testimony, *Minutes of the Canteen Committee Report,* C.E. Haygate at qq. 5431-38.

[111] Ibid., at xxiv, xxxvii.

[112] Wyndham, *The Queen's Service*, 98-99.

Recreation Room: This institute was a place "where soldiers can write letters and read the papers, and where amusements can be provided calculated to keep soldiers in barracks and away from the temptations of public-houses outside."[113] The large recreation rooms of garrisons and camps usually had theaters (admission price of 1d. to 2d.)[114] In effect, the recreation room was a "wholesome" club for privates and corporals whose members paid a monthly subscription of 2d. to 3d.[115] Though recreation rooms charged fees, most of their funding came from the wet canteen profits.[116] A major expense was purchase of athletic equipment: cricket bats and balls, foot balls, team uniforms, small boats, *etc.*

Libraries

Each garrison, camp, and station had a lending library. Membership was by subscription of 1d. to 2d. monthly. Soldiers' libraries were not institutes; they were operated directly by the army. The purpose of these libraries was "to encourage the soldiers to employ their leisure hours in a manner that shall combine amusement with the attainment of useful knowledge, and teach them the value of sober, regular, and moral habits."[117] Libraries were supervised by the Adjutant-General's Staff of the War Office.[118]

The libraries contained publications approved by the War Office and accepted books donated by the public. A "gentleman ranker" wrote in 1881 that only about 20% of soldiers used the library but there was a small clique of avid readers who borrowed books every other day. He noted that "books of travel, tales of emigration, books of military adventure, all had their fanciers ... but religious works were scarcely asked for at all."[119]

Soldiers' Homes

These were charitable organizations that offered enlisted men recreation rooms, reading rooms, meals, and short-term lodgings. Nearly all were operated as religious missions to servicemen and their offerings were served up with a

---

[113] *Canteen Committee Report,* at vi.

[114] Murray, *Six Months in the Ranks,* 95.

[115] Wyndham, *The Queen's Service,* 103-04, 142; Testimony, *Minutes of the Canteen Committee Report,* PVT Coughlin at qq. 350-52.

[116] Testimony, *Minutes of the Canteen Committee Report:* LTC Farmer at q. 62, COL Jeffries at q. 494; COL Crofton at q. 598.

[117] *Queen's Regulations, 1859,* Regulations for Troops in Barracks, §80.

[118] *King's Regulations, 1908,* appx. IV.

[119] Murray, *Six Months in the Ranks,* 327-328.

strong dose of religion.[120] Because of their sectarian nature, soldier's homes received no state funds.

The first soldiers' home was established in 1861 at Chatham by a Methodist organization. A year later Louisa Daniell, an unaffiliated evangelist, opened one at Aldershot. The first home in Ireland was established in 1877 at Tralee by Elise Sandes, another unaffiliated evangelist. By 1900, there were soldiers' homes operated by several religious bodies including the Church of England and the Salvation Army.[121]

Soldier's homes appealed only to a narrow segment of the enlisted ranks because the homes did not serve alcohol, stocked few books other than those with religious themes, carried few popular magazines, disallowed card-playing, and had no billiard tables. Overall, they were conducted more like places of worship rather than clubs. "Should a fresh arrival enter the reading-room he will certainly be made welcome. He will also have a tract pressed upon him."[122]

Married Life

Soldiers married either with their commanding officers' approval or without official sanction. Dependents of soldiers who married with permission were entered on the regiment's roll and received housing, subsidized food, and other benefits. For those families, the army was a provider of their basic maintenance, but in return they became subject to military discipline. Married women who broke the regimental rules risked being struck off-the-strength.[123] Off-the-strength dependents received nothing from the army and the War Office hardly recognized their existence.

Eligibility for Sanctioned Marriage

Parliament and the army establishment long considered marriage incompatible with a soldier's life and viewed wives as an expensive nuisance.[124] Accordingly, regulations required commanding officers to discourage soldiers from marrying.[125] Generals and War Office officials recognized; however, that state provision for military dependents was necessary to retain non-commissioned officers and attract high-quality, career-minded recruits. As a result, marriage was grudgingly authorized but effectively only for NCOs. That all sergeants could receive marriage permission was an inducement for promising corporals to extend their service. Permission to marry was at the discretion of a soldier's commanding officer.

---

[120] Wyndham, *The Queen's Service*, 84-85.

[121] Skelley, *The Victorian Army at Home*, 164.

[122] Wyndham, *The Queen's Service*, 84-85.

[123] *Trustram, Women of the Regiment*, 190.

[124] *Ibid.*, 30,39; Wyndham, *The Queen's Service*, 264-65.

[125] *Queen's Regulations, 1881*, Sec. VII, ¶182.

Army Life and Retirement: Other Ranks

To be eligible for marriage a soldier needed £5 in savings, two good conduct badges (six years blemish-free service), and seven years total service.[126] Warrant officers did not need permission to marry; their dependents automatically went on-the-strength.[127] The number of soldiers permitted to marry was determined by regimental grade structure. Each formation was authorized on-the-strength wives as follows: 100% for warrant officers and senior sergeants, 50 percent for other sergeants, 4 to 7 percent for corporals and privates, depending on corps. The percentage of rank and file "marriage billets" was highest in the Household Cavalry and Royal Army Medical Corps.[128] In a line infantry battalion, 57 soldiers were permitted to marry, calculated as follows:

Rank	Number Assigned	Wives Authorized
Warrant Officer	2	2 (100%)
Colour- & QM-Sergeants	9	9 (100%)
Sergeant	30	15 (50%)
Others	781	31 (4%)
Total	822	57 (6.8%)

Idealized Depiction of Army Family Life

---

[126] Ibid., ¶725.

[127] Ibid., ¶719.

[128] *Regulations Relating to the Issue of Army Allowances, 1884*, ¶98.

Until the turn of the twentieth century, permission to marry could not be retroactive. Many men married in secret and then when eligible for sanctioned marriage requested permission to wed. By Bloomsday, regulations allowed commanding officers to grant on-the-strength status to wives from a previously unauthorized marriage so long as no fully qualified applicant was awaiting permission to marry.[129]

Housing and Subsistence of Army Dependents

In the mid-nineteenth century soldiers and their families were housed in segregated barrack rooms or received "corner accommodation." A family barrack room was a large room occupied by several families. It was partitioned with curtains to form individual family dwelling-places. Corner accommodation was a curtained-off portion of the large room occupied by bachelor privates and corporals.[130]

By 1904, married soldiers and their families were housed in segregated, purpose-built, two-story apartment buildings. Family accommodation that met army standards were 2.5 room apartments with about 450 square feet (up to two children); 3.5 rooms with 675 square feet (three to five children); 4.5 rooms with 775 square feet (six or more children).[131] Family housing came with furniture, lamps, and bedding. In 1904, all family quarters did not meet standards and many barracks lacked enough family housing. Where family housing wasn't available soldiers received an annual housing allowance, including heat and light, of £14.4 for privates and corporals, £19.0 for sergeants, and up to £32.7 for higher ranks.[132] As lodging near barracks were invariably at a premium, a married soldier was unable to house his family decently unless he resided at great distance from his place of duty.

On-the-strength wives could do all their food shopping in barracks or camp. Wives obtained their husband's free meat and bread ration at the regimental cookhouse. Generally, the sergeant-cook made sure married men (nearly all NCOs) and the sergeants' mess received the best cuts of meat. The worst cuts, mostly bone, fat, and gristle, went into the privates' dinner.[133] Dependents could purchase other food items, plus meat and bread for themselves, at the regimental canteen. Many canteens gave families a 5% rebate on all such purchases, paid quarterly.[134]

Dependent Medical Care

All on-the-strength dependents were eligible for free medical care at army women and children hospitals, though they had to pay for hospital meals. There were few such hospitals, so the War Office paid for private-sector care. The general-officer-commanding contracted

---

[129] *King's Regulations, 1908*, ¶1349.

[130] Trustram, *Women of the Regiment*, 70-75.

[131] War Office, *Ground Plan of Buttevant Barracks, Co. Cork*, 1913; *Queen's Regulations, 1899*, ¶1055.

[132] Grierson, *The British Army*, 228.

[133] French, *Military Identities*, 121.

[134] *Canteen Committee Report*, 1903, [Cd. 1424], appx. 8.

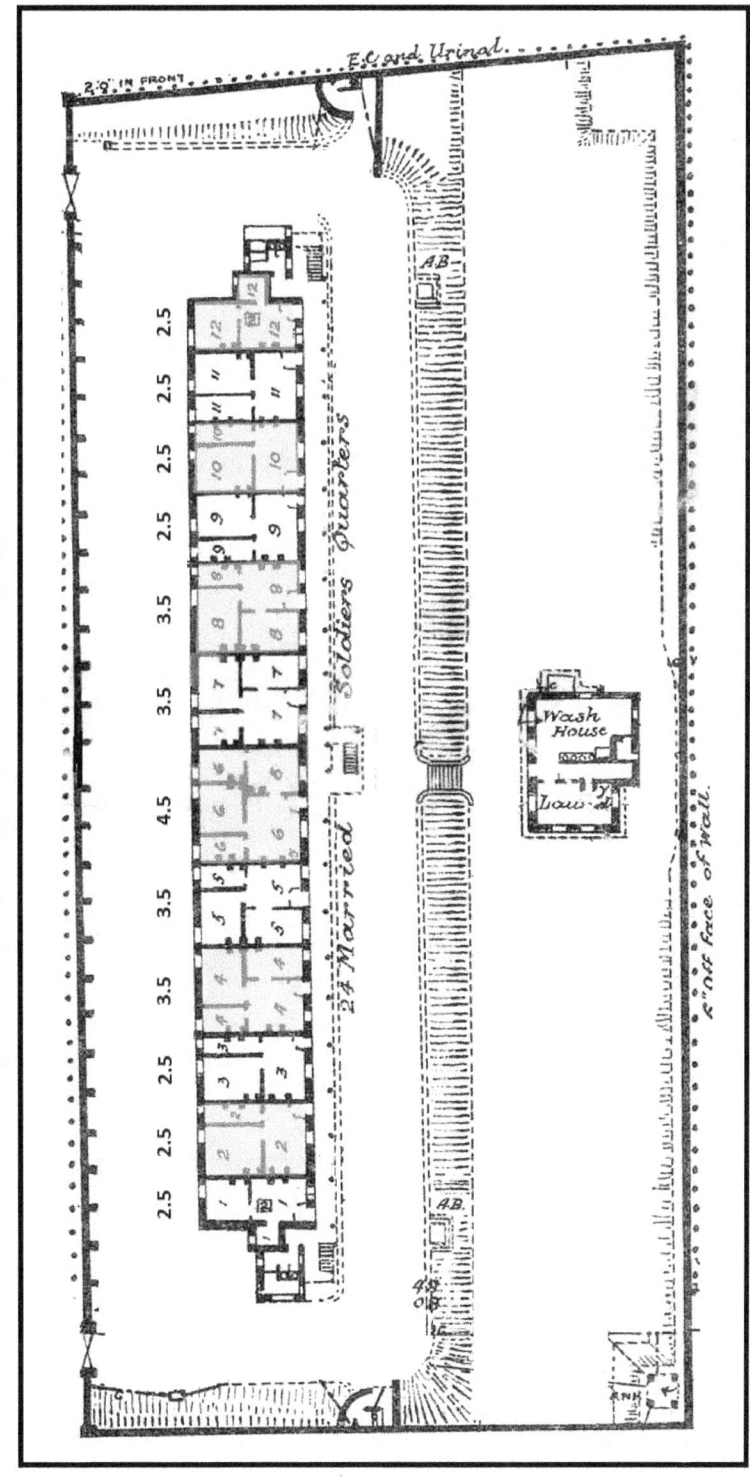

Family Quarters, 1896-97, Buttevant Barracks, County Cork

Military Archives, Defence Forces Ireland, IE/MA/MPD/ad134209-006, used with permission.

Numbers indicate rooms per apartment.

This was one of two apartment buildings that together housed fifty families at Buttevant Barracks, located midway between the cities of Limerick and Cork.

with medical practitioners to provide services to army dependents. Doctors could receive no more than £1 per day for treatment of army personnel, including on-the-strength dependents.[135]

Education of Children

The army provided low-cost, later free, primary education for dependent children. Secondary education was not offered as the law did not require such schooling in civil life. Nearly all dependent children attended either army regimental schools or War Office approved state schools. Some resided in army boarding schools.[136]

Regimental Schools

School attendance was compulsory by regulation. During most of the nineteenth century, soldiers were charged school fees of 2d. per month for one child and 1d. for each additional child. Fees were abolished in 1871.[137] By 1904, mandatory schooling encompassed children aged four to fourteen, as did compulsory education laws applicable to civilians.[138] Pupils were instructed by army schoolmasters and civilian schoolmistresses. Such professionals were assisted by soldiers and on-the-strength wives and daughters. The curriculum conformed to the Day School Code of the Council on Education for England and Wales, though the only optional, advanced subjects offered were algebra and geometry. In the Edwardian Era, a regimental school education prepared children for clerical and needlework employment, not secondary schooling.[139]

State Schools

When there was no army school within two miles of a family's residence, the children attended civil schools approved by the War Office. If there was not a nearby, free civil school, the army paid for attendance at a private school and provided an annual textbook allowance of 6s. per child.[140]

Military Boarding Schools

As noted previously, the War Office maintained two boarding schools for army orphans and dependent children: The Royal Hibernian Military School, Dublin and the Duke of York's Military School in Chelsea, London. Prior to 1893, the

---

[135] *Royal Warrant for the Pay, 1899*, 1906, Art. 375.

[136] 410 boys at the Hibernian School, 570 boys at the Duke of York's School, 400 boys and 300 girls at the Lawrence Asylums in India.

[137] Skelley, *The Victorian Army at Home*, 102.

[138] Grierson, *The British Army*, 215.

[139] Williams, *Tommy Atkins' Children*, 103-07.

[140] *Royal Warrant for the Pay, 1899*, Art. 1275.

London school was titled "the Royal Military Asylum." A Dublin charitable organization established the Hibernian School in 1769; Parliament established the London school in 1801 and it received its first pupils in 1803. In 1816, the government opened a third military boarding school at a disused cavalry barracks in Southampton as a branch of the London school. Orphaned infants, at the time cared for at a private, charitable institution on the Isle of Wight, were also transferred to the Southampton school.[141]

At first, the schools were co-educational but in 1825 the government segregated the two English schools; London for boys, Southampton for girls and all orphans under age five. In November 1840, the War Office closed the Southampton school and transferred its 54 boarders to London for an annual savings of £1,600.[142] No further girls were admitted to the London school. The governors of the Hibernian School ended admission of girls in 1848 and by the time of the Crimean War, no girls were in the Dublin school.[143]

The London school was open to boys aged ten to twelve; the Dublin school aged seven to twelve. The schools operated in a regimental manner under the supervision of army officers and NCOs. Pupils were assigned fatigues such as scrubbing floors and tables. Discipline was mindlessly strict and until the 1870s, punishments for infractions included flogging and solitary confinement with a bread and water diet. By the 1890s, severity of punishment was lessened to that found in elite, British public schools.

In 1882, two-thirds of the children nominated to these schools were accepted for enrollment. Orphans were given admission preference; their acceptance rate was 90%. Five hours of academic instruction were provided daily by seven schoolmasters and twenty-one student-teachers (about one instructor for every thirty-five pupils). Pupils also received two to three hours of instruction by army personnel in music, carpentry, shoemaking, and tailoring. About one-fourth of pupils failed to reach the civil standard for ten-year olds with three years of primary schooling. With few exceptions, the schools released boys when they reached age fourteen.[144] Staff pressured pupils, and their parents, to consent to enlistment as boy recruits. Boys who did not enlist in the armed forces were placed by the schools into civilian apprenticeships.[145] Throughout the 1880s the quality of

---

[141] Clarke, "The Royal Hibernian School in Dublin;" Cockerill, *The Charity of Mars*, 21-23, 37, 130-32.

[142] Cockerill, *The Charity of Mars*, 143; *Hampshire Advertiser*, December 5, 1840; Thomas Macaulay, Secretary at War, 56 Parl. Deb. (3d ser.) (1841) 1367.

[143] *The Times*, June 11, 1856; *The Evening Freeman*, September 26, 1856.

[144] *Report of the Committee Appointed to Inquire into the Royal Hospitals at Chelsea and Kilmainham, and The Royal Military Asylum, Chelsea and the Royal Hibernian Military School*, 1883, [C. 3679]. Boys in the band could remain at the schools until age 16.

[145] Cockerill, *Sons of the Brave*, 95-96; Skelley, *The Victorian Army at Home*, 108. A prerequisite for admission was parental consent to placement in apprenticeships. *Royal Warrant for the Pay, 1899*, Art. 1272.

education at the boarding schools declined. In 1889, only 20% of pupils met the civil academic standard for their age; 17% were one level below standard; 63% two or more levels below standard. About two-thirds of the boys who left these institutions at age fourteen entered the army.[146]

In the 1890s, the War Office compelled the boarding schools to increase the amount of academic instruction, reduce time spent on army drill and vocational training, and make the academic curriculum more rigorous. Pupils were also relieved of the heavier fatigue duties.[147] Within a few years, student achievement improved greatly. In 1896, half the pupils were at or above the civilian standard for their age, compared to one-fifth ten years earlier.[148] Though the quality of education had improved since the mid-1880s, in the early twentieth-century the military boarding schools, overall, were inferior to comparable, state primary schools.

In addition to the military boarding schools in the United Kingdom, there were four similar schools in India for children of British soldiers of the Indian Army and British Army units stationed in India. The Lawrence Military Asylums in India were private charities that received some state funding.[149] They also charged fees based on the soldier's ability to pay. The schools were open to girls, were less military in nature than the UK schools, and few of the boys enlisted upon leaving.[150]

Survivor Benefits

Before the Boer War, widows of enlisted men, unlike those of officers, were not authorized pensions. The state supervised Royal Patriotic Fund, founded in 1854, was the kingdom's principal charity for military widows and orphans.[151] If that organization did not

---

[146] *Annual Report of the Director-General of Military Education*, 1889, [C. 5805].

[147] Williams, *Tommy Atkins' Children*, 105.

[148] Skelley, *The Victorian Army at Home*, 111; *Annual Report of the Director-General of Military Education*, 1896, [C. 8421].

[149] Cockerill, *Sons of the Brave*, 110. The schools were the Lawrence Military Asylums at Sanawar (500 pupils), Ghora Gali (170 pupils), Mount Abu (100 pupils), and Lovedale (500 pupils).

[150] For Sanawar, of the 44 boys who left in 1898 on reaching leaving age, only 10 entered the army. *Annual Report of the Lawrence Military Asylum, Sanawar, 1897-98* (Lahore: Punjab Govt. Press, 1898).

[151] The Royal Patriotic Fund was administered by a government appointed body but received no state funding. One of its programs was the provision of boarding school education for orphans of soldiers, sailors, and officers. The Fund placed children into private boarding schools as well as its own school for girls, the Royal Victoria Patriotic School in London (250 girls). Orphans of officers received a classical, academic education at a public school, most often Wellington College (60 boys). The Royal Patriotic Fund, *Catalogue Description*, UK National Archives. In 1903, Parliament reorganized the Patriotic Fund as a Crown corporation. 3 Edw. 7, c. 20.

provide assistance, the War Office paid a lump sum to the widow as follows: one-year's base pay plus one-third of annual base pay for each child under age sixteen. Such benefit was only paid if the deceased was killed-in-action or died of wounds, and was only for on-the-strength dependents.[152] In 1901, the government relented to public pressure that arose during the Boer War and authorized pensions for widows who had been on-the-strength wives. For most survivors, their pensions, under £30 per year, could barely support a penurious existence. They relied on family and charity just to achieve a standard of living a notch or two above that of the workhouse.

For widows of soldiers below warrant officer rank, pensions were paid if the deceased died as a result of combat or a service-related injury or disease. The War Office granted pensions only to widows it found "worthy" and the amounts were determined by the deceased soldiers' rank.[153] Annual payments were from £13 to wives of privates to £26 to wives of quartermaster-sergeants. An additional £3 18s. to £5 4s. was paid for each boy under age fourteen and each girl under age sixteen. Only legitimate children qualified for the additional payment. Pensions could be reduced if the survivors received public assistance or private charity and terminated on the widow's remarriage. The War Office could suspend or terminate payment if it determined that the widow was no longer worthy.

The Patriotic Fund sought to place widows into employment and to that end provided child care services. Where employment was impracticable, it provided weekly cash benefits, the amount of which depended on the deceased husband's rank and the number of dependent children. Generally, widows with infants were not expected to work. Widows of privates received annually £18.2; widows of senior NCOs £23.4. At age 60 all widows received a £2 annual increase; at age 70 another £1.[154] In 1895, the Fund assisted 2,154 servicemen's widows and cared for 414 orphans.[155] In the years following the Boer War, the Fund assisted annually about 3,700 families and 550 orphans.[156] Approximately two-thirds of beneficiaries were former army dependents; the remaining third former navy and marine dependents.

At the turn of the twentieth century, John Lumsden, chief medical officer of the Guinness brewery, determined the annual income necessary to support a family in Dublin at the subsistence level. His findings came from interviews and surveys of hundreds of Guinness employees. Lumsden found that a single adult needed an annual income of £18 to live without poor relief or charitable assistance. Each child required an additional £8.[157] Accord-

---

Other charitable institutions open to soldiers' children were the Royal Caledonian Asylum (160 children), the Drummond Institution (60 girls), the Guards' Industrial Home (58 girls), and the Royal British Female Orphan Asylum (150 girls).

[152] *Royal Warrant for the Pay, 1899*, Art. 1262.

[153] Ibid., Arts. 1262, 1262A-E.

[154] Burn and McDonald, "An Investigation into the Rates of Re-Marriage - Patriotic Fund."

[155] *Report from the Select Committee on the Royal Patriotic Fund*, 1896, H.C. Accounts & Papers, No. 368.

[156] Trustram, *Women of the Regiment*, 177.

[157] Corcoran, *Guinness: the Greatest Brewery on Earth*, 43.

ingly, to live at the subsistence level in Dublin, a woman and three children required an annual income of £42. The following graphic shows survivor pensions in relation to Lumsden's minimum income findings.

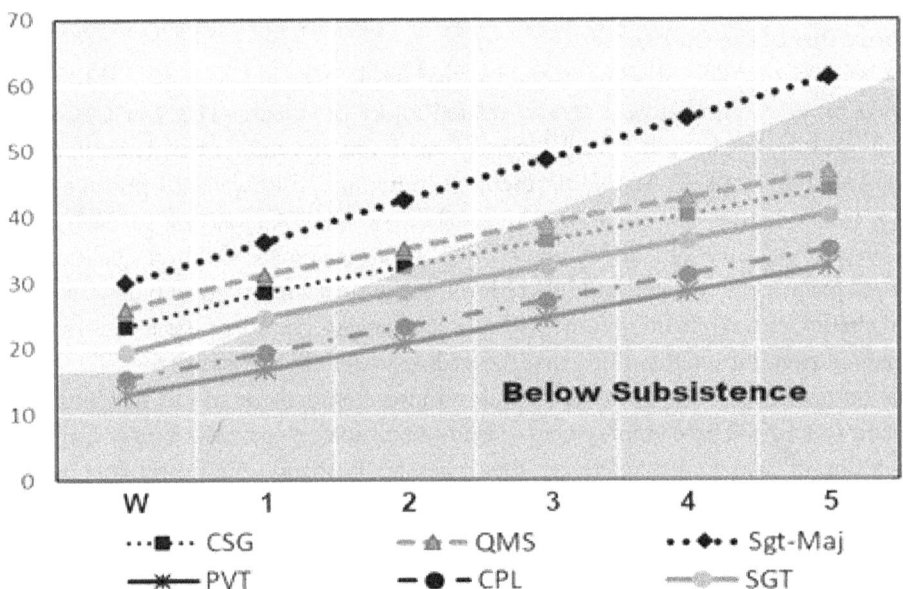

Widows of warrant officers (sergeant-majors, bandmasters, senior school-masters, *etc.*) received pensions under the same conditions as other soldiers; however, the amounts were somewhat higher and the basis for qualification was broader.[158] Widows received £30 annually plus £6 5s. for each qualifying child. Unlike widows of lower-ranked enlisted men, these widows could receive pensions if their husbands died of any cause after five years in rank. The annual pension for non-service related deaths was £20 plus £5 for each qualifying child. A widow in such situation with two qualifying children, would receive less than the earnings of an unskilled, Dublin laborer.

For a war widow with three qualifying children, the annual pension was from £24 14s. (privates) to £48 15s. (warrant officers). For orphans, the annual payment was £10 to £13 for each child of a warrant officer; of other soldiers from £7 16s. to £10 8s. depending on the father's rank. The War Office could increase the orphan benefit to allow placement of a child in a "benevolent institution." In 1904, widows of warrant officers received compassionate allowances which averaged £5.5s. per child.[159] There were no such allowances for widows of lower-ranked soldiers.

---

[158] *Royal Warrant for the Pay, 1906*, Arts. 738-40.

[159] War Office, *Army Estimates for the Year 1904-05*, 1904, H.C. Accounts & Papers, No. 73, Vote 15.

State Employment Opportunities for Dependents

There were several army civilian jobs for which the War Office gave absolute hiring preference to on-the-strength female dependents. The predominant position for employed wives and daughters was regimental laundress. A woman earned annually about £11 as an officers' laundress; £4 to £8 as a rankers' laundress.[160] Military etiquette forbade NCO's wives from doing laundry for privates. Another commonly held position was that of an officers' "daily." For cleaning an officer's quarters, a woman received about 5s. monthly; £3 annually.[161]

A wife of at least twenty years of age who had a primary school education could earn £8 to £18 annually as a part-time, assistant schoolmistress in a regimental school. Such women were assigned to the "infants" classes for children under age seven. A teenaged daughter who had a primary school education could earn £4 to £5 as a classroom "monitress."[162] Such girls taught mechanically from detailed lesson plans provided by schoolmasters. Upon reaching age twenty, they were eligible for the schoolmistress training course at the Army Model School, Aldershot.[163]

Wives who held midwifery credentials were engaged regularly by the War Office for such services. Wives and older daughters also obtained employment as seamstresses, nursing aides, domestic servants of officers' families, and canteen assistants.[164] Tailoring work for enlisted men had to be done surreptitiously as such services infringed on the monopoly rights of the regimental sergeant master-tailor and his soldier assistants and apprentices.

Overseas Deployment and Separation Allowances

Soldiers spent half of their careers abroad through a series of multi-year, overseas postings. Due to insufficient family accommodation at certain colonial stations, not all on-the-strength families could accompany the soldier abroad. Also, ever since the Crimean War, the army did not send military dependents abroad with soldiers on campaign. Families left at home usually had to vacate their army housing and make do on a separation remittance. This sum, paid monthly in advance, consisted of a stoppage from the soldier's pay and a supplemental allowance provided by the state. An allowance was made, and a stoppage required, for a soldier's wife and each son under age 14 and daughter under age 16. The allowance and minimum stoppage amounts depended on rank. Soldiers could voluntarily have greater amounts sent home, not exceeding three-fourths of base pay.[165]

---

[160] Women received £1 monthly washing for officers, 4d. to 8d. daily washing for enlisted men. Trustram, *Women of the Regiment*, 109-10; Grierson, *The British Army*, 241.

[161] Trustram, *Women of the Regiment*, 110-11.

[162] *Royal Warrant for the Pay*, 1906, Art. 1077; Williams, *Tommy Atkins Children*, 101.

[163] Williams, *Tommy Atkins Children*, 101, 118.

[164] Trustram, *Women of the Regiment*, 111-14; Buxton, *The Elements of Military Administration*, 406.

[165] Army Orders, 1905, No. 205.

Army Life and Retirement: Other Ranks

The table and graphic below show the allowances and stoppages by rank and the total remittance in relation to Lumsden's subsistence requirement. Note that the remitted amount was rarely enough to provide the left behind dependents a "respectable" working class standard of living.

Table 31.
Annual Family Separation Remittance, 1904
(Allowance and Stoppage)

	Allowance		Stoppage		Total Sent	
	Wife	Child	Wife	Child	Wife	Child
Warrant Officer	£41.1	£3.0	£15.2	£3.0	£56.3	£6.0
Quartermaster-	38.0	3.0	15.2	3.0	53.2	6.0
Colour-Sergeant	24.3	3.0	15.2	3.0	39.5	6.0
Sergeant	19.8	3.0	15.2	3.0	35.0	6.0
Corporal	19.8	3.0	9.1	1.5	28.9	4.5
Private	19.8	3.0	9.1	1.5	28.9	4.5

Sources: *Royal Warrant for the Pay, 1906*, Art. 955; *Regulations for the Allowances of the Army, 1903*, ¶225.

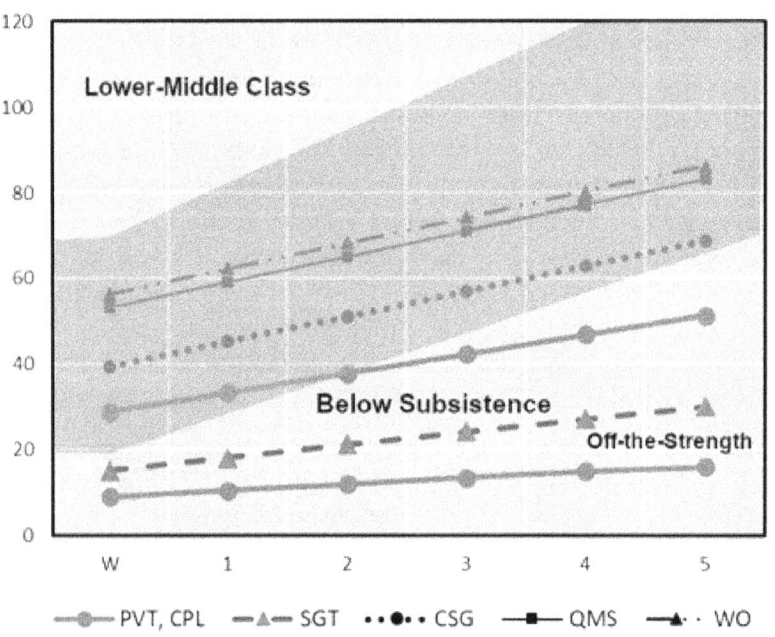

Most separated wives receiving the minimum allowance relied on family or charity, such as that provided by the Soldiers' and Sailors' Families Association. This private charity, founded in 1885, operated like the Patriotic Fund. The SASFA helped find employment for wives, provided child daycare, and home nursing. "Lady visitors" from the society also provided practical guidance for the now independent army wife as well as emotional support. When necessary, the SASFA provided cash payments.[166] Like the Patriotic Fund, aid was provided to off-the-strength families. In 1895, the SASFA assisted 3,380 families, three-fourths of which were army families.[167] During the Boer War, the SASFA assisted 206,438 service families and distributed £1.25 million in direct relief.[168]

Off-the-strength dependents received only the stoppage from the soldier's pay, which for the wife and two children of a private or corporal came to £12.1 per year. The War Office mandated the stoppages for such dependents as by 1904, the law required soldiers to support their families.[169] Note that a soldier only had to support illegitimate children if he acknowledged paternity or the mother obtained a judicial paternity decree.[170] Stoppages alone were insufficient to support a family. The graphic on the previous page shows separation allowances in relation to Dr. Lumsden's minimum income requirements for Dublin. Note that the separation remittance for large, on-the-strength families of corporals and privates was less than that required to live at the subsistence level. The remittance for all off-the-strength families was below the subsistence level.

In 1904, though half of all soldiers were stationed overseas, only about thirty percent of on-the-strength families were abroad. Family quarters were available at most foreign stations; however, there was insufficient accommodation in Egypt and South Africa, which together quartered about 26,000 troops. Other foreign stations that lacked enough family quarters were Crete, Mauritius, Ceylon, North China, and the Straits Settlements.[171] Additionally, families invalided home received a separation allowance for up to two years and a pregnant wife could not embark for a foreign station until she gave birth and was medically fit to travel.[172] In India, army families were usually housed in bungalows located in cantonments; in other overseas stations, such as Gibraltar, in married soldiers' apartments. Families abroad, unlike at home, received a free meat and bread ration; half a soldier's ration for wives, one-fourth for children under age 14. A soldier's ration overseas was one pound of meat and one pound of bread.[173] Meat was weighed "bone-in" and like at home, dependents and the sergeants' mess received the best cuts. Also, like in the UK, dependents could buy groceries

---

[166] Trustram, *Women of the Regiment*, 179-81.

[167] *Report from the Select Committee on the Royal Patriotic Fund*, 1896, H.C. Accounts & Papers, No. 368.

[168] Barnes, "History of SSAFA, Part 1."

[169] Trustram, *Women of the Regiment*, 66-67. In 1904 a soldier who abandoned his family and was subject to a maintenance order could have stopped from his daily pay up to 1s. for the wife, 6d. for each legitimate child, and 3d. for each bastard. Annual Army Act, 1904, 4 Edw. 7, c. 5.

[170] The War Office defined a soldier's family to include only legitimate children and step-children. *Regulations Relating to the Issue of Army Allowances, 1884*, ¶94.

[171] *Army Medical Department Report, 1904*, 1906, [Cd. 2700].

[172] *Regulations Relating to the Issue of Army Allowances, 1884*, ¶¶120, 124.

[173] Ibid., ¶3.

from the canteen at somewhat below market prices. Following is a schematic diagram of unit overseas rotations for the report year that included Bloomsday.

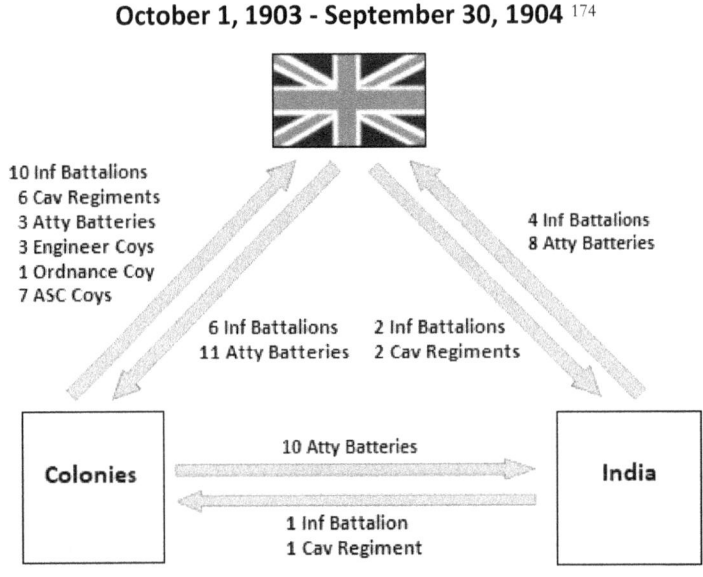

The usual period for service abroad, excluding India, was five years. For the West Indies, Ceylon, Hong Kong, Mauritius, and the Straits Settlements (Singapore and Malaya) it was three years. Those postings were of shorter duration due to the higher incidence of disease in the tropics. For Bermuda and St. Helena, the tour of duty was also three years as they were hardship stations (small, isolated islands). Service in West Africa was limited to one year.[175] Only a few British Army soldiers were stationed in West Africa, and their dependents were not relocated there.[176] In 1904, there were only 228 European military personnel in the West African stations. That part of the empire was garrisoned by a battalion of the West Indies Regiment and locally recruited units of the Colonial Corps.[177] There was no regulation duration of postings to British India, which included Burma. Nearly all were lengthy and infantry regiments remained there from seven to fourteen years with ten years the average. Within India, units had frequent changes of stations.[178] Except for postings to Gibraltar, foot

---

[174] *Annual Army Report, 1904*, 1905, [Cd. 2268]. The illustrated rotation was atypical for the British Army in peacetime as it includes the return home of units sent to South Africa for the Boer War.

[175] *Queen's Regulations, 1899*, ¶1465A.

[176] The army could not withhold separation allowances from families that refused to relocate to West Africa. *Regulations Relating to the Issue of Army Allowances, 1884*, ¶127.

[177] West African units of the Colonial Corps were a battalion of infantry, a company of fortress engineers, and a company of garrison artillery. *Monthly Army List, July 1904*.

[178] *Hart's Annual Army List, 1905*.

guards and household cavalry were sent abroad only in wartime.[179] The Indian Army had its own corps of engineers and support departments such as medical and ordnance. Accordingly, only line infantrymen, line cavalry troopers, and artillery gunners were subject to Indian postings (83.0% of enlisted personnel.)[180]

Death and Disability

Unlike civilians, soldiers live with the risk of combat death or disability. In common with civilians, soldiers when not engaged in hostilities, have a risk of death or disability from disease and accident. British soldiers' mortality and disability rates not related to war; however, varied greatly from those for their civilian counterparts (males age twenty through forty-four).

Mortality Rates

Shortly after the Napoleonic Wars ended, the annual mortality rate for soldiers stationed in the United Kingdom was 40% higher than for their civilian counterparts. This difference then narrowed progressively until the end of the Crimean War when parity was achieved. Afterwards, the mortality rate for soldiers declined in relationship to that for civilians until by Bloomsday the annual mortality rate for soldiers at home was 40% lower than their civilian counterparts.

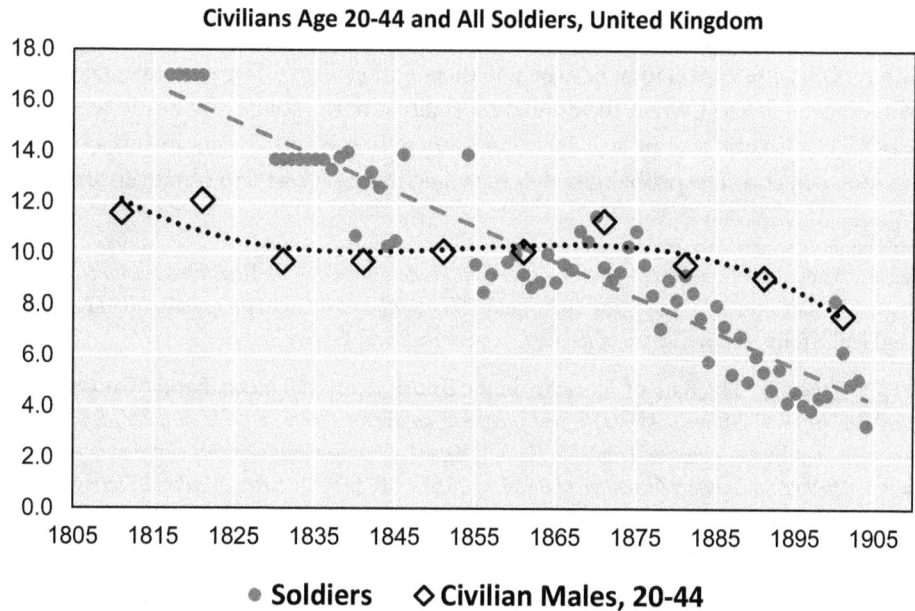

---

[179] Cairnes, *Social Life in the British Army*, 25.

[180] *Annual Army Report, 1904*, 1905, [Cd. 2268].

In the preceding graphic the broken lines represent ordinary least square data fits; straight line for soldiers and cubic polynomial for civilian. Note that for forty-five years, 1830 to 1875, the mortality rate for civilians was nearly constant.[181]

For the typical recruit, the disparity in mortality rates was often greater than for the army as a whole. In the early part of the nineteenth century, most recruits came from rural areas where annual, overall mortality rates were lowest: 2.2% for agricultural districts, 2.4% for all England & Wales, 3.3% for the five largest urban areas. Accordingly, the mortality rate for young men from the countryside likely doubled upon enlistment. At the end of the nineteenth century, the typical army recruit was an urban slum-dweller. In 1904, the overall annual mortality rate for urban counties was 1.7%; rural counties 1.2%. Enlistment at that time likely halved the annual mortality rate for young urban men.[182]

The lack of decline in mortality for the civilian population during most of the nineteenth century was due to the shift in population from rural areas to the slums of industrial cities. Note that the civilian death rates through 1851 are for England and Wales only; 1861 for Great Britain. For 1871 and later the civilian death rates are for the entire United Kingdom as beginning in 1864, the law required death registration in all three constituent nations.

Few soldiers of the Victorian Era army spent their entire careers in the United Kingdom. On average, a soldier was abroad for half his military service.[183] For most foreign stations mortality rates were higher than at home and much higher in India, the West Indies, and West Africa. Table 32 on the following page shows army mortality rates for the UK, India and two colonial stations from the end of the Napoleonic Wars to shortly before the First World War. Note from Table 32 that at the beginning of the nineteenth century the mortality rate for British soldiers in India was 5.1 times that for soldiers in the United Kingdom but by the early twentieth century that difference had narrowed to a factor of 2.6.

The disparity in annual mortality in the early nineteenth century had a dramatic effect on the difference in life expectancy between soldiers and civilians. For example, take a 20-year old who enlisted in 1811 when the annual mortality rate for soldiers at home was 1.7% and in India 8.7%. Assume that he will serve 14 years at home and 7 years in India before he is eligible for a pension. The probability that he would live to age 41 and obtain an army pension

---

[181] Civilian: Annual civilian deaths prior to the *First Annual Report of the Registrar-General of England (1838)* are estimates from Wrigley and Schofield, *The Population History of England, 130-42*. Data from 1841 and later from *Annual Reports of the Registrar-Generals* and *Census Reports*.

Military: Chaplin, "The Rate of Mortality in the British Army 100 Years Ago;" War Office: *Statistical Reports on the Sickness, Mortality, and Invaliding among the Troops in the United Kingdom, the Mediterranean, and British America*, 1852-53, [1639]; *Report of the Commissioners appointed to inquire into the Regulations Affecting the Sanitary Condition of the Army*, 1857-58, [2318]; Army Medical Department: *Statistical, Sanitary, and Medical Reports for the year 1860*, 1862 [3051]; *Army Medical Department Reports - for 1885*, 1887 [C. 5128], *for 1904*, 1906 [Cd. 2700].

[182] *Annual Reports of the Registrar-General for England*, 1838, 1904; Spiers, *The Army and Society*, 48-49, 59-60.

[183] Each year, 1879 through 1904, excluding the Boer War years, about half the army was stationed abroad. *Annual Army Reports: for 1898*, 1899, [C. 9426]; *for 1903*, 1904 [Cd. 1904]; *for 1904*, 1905, [Cd. 2268].

was only 41.6%.[184] In 1811, a 20-year old, male English civilian had a 78.3% probability of living to age 41.[185] In 1901, under the same assumptions, a 20-year old recruit had an 86.3% probability of living to age 41 (twice what it was in 1811); the civilian an 85.2% probability.[186]

Table 32.
British Army Annual Mortality Rates
At Home and Selected Stations Abroad

	1817-1821		1859		1876-1885		1904	
	Pct.	X UK	Pct.	X UK	Pct.	X UK	Pct.	X UK
UK	1.7		1.0		0.7		0.3	
India	8.7	5.1	3.5	3.5	1.6	2.3	1.1	3.7
Gibraltar	1.8	1.1	0.8	0.8	0.7	1.0	0.3	1.0
Malta	2.5	1.5	1.9	1.9	1.0	1.4	0.6	2.0
Egypt	NA	NA	NA	NA	* 1.1	* 1.6	0.6	2.0

* For 1885 only.

Sources: See note 181.

Medical Discharge Rates and Disability Pensions

In the British Army of the Late-Victorian Era, during periods of no major hostilities, 1.5% to 2.0% of enlisted men were invalided out of the service annually. No more than ten percent of such soldiers received permanent pensions, while about one-third to one-half received temporary pensions of from 18 to 39 months. About half of invalided soldiers received nothing from the army or a lump sum "gratuity" of up to £30. Table 33 shows the number of soldiers invalided-out annually, and the number and nature of disability pension.

Table 33.
Enlisted Men Invalided Out of the Army
1864, 1874, 1884, 1894, 1904

	Strength	Invalided Out	%	Perm. Pension Number	%	Temp. Pension Number	%	Gratuity / Nothing Number	%
1904	279,303	4,973	1.8	494	9.9	2,914	58.6	1,565	31.5
1894	210,863	3,152	1.5	76	2.4	1,184	37.6	1,892	60.0
1884	181,297	2,962	1.6	180	6.1	1,197	40.4	1,585	53.5
1874	188,379	3,996	2.1	260	6.5	1,910	47.8	1,826	45.7
1864	206,014	4,788	2.3	NA	NA	3,417	71.4	NA	NA

War Office Sources: *Return of the Number of Men who Retired from the Army*, 1905, H.C. Accounts & Papers, No. 318; *Annual Army Reports - 1904*, 1905, [Cd. 2268], *1894*, 1895 [C. 7885], *1884*, 1885 [C. 4570], *1874*, 1875 [C. 1323]. Permanent pensions for 1874 and 1884 are estimates.

---

[184] $(1-.017)^{14} \times (1-.087)^7 = 0.416$.

[185] $(1-.0116)^{21} = 0.7827$.

[186] $(1-.005)^{14} \times (1-.011)^7 = 0.8628$ for the soldier; $(1-.0076)^{21} = 0.8520$ for the civilian.

Warrant officers (sergeant-majors, bandmasters, *etc.*) received permanent disability pensions regardless of the cause of the infirmity (combat, service-connected, and non-service connected). For all other enlisted men, permanent disability pensions were awarded only for

(1) combat-connected disabilities (regardless of a soldier's years of service), and

(2) non-combat, service-connected disabilities for a soldier with 14-years' service.

The regulations defined combat casualties to include diseases contracted in connection with war operations. Table 34 summarizes the disability pension scheme for enlisted personnel. The pension amount was determined by the War Office based on the cause and extent of the disability. The maximum benefit went to those the Medical Corps deemed unfit for any employment. Warrant officers with at least five years in rank received £64 annually regardless of the disability.

Table 34.
Annual Disability Pensions for Other Ranks

	Combat		Line-of-Duty		Other	
	Min.	Max.	Min.	Max.	Min.	Max.
WOs 5+ years	£ 64	£ 64	£ 64	£ 64	£ 64	£ 64
WOs <5 years	18	64	12	27	12	18
All Sergeants	18	64	12	27	0	18
Corporals	14	55	9	23	0	18
Privates	9	46	9	18	0	18
Term	Life		1.5 yrs.	Life	Life	

Source: *Royal Warrant for the Pay, 1907*, Arts. 714, 1085-90.

Soldiers with fewer than fourteen-years' service who were invalided-out for non-combat, service-connected disability received temporary pensions as follows: 18 months if fewer than six years service plus an additional 3 months for each year of service over six years (maximum 39 months). When a veteran's temporary pension was soon to expire, an examining military physician could extend the pension for a second term. Near the end of the second term, the War Office had discretionary power to make the temporary pension permanent for ex-soldiers of "good character."

The War Office, with Treasury concurrence, could award special pensions to disabled veterans not in receipt of an army pension. The maximum annual benefit was £18.[187] This special pension provision applied to temporary pensioners whose benefit term expired and to soldiers invalided-out for non-service connected disabilities.

Except for the relative handful of veterans with combat-related disabilities, pension amounts for invalided soldiers were inadequate.[188] The state did not provide veterans with

---

[187] *Royal Warrant for the Pay, 1907*, Art. 1090.

[188] Skelley, *The Victorian Army at Home*, 52.

medical care so former soldiers with chronic ailments either had to pay for care or receive whatever treatment was available for paupers. Soldiers of all ranks (other than warrant officers) with a non-service-connected disability could receive no more than £18 annually, which was also the maximum for privates with a non-combat, service-connected disability. Generally, disabled veterans incapable of working relied on family, charity, and begging to support themselves. Many disabled ex-soldiers became permanent patients in workhouse hospitals.[189]

The Health Risk of Army Service

A good indicator of the additional health risk a young man incurred upon enlistment is a comparison of the death and disability rates of the peacetime Regular Army and the Army Reserve. Both groups are young and entry into both organizations requires good health (soldiers underwent a medical inspection prior to transfer to the reserves). For the twelve months ended September 30, 1904, the Regular Army lost 2.4% of its strength to death and disability (mortality 0.6%, disability 1.8%). The Army Reserve lost only 0.7% (mortality 0.2%, disability 0.5%). As reservists hold civilian occupations, this data probably gives the best indication of the additional health hazard faced by soldiers.

Some of the disparity in attrition between regulars and reservists may be due to the difference in age distribution among the two groups. In 1904, about one-fourth of regular soldiers were below age twenty-one; no reservists were that young. Following is the age distribution for the two forces in 1904:

	Under 21	21 - 29	30 - 34	35 - 39	40 +
Regulars	25.8%	60.4%	8.8%	4.0%	1.0%
Reservists	0	67.1	30.3	2.4	0.2

Time Off: Passes, Leave, and Furloughs

Soldiers were required to remain in barracks, camp, or garrison unless on furlough, on leave, or in receipt of a pass. Note that "barracks" was an urban or town army facility and not an individual building. A dormitory was known as a "barrack," the singular of barracks. Confinement to a barrack was one of several punishments for minor infractions of rules.

By custom, all soldiers had passes to be out of barracks on Saturdays after end of the duty-day, and Sundays after church parade. Soldiers with the rank of sergeant and higher had passes for weekdays.[190] Corporals and privates could receive weekday passes at the discretion of the commanding officer. Corporals and privates out on pass had to return to barracks in time for the 9:30 pm roll call; sergeants and higher could remain out until midnight. When out of barracks on pass, all soldiers had to remain in station (the town or city in which their

---

[189] Ibid.

[190] *King's Regulations, 1908*, ¶1818.

unit was quartered).[191] During the Boer War, Maud Gonne, feminist, Irish nationalist, and anti-recruiting activist, complained vociferously of the Dublin Garrison policy to grant weekday passes to all soldiers. She viewed the liberal pass policy as a plague inflicted by the British on Dublin's residents, especially young women.[192]

Authorization to be out of barracks overnight was known as "leave" which a commanding officer could grant for a maximum of six consecutive nights.[193] Leave was rare and usually granted as a reward for outstanding service or for compassionate reasons (death in the family, urgent family legal matters, *etc.*).

Extended time away from barracks was known as "furlough." After the Boer War, the army reduced the furlough season from five months to four, October 1st through January 31st. The regulations did not specify the length of annual furlough; they only limited the number of men on furlough to 25% of each rank.[194] Furloughs were from four to six weeks.[195] The only restriction on a soldier's right to annual furlough was that he had to be out of debt to the War Office, regimental institutes, and regimental tradesmen.[196] Soldiers stationed abroad for six consecutive years received a furlough of about two months plus passage home and back.[197]

On leave and furlough, all enlisted men could wear civilian clothes. When out of barracks on pass, only soldiers of rank colour-sergeant and higher could be out of uniform.[198] Worn by soldiers when off-post was the "review" uniform of rust-red tunic, dark trousers, and cap. Though not mandated by army regulations, regimental rules and custom dictated that enlisted men, when off post, carry a swagger stick. This was an oddity in that on-duty, only commissioned officers, warrant officers, and company sergeant-majors carried swagger sticks. Joyce was aware of the swagger stick custom as in *Ulysses* he describes Privates Carr and Compton as carrying such accoutrement.[199]

---

[191] Ibid.

[192] U (Gabler) Lotus Eaters 5:70; Gifford and Seidman, *Ulysses Annotated*, 86. In an open letter to Gonne, Alfred Webb praised her efforts to "free some of our principal thoroughfares from being occupied to such a large extent by military, and their too often foolish companions, or worse, in the evening, and far on into the night." *Freeman's Journal*, June 3, 1904.

[193] *King's Regulations, 1908*, ¶1819.

[194] *Queen's Regulations, 1899*, ¶1935; *King's Regulations, 1908*, ¶1296.

[195] Murray, *Six Months in the Ranks*, 124; Wyndham, *The Queen's Service*, 77.

[196] *King's Regulations, 1908*, ¶1301.

[197] Ibid., ¶1308.

[198] Ibid., ¶1694.

[199] "Private Carr and Private Compton, swaggersticks tight in their oxters …" U (Gabler) Circe 15:48-49. Oxter is a term for armpit "mostly North of England and Scottish but used occasionally in other parts by obscurantists." Norman Schur, *British English A to Zed* (New York: Skyhorse, 2013), 256.

Privates in Review Uniform with Swagger Sticks

By Mayger (London), c. 1898   By W.G. Wise (Wiltshire), c. 1903

The Reserves

The Old Army Reserve (1859-1869)

The Army Reserve was established by Parliament in 1859 and consisted initially of two classes.[200] Reservists of the 1st Class were soldiers released by the army prior to completion of their twelve-year engagement. Reservists of the 2nd Class were veterans who completed their initial engagement and agreed to be subject to recall. This class included pensioners. 1st Class reservists received 2d. daily plus an annual allowance of £1. If recalled to service, they could be deployed worldwide. 2nd Class reservists received ½d. daily plus the same annual allowance. If mobilized, such reservists could only serve in the United Kingdom.

When the reserve scheme was implemented, 1st Class reservists trained 40 days annually; 2nd Class 12 days[201] By 1870, training was no longer authorized for 1st Class reservists while training was voluntary for 2nd Class reservists and about half did so each year. Reserve service

---

[200] Reserve Forces Act, 1859, 22 & 23 Vict, c. 42; amended by the Reserve Forces Act, 1867, 30 & 31 Vict., c. 110.

[201] War Office, *Army Estimates of Effective and Non-Effective Services for 1870-71*, 1870, H.C. Accounts & Papers, No. 36.

by non-pensioners in the 2nd Class reckoned towards pension eligibility; two years reserve time counting as one year of regular service. Accordingly, reservists of this class would be eligible for a pension after 18 years in the reserve. As reservists did not serve for fixed terms, and there was no age limit on reserve enrollment, many veterans remained reservists well into their 60s. In 1890, about one-in-seven of all longevity pensions were awarded to 2nd Class reservists.[202]

Reservists could be mobilized only by Royal Proclamation; in effect, by order of Parliament. The 1st Class had to be recalled fully before the 2nd Class could be mobilized. At the beginning of 1870, the 1st Class Reserve had 1,939 men, the 2nd Class 18,578 (3,885 single-engagement men and 14,693 pensioners).[203] If recalled to service, a reservist could be assigned to any unit within his corps or department.

The Reserve Forces Acts of 1859 and 1867 failed to create a large pool of trained veterans to supplement the Regular Army during time of war. The War Office granted early release to very few men. Those released were typically unsuitable for military service, but their shortcomings were not egregious enough to warrant discharge.

The Reorganized Army Reserve (1870-1913)

The Army Reserve was reorganized significantly under the Army Enlistment Act of 1870.[204] Under authority of the Act, the War Office divided the 1st Class Reserve into four sections, each for a specified type of reservist.[205]

Section A: These were short-service men who volunteered to be recalled at the discretion of the War Office during their first year of reserve service. Such reservists received retainer pay of 1s. daily. At the expiration of their one-year term, these reservists were transferred to Section B. Section A enrollment was limited to 5,000 men.

Section B: Consisted of the bulk of short-service men. They remained in reserve status until expiration of their twelve-year, total engagement. Section B reservists received retainer pay of 6d. daily.

Section C: These were full-service men released from regular service prior to expiration of their twelve-year engagement. Like Section B reservists, they received 6d. daily. In 1904, the War Office abolished Section C. Soldiers released early from regular service were classified as Section B reservists.[206]

---

[202] Review of 230 pension records, July-September, 1890. UK National Archives, WO 117/44/2.

[203] War Office: *Detailed Statement of the present Strength of the Army Reserve*, 1870, H.C. Accounts & Papers, No. 104.

[204] 33 & 34 Vict., c. 67.

[205] Grierson, *The British Army*, 26; *Royal Warrant for the Pay, 1906*, Art. 1289.

[206] *Annual Army Report, 1904,* 1905, [Cd. 2268].

Section D: These were men who volunteered for further service after completion of their twelve-year engagement, either with the Army Reserve or the Regular Army. Initial service in Section D was under a four-year enlistment. Re-enlistment was at the discretion of the War Office and allowed only for specialists. Section D reservists received 6d. per day and could not serve past age 50.

At the outbreak of the Boer War, the 1st Class Reserve totaled 78,839 men. On September 30, 1903 after the Boer War demobilization, there were 69,148 reservists.[207] From then on, the reserve increased progressively until it totaled 146,756 men at the end of the reporting year immediately prior to the First World War (September 30, 1913).[208]

When the War Office reorganized the 1st Class Army Reserve, it barred non-pensioner enrollment in the 2nd Class Army Reserve. From 1870 onwards, soldiers who completed twelve-years' service with the regulars and wished to serve in the reserve, had to enlist in Section D of the 1st Class Reserve. In 1881, the War Office made all pensioners subject to recall and accordingly, barred new enrollment in the 2nd Class Reserve.[209] On January 1, 1896, the 2nd Class Reserve had only 111 men and by Bloomsday it was extinct.[210]

Reserve Service

In 1904, reserve service was governed by the Reserve Forces Act of 1882.[211] Reservists were subject to recall for military service by Parliament and call-out in aid of the civil power by a magistrate (the Lord Lieutenant in Ireland). The Army Reserve was mobilized three times prior to Bloomsday: 1878 for an expected war with Russia, 1882 for the Egyptian Expedition, 1899 for the 2nd Boer War. Reservists were also subject to periodic medical inspection and training. After having spent eighteen-months in reserve status, a reservist could be called-out annually for twelve-days training with the army or militia, or ordered to attend 20 drills (2 to 4 hours each) with a volunteer force unit. For annual training, reservists were paid at the army rates; for each drill attended they received 4d. If called upon to aid the civil authority, reservists received army pay plus a bonus of 6d. daily.[212]

When initially constituted, reservists were required to train for up to 40 days annually. Successive governments progressively lowered the amount of mandatory training until by 1870, training was voluntary. Usually no training was available due to lack of a Parliamentary appropriation. The reserve effectively became dormant because governments did not want to inconvenience reservists' employers and feared ex-soldiers would not be able to obtain

---

[207] Ibid.

[208] Board of Trade, *Statistical Abstract for the United Kingdom: 1901-1915*, 1917, [Cd. 8448].

[209] *Army Circulars for 1881*, Clause 68.

[210] *Annual Army Report, 1898,* 1899, [C. 9426]; *Annual Army Report, 1904,* 1905, [Cd. 2268].

[211] 45 & 46 Vict., c. 48.

[212] *Royal Warrant for the Pay, 1906*, Arts. 1295-97.

civil employment if they were routinely called up for training. This concern that periodic reserve training would adversely affect recruiting (prospective soldiers did not want to be unemployable upon release from full-time service) dates to the 1860s.[213]

In 1904, there was a limited compulsory training requirement for infantry reservists of Section B. For those who had fewer than three-years' full-time service, there was a one-day musketry drill in the 5th, 7th, 9th, and 11th year of the initial engagement. For others, this training was required only in the 10th year of their 12-year engagement. The musketry drill consisted of one hour of instruction and the firing of 28 rounds of ammunition. There was also a voluntary training program for infantry reservists of six days per year. No more than 30 Section B reservists could train for six days annually with each infantry militia battalion.[214] As there were 124 militia infantry battalions, a maximum of 3,720 reservists could undertake voluntary, annual training (7% of 53,166 infantry reservists).[215]

For many veterans released recently from full-time service, reserve pay was their only source of regular income. In the 1901 Census of Ireland, 4,151 respondents listed "Army Reserve" or "Reservist" as their occupation; 699 in County Dublin, 617 in Counties Antrim and Down (Belfast). In the 1911 census, 4,487 Irish respondents gave reservist as their occupation; 1,179 in County Dublin, 642 in Counties Antrim and Down.

Army Reserve Reported as Occupation

At the outbreak of the First World War in 1914, 17,804 reservists in Ireland were recalled to service; 12.25% of total army reservists.[216] Three years before the war, the Army Reserve totaled 137,682 men which indicates about 17,000 were Irish residents.[217] Accordingly, in

---

[213] House of Lords speech by Field Marshal Hugh Rose (1st Baron Strathnairn), who served 50 years with the army, and response by former War Minister Edward Cardwell (1st Viscount Cardwell). 231 Parl. Deb. (3d ser.) (1876) 670-88.

[214] Army Orders, 1904, No. 64.

[215] *Annual Army Report, 1904,* 1905, [Cd. 2268].

[216] Dublin Castle, *Report on Recruiting in Ireland*, 1916, [Cd. 8168]; War Office, *Statistics of the Military Effort of the British Empire During the Great War* (London: HMSO, 1922).

[217] Board of Trade, *Statistical Abstract for the United Kingdom: 1901-1915*, 1917, [Cd. 8448].

1911 about 27% of Irish reservists were without regular employment. The indicated unemployment rate for May 1901, is 44%; however, that figure includes thousands of reservists recently released from Boer War service. The illustration on the previous page is of a census return for a household of three young siblings and their cousin who was a 26-year-old army reservist.

Reservists were managed by the regimental territorial district for the area in which they resided. They had to report to the district staff quarterly and failure to do so for two consecutive quarters caused forfeiture of six-months' pay.[218] Reservists of Section A received about £18 annually; reservists of the other sections £9. All reservists were given a travel allowance if ordered to report for training or inspection more than five miles from their registered place of abode. Reservists also received a travel allowance to enlist or re-enlist in Section D.[219]

In general, reservists had to remain in the United Kingdom. With permission of the territorial district commanding officer, a reservist could be employed at sea or travel abroad. With War Office permission, he could reside in India or any colony with a British garrison.[220] In 1910, 7,198 of the then 131,267 army reservists resided abroad (5.5%).[221]

Time with the reserve counted towards pension eligibility but only for recalled or re-enlisted reservists.[222] After three years of full-time service, reservists who had returned to the colours could qualify for a reduced pension (£12 annually if a private, £23 a corporal, £27 a sergeant).[223]

Reservists with at least fourteen years' total service (regular and reserve) who were discharged as medically unfit received gratuities (lump sum payments) of from £3 to £6 depending on years of service.[224]

Termination of Service and Pensions

A soldier's military life ended for a variety of reasons other than retirement, time expiration, and medical unfitness. Soldiers could be released from service by purchase, compassionate discharge, and for misconduct. Men were also discharged if their commanding officer found them "unlikely to become efficient soldiers." Finally, soldiers could unlawfully leave the army through desertion. Desertion was somewhat common; however, about two-thirds of deserters returned to the army voluntarily.

---

[218] *Royal Warrant for the Pay, 1906*, Art. 1298A.

[219] *Regulations for the Army Reserve, 1904*, §§94, 96.

[220] Ibid., §§54, 92.

[221] Statement of Bron Herbert (Lord Lucas), Under-Secretary of State for War, 6 H.L. Deb. (5th ser.) (1910) 677-80.

[222] *Royal Warrant for the Pay, 1906*, Art. 1146.

[223] Table 35, *infra*.

[224] *Royal Warrant for the Pay, 1906*, Art. 1200.

Army Life and Retirement: Other Ranks

In 1904 only about half the men who left the Regular Army did so because of expiration of their service obligation, while about one-in-seven died or were invalided out, and about one-in-ten left through misconduct (dismissals and net desertions). That same reporting year, 10,616 reservists fulfilled their total military obligation and were accordingly discharged (119 as invalids).[225]

Decrease in Regular Army Soldiers, 1904

	Number	Pct. of Releases
Completion of Service [226]	24,393	54.2
Purchased Discharge	2,456	5.4
Compassionate Discharge	385	0.8
Other Releases, Short-Service	492	1.1
Other Releases, Long-Service	1,505	3.3
Sub-Total, Early Termination	4,838	10.6
Death	1,699	3.8
Invalided-Out	4,973	11.0
Sub-Total, Death and Disability	6,672	14.8
Dismissal, Lack of Aptitude	1,373	3.1
Dismissal, Misconduct	3,830	8.5
Desertion	3,959	8.8
Returned from Desertion	(2,673)	
Sub-Total, Soldier Shortcomings	9,162	14.4
Total Released from Regular Army	45,065	

Discharge by purchase was available to both regulars and reservists. During peacetime, most soldiers who wished to buy out of the army needed only their commanding officer's approval. Approval was given liberally. For soldiers with special skills and orders for assignment abroad, approval of the general-officer-commanding was required. Such approval was not perfunctory and frequently not given. For nearly all soldiers, the discharge purchase price was £18. The War Office charged highly skilled machinery artificers £30 for their discharge.[227] Reservists with only three-years' full-time service had to pay £25.

---

[225] *Annual Army Report, 1904,* 1905, [Cd. 2268].

[226] Transfer to reserve, longevity pension, and completion of 12-years' full-time service.

[227] *King's Regulations, 1908,* ¶¶391-96; *Royal Warrant for the Pay, 1908,* Arts. 1056-60.

Terminal Furlough and Discharge Gratuities

Two months before release from regular service, nearly all time-expiring soldiers received terminal furlough. As soldiers received furlough pay in advance, this early release effectively provided some invalided reservists and almost all regulars with a discharge bonus.[228] In addition to this extra remuneration, released soldiers could receive a gratuity. Following is a schedule of such "gratuitous" payments:[229]

Reservists found medically unfit after 14 or more years service:	£3 - £ 6
Regulars who served 3 or more years, £1 per year, max. of £12:	£3 - £12
Regulars who served fewer than 3 years:	£2

Accordingly, an infantry corporal who was released from regular service after seven years received £14 15s. (furlough pay and rations allowance plus release gratuity). Privates after three years received £7 10s.

Longevity Pensions

Soldiers with 21-years' service had the right to retire; those with 18-years' service could retire with War Office permission. Soldiers who were discharged involuntarily due to a reduction-in-force received a reduced pension if they had 14-years' service.[230] Very few soldiers received a service longevity pension. After the introduction of short-service, most recruits had no intention to make the army a career. For the remainder, death and disability precluded 21-years' service. Also, many soldiers, upon completion of their initial engagement of twelve-years did not receive permission to re-engage. This was usually due to budgetary constraints, but commanding officers sometimes withheld permission to remain in service for soldiers who were disciplinary problems. During the Late Victorian Era, only about 6% of enlistees remained with the army long enough to receive a pension.[231]

Table 35 on the following page shows pension amounts for soldiers retiring on or about June 16, 1904. Minimum amounts are for soldiers with 21-years' total service of which 3-years were immediately prior to discharge. Soldiers with 30-years' service received the maximum amount for their rank. Generally, only warrant officers (sergeant-majors in the infantry) remained with the army for 30 or more years.

---

[228] *King's Regulations, 1908*, ¶1310.

[229] *Royal Warrant for the Pay, 1908*, Arts. 1118, 1046-47, 1152.

[230] After 18 years a soldier could be pensioned off "for the benefit of the public service." Ibid., Art. 1067.

[231] For the years 1882 through 1884, annual enlistments averaged 30,850; 1,784 longevity pensions were awarded in 1904. For the years 1872 through 1874 and 1894, the respective numbers were 18,542 and 1,243. War Office: *Return of the Number of Officers, Non-Commissioned Officers, and Men who Retired from the Army (time expired only)*, 1905, H.C. Accounts & Papers, No. 318; *Annual Army Report, 1884*, 1885, [C. 4570].

Table 35.
Annual Pension Amounts
Soldiers Retiring in 1904-1905, Infantry Rank Titles

	Pct.	Minimum	Maximum	Typical
Sergeant-Major	9.6	£ 64	£ 82	£ 68
Quartermaster-Sgt.	12.4	36	64	50
Colour-Sergeant	31.6	32	59	45
Sergeant	16.8	27	55	30
Corporal	10.8	23	44	25
Private	18.8	12	33	18

Sources: *Royal Warrant for the Pay, 1906*; UK National Archives, WO 117/59.

Retirement for a soldier, unless he obtained employment, meant a decline in his standard of living. As indicated by Table 35, only the unmarried, senior sergeants retired with pensions that could provide a spartan, working class standard of living (about half of retirees). For soldiers with dependents, only those with the rank of quartermaster-sergeant or sergeant-major would have that standard of living (about one-fifth of retirees). Many retired corporals and privates resorted to charity, public relief, or begging simply to get by. Following is a comparison of annual army pensions, in pounds sterling, to selected working class wages for Dublin on Bloomsday.[232]

	Dublin Wages	Typical Army Pension	
Construction Tradesmen	75-110		
Police Constables	58 - 75	68	Sergeant-Major
Factory Workers	40 - 70	45/50	Qtrmstr.-Sgt. / Clr.-Sgt.
Laborers	30 - 55	30	Sergeant
		25	Corporal
		18	Private

Note that sergeant-majors received in retirement less money than that earned by a lower-paid, journeyman carpenter.

The Royal Military "Hospitals"

The War Office maintained two retirement homes for destitute, elderly army pensioners: The Royal Military Hospital, Chelsea (London) and The Royal Military Hospital, Kilmainham (Dublin). Both were founded by Charles II and modeled on *Les Invalides* in Paris. The Dublin

---

[232] *Reports of an Enquiry by the Board of Trade into the Earnings and Hours of Labour of Workpeople of the United Kingdom in 1906,* 1910, [Cd. 4844, 5086, 5196], 1912 [Cd. 6053], 1913 [Cd. 6556]; *Estimates for Civil Services for the Year Ending 31 March 1907*, 1906, H.C. Accounts & Papers Nos. 71-II, 272-VIIII.

facility opened in 1684, the London facility in 1686. Residents forfeited their pensions but received pocket money which in 1904 was 1d. per day.[233]

In 1904, Chelsea Hospital had 558 residents, a staff of 85, and an annual budget of £32,940 (£59 per resident); Kilmainham 137 residents, staff of 20, and budget of £6,760 (£49 per resident).[234]

Upon creation of the Irish Free State, the War Office relocated the Kilmainham Hospital residents to Chelsea Hospital. The building and grounds are currently owned by the Office of Public Works, Republic of Ireland, and house a Royal Hospital museum, a park, an event center, banquet rooms, and the National Museum of Modern Art.[235] The Chelsea Hospital still functions as an army retirement home and in 2020 had about 300 residents.[236]

---

[233] Skelley, *The Victorian Army at Home*, 208.

[234] War Office, *Army Estimates for the Year 1904-05*, 1904, H.C. Accounts & Papers, No. 73.

[235] Website of the Royal Hospital Kilmainham, www.rhk.ie.

[236] Website of the Royal Hospital Chelsea, www.chelsea-pensioners.co.uk.

Chapter Bibliography

Arnold-Foster, H.O. *The Army in 1906*. London: Murray, 1906.

Barnes, Alison. "History of SSAFA." Soldiers, Sailors, and Airmen Family Association, www.ssafa.org.uk.

Bowman, Timothy and Mark Connelly. *The Edwardian Army: Recruiting, Training, and Deploying the British Army, 1902-1914*. Oxford: Oxford Univ. Press, 2012.

Burn, J. and J. McDonald, "An Investigation into the Rates of Re-Marriage and Mortality amongst Widows in receipt of relief from the Patriotic Fund." *Journal of the Institute of Actuaries* 38, no. 5 (July 1904):433-501.

Buxton, J.W. *The Elements of Military Administration*, Part I. London: Keegan, Paul, Trench, 1883.

[Cairnes, William E.]. *Social Life in the British Army*. New York: Harper, 1899.

—— *The Army from Within*. London: Sands, 1901.

Chaplin, Arnold. "The Rate of Mortality in the British Army 100 Years Ago," *Proceedings of the Royal Society of Medicine* 9 (1916): 89-99;

Clarke, Howard. "The Royal Hibernian Military School in Dublin," *Journal of the Genealogical Society of Ireland* 13 (2012): 16-29.

—— Note 1819, The Royal Hibernian Military School, *Journal of the Society for Army Historical Research* 85, no. 341 (Spring 2007): 85-87.

Cockerill, Arthur W. *Sons of the Brave*. London: Cooper and Martin, Secker, Warburg, 1984.

—— *The Charity of Mars – a history of the Royal Military Asylum*. Coburg, Canada: Black Cat, 2002.

Corcoran, Tony. *Guinness: the Greatest Brewery on Earth - Its History, People, and Beer*. New York: Skyhorse, 2009.

Edmondson, Robert. *John Bull's Army from Within*. London: Griffiths, 1907.

French, David. *Military Identities: The Regimental System, the British Army, and the British People c. 1870-2000*. New York: Oxford Univ. Press, 2005.

Gifford, Don with Robert J. Seidman. *Ulysses Annotated*. Berkeley: Univ. of Calif. Press, 1988.

[Grierson, James Moncrieff]. *The British Army*. London: Sampson, Low, Marston, 1899.

Leeson, D.M. "Playing at War: The British Military Manoeuvres of 1898," *War in History* 15, no. 4 (2008): 432-61.

Mansfield, Nick. *Soldiers as Workers*. Liverpool: Liverpool Univ. Press, 2016.

[Murray, Grenville]. *Six Months in the Ranks*. London: Smith, Elder, 1881.

Rosenbaum, S. "More than a Century of Army Medical Statistics," *Journal of the Royal Society of Medicine* 83 (July 1990): 456-63.

Skelley, Alan Ramsay. *The Victorian Army at Home*. London: Croom Helm, 1977.

Spiers, Edward M. *The Army and Society 1815-1914*. New York: Longman, 1980.

—— *The Late Victorian Army, 1868-1902*. Manchester: Univ. of Manch. Press, 1992.

Trustram, Myna. *Women of the Regiment*. Cambridge: Cambridge Univ. Press, 1984.

Ward, B.R. *The School of Military Engineering, 1812-1909*. Chatham, UK: Royal Engineers Institute, 1909.

Wheeler, Owen. *The War Office Past and Present*. London: Methuen, 1914.

Williams, N.T. St. John. *Tommy Atkins' Children*. London: HMSO, 1971.

Wrigley, E.A. and Roger Schofield. *The Population History of England, 1541-1871: A Reconstruction*. Cambridge: Harvard Univ. Press, 1981.

Wyndham, Horace. *The Queen's Service*. London: Heinemann, 1899.

Chapter 8
Officers and Soldiers of the Auxiliary Forces

Prior to 1908, there were three part-time military forces in the UK: the militia, imperial yeomanry, and the volunteer force. In Ireland, there were no volunteer force units and only two imperial yeomanry regiments, each with an authorized strength of 488, all ranks. Accordingly, this chapter will concentrate on the militia. In 1904 the militia in Ireland had 18,050 enlisted men and 519 officers.[1]

The primary war-time role of the militia was replacement of home-based regular army units that were sent abroad. Its secondary military role was to reinforce the regular army in the unlikely event of imminent invasion. During peacetime, the militia stood available to quell disturbances and reinforce the regular army in case of rebellion. In 1904 a royal commission to examine the state of the militia and volunteer force (Norfolk Commission) concluded that "the Militia, in its existing condition, is unfit to take the field for the defence of this country" and that the volunteer force was unable "to face, with prospect of success, the troops of a Continental army."[2]

Enrollment in the militia ended in 1908 when the auxiliary forces were reorganized by the Liberal war minister, Richard Haldane, pursuant to the Territorial and Reserve Forces Act, 1907.[3] That statute created the Special Reserve into which most militiamen transferred, and merged the yeomanry and volunteer force to form the new Territorial Army. Unlike their predecessors, the new auxiliary forces were liable for service abroad.

In 1904, the militia operated under the Militia Act, 1882 as amended by the Reserve Forces and Militia Act of 1898 and the Militia and Yeomanry Acts of 1901 and 1902.[4] Militiamen, other than those who volunteered for the misnamed Militia Reserve, could not be compelled to serve abroad. Militia officers were at all times subject to military law; enlisted personnel only when embodied (called to full-time service) or in training.[5] Both officers and rankers were paid under the Regular Army pay schedules set forth in the Royal Warrants for the Pay of the Army.

---

[1] War Office, *Return showing the Establishment of Each Unit of Militia in the United Kingdom and the Numbers Present, Absent, and Wanting to Complete at the Training of 1904*, 1905, [Cd. 2432]. Hereafter cited as *Annual Militia Return*.

[2] *Report of the Royal Commission on the Militia and Volunteers*, 1904, [Cd. 2061], at 6-7. Hereafter cited as *Norfolk Commission Report*. The commission was headed by Henry Fitzalan-Howard, the 15th Duke of Norfolk. Norfolk served in the Boer War with the Imperial Yeomanry and was wounded in action.

[3] 7 Edw. 7, c. 9.

[4] 45 & 46 Vict, c. 49; 61 & 62 Vict., c. 9; 1 Edw. 7, c. 14; 2 Edw. 7, c. 39.

[5] Annual Army Acts. For 1904, 4 Edw. 7, c. 5, §§175-76.

Militia Officers

Commissioning

As in the Regular Army, militia officers were commissioned into a "regiment" which in Ireland were infantry battalions, garrison artillery brigades, and the militia of the Royal Army Medical Corps. In England, there were also "divisions" of harbor mining engineers (seven), "corps" of fortress engineers (two), and one militia field artillery brigade. A young gentleman who sought a militia commission required nomination by the lord lieutenant of the county in which his chosen militia unit was located.[6] Nominees required approval of the unit's commanding officer and the general officer commanding the military district in which the unit was located. The final appointment decision was made by the Office of Inspector-General for Auxiliary Forces.[7]

Militia Officers, the Landed Classes, and Officer Strength

Though property requirements for militia commissions were abolished by the War Office in 1869, commanding officers favored for appointment young men of the gentry and aristocracy.[8] A survey by the Norfolk Commission revealed that "gentlemen of independent means" accounted for three-fourths of officers excluding candidates for Regular Army commissions.[9] The questionnaire did not ask for the source of their income; however, it did ask for the number of officers that resided in the unit's recruiting territory. Only one-third of officers were local to their unit which indicates that at most, two-fifths of militia-intended officers were of the landed classes.[10]

From 1870 to the militia's demise in 1908, the senior ranks remained the domain of the landed classes. The social background of the captains and lieutenants; however, changed dramatically during that period.[11] Agriculture in the United Kingdom experienced an economic depression in the late nineteenth century as grain imports from North America progressively reduced domestic crop prices. As a result, the pool of landed families that could provide their sons with independent incomes decreased.[12] Accordingly, by the turn of the

---

[6] Militia Act, 1882. 45 & 46 Vict., c. 49, §6.

[7] Various Testimony, *Norfolk Commission Report, Minutes of Evidence*, 1904, [Cd. 2062, Cd. 2063].

[8] Abolition of the property requirement for militia commissions was one of the many reforms instituted by Edward Cardwell during his tenure as Secretary of State for War. French, *Military Identities*, 206-08.

[9] Summary of Answers to Circular of Questions, *Norfolk Commission Report, Appendices*, 1904, [Cd. 2064]. Responses were provided by all commanding officers of all 32 artillery, 9 engineer, and 124 infantry militia units.

[10] Army candidates accounted for 25.0% of officers and presumably were not of the 31.1% who were local to their regiment. Accordingly, at most 31/75 of officers were landed and intent upon a militia "career."

[11] Stoneman, "The Reformed British Militia," 85-86.

[12] French, *Military Identities*, 208.

twentieth century there were few young, idle country gentlemen of independent means interested in a militia commission. Reluctantly, county lieutenants and militia colonels turned to the middle class to fill the increasing number of vacancies in the officer ranks. By Bloomsday, the proportion of junior officers who were of the landed classes was half what it was in 1870.[13] Note that in nineteenth century Ireland, 7% to 8% of landed gentlemen took militia commissions. In the first decade of the twentieth century that proportion was 3%.[14]

Despite the militia's outreach to middle class young men, it was unable to fill its authorized, lieutenant positions. Though not in the public discourse of the time, the recent Boer War experience must have dissuaded many young gentlemen from entering the militia. During the war, all militia regiments were embodied, and few units remained in their home county. Not very many young men of means wanted to spend five months or more in an isolated garrison town, an industrial city center, or a small Mediterranean colony. Additionally, some units volunteered for service in the enteric fever laden, war zone of South Africa, a strong disincentive for acceptance of a militia commission. In 1904, only about one-half of infantry lieutenant positions were filled; three fourths of such artillery positions. Table 36 shows the shortfall in militia captains and lieutenants in 1904.

Table 36.
Militia Infantry and Artillery Junior Officers
Authorized and Actual, 1904

	Authorized	Actual	Pct.
England & Wales	1,612	1,109	68.8
Ireland	482	369	76.6
Scotland	279	173	62.0
Total for 123 Infantry Battalions	2,373	1,651	69.6
England & Wales	206	169	82.0
Ireland	177	136	76.8
Scotland	23	64	80.9
Total for 32 Artillery Brigades	456	369	80.9
Grand Total, Infantry & Artillery	2,829	2,020	71.4

War Office Sources: *Annual Militia Return, 1904*, 1905, [Cd. 2432]; *Army Estimates for the Year 1904-05*, 1904, H.C. Accounts & Papers, No. 73h. Review of all militia units in Ireland, *Monthly Army List, December 1904*.

Though the militia could not fill its junior officer positions, neither could the volunteer force and yeomanry. In that the volunteer force never had a nexus with county society, and few of its officers were of independent means, indicates that the economic decline of the

---

[13] Stoneman, "The Reformed British Militia," 87.

[14] Perry, "The Irish Landed Class and the British Army.".

landed classes may have had little to do with the militia's shortage of junior officers. Following is the infantry and artillery junior officer strength of the militia, volunteers, and the yeomanry, the volunteer-like cavalry auxiliary force.[15]

	Infantry	Artillery	Total
Volunteers	74.9	72.7	74.4
Militia	69.6	80.9	71.4
Yeomanry	NA	NA	74.1

Most young men who sought militia commissions planned to use the auxiliary force as a steppingstone to a Regular Army commission. The militia was the "backdoor" to the officer corps for young gentlemen who failed to gain entry to Sandhurst or Woolwich. Annually, the War Office awarded about 275 commissions to militia officers.[16] Selection was by semi-annual, competitive examination. Candidates had to be unmarried, under age 22 at the start of the examination year and have completed two annual trainings; under age 23 if completed three annual trainings. Candidates could sit for the examination three times.[17] Militia officers with "scarlet fever" who failed to be selected usually resigned their militia commissions.

Few regular commissions were available to young officers of the yeomanry and volunteer force. Had the militia not been the backdoor to the officer corps of the Regular Army, it would have had far fewer junior officers than it did. In the militia at the turn of the twentieth century, about one-half of infantry lieutenants and two-thirds of artillery lieutenants openly proclaimed their intention to enter the Regular Army.[18]

Attrition among junior militia officers was high and in Ireland, about 40% of the militia lieutenants on the regimental rolls on December 1902, were no longer there three years later. Note from the following table that attrition rates varied widely among units. In the infantry, there was even a substantial difference among battalions of the same regiment. For example, within the Royal Irish Regiment its 3rd Battalion (Wexford) had a 41.7% loss of lieutenants, its 4th Battalion (Clonmel) 20.0%, and its 5th Battalion (Kilkenny) 60.0%, three times the rate for the 4th Battalion. The recruiting area for that regiment, Army Territorial District 18, consisted of Counties Kilkenny, Tipperary, Wexford, and Waterford.

---

[15] Volunteer Infantry: 8,115 (711 senior) officers authorized, 6,254 actual; all senior positions filled. *The Annual Return of the Volunteers Corps of Great Britain, 1904*, 1905, [Cd. 2438].

Volunteer Artillery: 2,178 (203 senior) officers authorized, 1,639 actual; all senior positions filled. *Imperial Yeomanry Training Return, 1904*, 1905, [Cd. 2267].

Imperial Yeomanry: 1,538 (352 senior) officers authorized, 1,231 actual; all senior positions filled. *Army Estimates for the Year 1904-05*, 1904, H.C. Accounts & Papers, No. 73.

[16] MG A. Turner, Inspector-General of Auxiliary Forces, *Norfolk Commission Report, Minutes of Evidence*, 1904, [Cd. 2062], at q. 3750.

[17] *Regulations Under Which Commissions in the Army may be Obtained by Officers of the Militia, 1899. Norfolk Commission Report, Appendices*, 1904, [Cd. 2064], no. 8.

[18] Summary of Answers to Circular of Questions, *Norfolk Commission Report, Appendices*, 1904, [Cd. 2064].

Table 37.
Attrition Rates for Militia Lieutenants, Ireland
December 1902 to December 1905

	Dec 1902 LTs	Dec. 1905 LTs	CPTs	Resigned	Pct.
3/ Royal Irish Regiment	12	6	1	5	41.7
4/	5	2	2	1	20.0
5/	5	2	0	3	60.0
3/ Royal Inniskilling Fusiliers	9	5	2	2	22.2
4/	7	2	1	4	57.1
5/	8	4	0	4	50.0
8/ Royal Rifle Corps	6	1	0	5	83.3
9/	12	7	4	1	8.3
3/ Royal Irish Rifles	7	5	0	2	28.6
4/	6	1	3	2	33.3
5/	4	3	0	1	25.0
6/	6	3	1	2	33.3
3/ Royal Irish Fusiliers	11	2	3	6	54.6
4/	8	3	1	4	50.0
5/	8	2	1	5	62.5
3/ Connaught Rangers	6	1	2	3	50.0
4/	6	0	2	4	66.7
5/	11	6	2	3	27.3
3/ Leinster Regiment	6	4	1	1	16.7
4/	7	4	2	1	14.3
5/	9	6	0	3	33.3
3/ Royal Munster Fusiliers	8	6	1	1	12.5
4/	8	4	0	4	50.0
5/	8	4	0	4	50.0
3/ Royal Dublin Fusiliers	8	5	2	1	12.5
4/	11	5	0	6	54.6
5/	8	2	0	6	75.0
6/ The Rifle Brigade	11	4	3	4	36.4
Total for 28 Infantry Battalions	221	99	34	88	39.8
Antrim Artillery	8	6	1	1	12.5
Clare Artillery	8	4	2	4	50.0
Cork Artillery	8	5	1	2	25.0
Donegal Artillery	12	6	1	5	41.7
Dublin City Artillery	9	2	2	5	55.6
Limerick City Artillery	9	2	1	6	66.7
Londonderry Artillery	4	0	1	3	75.0
Mid-Ulster Artillery	5	2	2	1	20.0
Sligo Artillery	4	0	1	3	75.0
Tipperary Artillery	4	0	2	2	50.0
Waterford Artillery	5	1	1	3	60.0
Wicklow Artillery	8	4	0	4	50.0
Total for 12 Artillery Brigades	84	32	15	39	46.4
Total, Infantry & Artillery	**305**	**131**	**49**	**127**	**41.6**

Sources: *Hart's Annual Army List 1903, 1906*

Officer Training and Commissioned Service

Newly commissioned officers served full-time for five months during which they received their initial training. For infantry officers all training was informal; artillery officers had to spend two or three months on a gunnery course.[19]

Novice infantry officers trained through one of two programs: Preliminary drill or attachment to a regular infantry battalion. The preliminary drill program consisted of 49 days general training by the militia battalion's cadre, 14 days recruit musketry training, 27 days at the unit's regular annual training, and finally 60 days of on-the-job training with a regular infantry battalion. The attachment program consisted of 123 days at the regimental depot or the barracks of a regular infantry battalion plus the 27 days at the militia unit's annual training.

Like their infantry colleagues, new artillery officers trained through one of two programs. Their preliminary drill program consisted of 49 days general training by the militia unit's cadre, 14 days recruit gunnery training, 27 days at the unit's annual training, and finally 60 days on a gunnery course. The "schooling" program consisted of 30 days with a regular artillery brigade, the 27 days of the militia unit's annual training, and then 93 days on a gunnery course. The gunnery course was undertaken at branches of the School of Gunnery: Isle of Wight, Plymouth, and Sheerness on the River Medway, all fortified positions.[20]

After completion of initial training, no further training was required for infantry officers; however, lieutenants who desired promotion to captain had to complete the four-week officer's course at the School of Musketry. In 1903, 313 infantry lieutenants competed for the school's 121 places reserved for militia officers.[21]

Beginning in 1904 all militia artillery officers were required to take a short course every three years at the Woolwich branch of the School of Gunnery.[22]

Like enlisted militiamen, officers were required by law to attend their units' annual, 27-day training held during the summer months. Artillery officers, like enlisted gunners, trained for 41 days every third year at their unit's mobilization fortification. Unlike enlisted militiamen, officers were required by mess tradition to engage in regimental social functions throughout the year.

As was the case for their Regular Army counterparts, the pay and allowances of junior militia officers did not cover their military-related expenses. The expenses of a militia officer were maintenance of uniforms and equipment, support of the officers' mess, and funding the regimental band. Many officers also incurred travel expense for annual training.[23] These

---

[19] Deficiencies in Officer Ranks, *Norfolk Commission Report, Appendices*, 1904, [Cd. 2064], no. 37; Instruction of Officers, Ibid., no. 38; Various Testimony, *Norfolk Commission Report, Minutes of Evidence*, 1904, [Cd. 2062, Cd. 2063].

[20] Grierson, *The British Army*, 219.

[21] Number of Applications to Undergo a Course of Musketry, *Norfolk Commission Report, Appendices*, 1904, [Cd. 2064], no. 25.

[22] Cavenagh-Mainwaring, *The Royal Miners*, 108.

[23] There was no army allowance for travel expense of militia officers who resided outside their unit's recruitment county. For Irish regiments, about half the officers lived in England.

Officers and Soldiers of the Auxiliary Forces

expenses were significant and precluded commissioned service by all but the financially well-off.[24]

Initial Costs:	
Uniform	£40 to £65
Mess Initiation	8
	£48 to £73
Annual Costs:	
Mess Subscription	£3.5 to £5.0
Extra Mess Expense	varied
Meals	5.0 to 10.0
Uniform Maintenance	varied
Uniform Upgrades	3.5 to 5.3
Band Subscription	1.0 to 3.0
Travel	0.0 to 9.9+
	£30 to £35

The pay of junior militia officers for 27-days' training was £7.1 for a second lieutenant, £8.8 for a junior first lieutenant, and £10.1 for a first lieutenant with seven-years' service.[25] Overall, the out-of-pocket annual expense to serve as a militia lieutenant was about £25.[26]

Officers of 4/Royal Inniskilling Fusiliers (Tyrone Militia), 1903

The Inniskillings Museum, www.inniskillingsmuseum.com.

---

[24] *Report of the Committee on Expenses Incurred by Officers of the Army*, 1903, [Cd. 1421]; Various Testimony, *Norfolk Commission Report, Minutes of Evidence*, 1904, [Cd. 2062, Cd. 2063]; Bowman and Connelly, *The Edwardian Army*, 115; Trevor Herbert and Helen Barlow, *Music and the British Military in the Long Nineteenth Century* (Oxford: Oxford Univ. Press, 2013); Cavenagh-Mainwaring, *The Royal Miners*, 87, 92-93, 113.

[25] *Royal Warrant for the Pay*, 1906.

[26] Testimony LTG Kelley-Kenny, Adjutant-General, plus that of several other witnesses. *Norfolk Commission Report, Minutes of Evidence*, 1904, [Cd. 2062, Cd. 2063].

Annual training was conducted typically on rented land or army camps, and officers lived in tents, or if available, rooms at nearby inns and hotels. If there was a restaurant or hotel close to camp, the officers' mess would be established in a hired banquet room; otherwise, the mess was set up "under canvas." During the year, some regiments had an occasional mess event and the colonel expected the officers to attend.[27] The extent of this extra-curricular activity was dictated by regimental tradition. For officers who lived a great distance from their unit's barracks, such as Englishmen serving in Irish regiments, travel for social activities could be expensive.

Militiamen

Like the Regular Army, the militia's rank-and-file was overwhelmingly of the lower strata of the working class.[28] As historically the militia was a county force, agricultural laborers were over-represented in its enlisted ranks. In 1903, when only about 16% of British employed males engaged in agricultural work, 22% of militiamen were agricultural laborers. In Ireland, where 55% of employed males engaged in agricultural pursuits, 60% of militiamen were farm and field workers.[29] The rural complexion of the militia had faded notably over the course of the Boer War. In 1898, the year before outbreak of hostilities in South Africa, 31% of all militiamen were agricultural laborers (two-thirds of Irish militiamen).[30] The militia attracted fewer skilled workers than did the regular army. There were only a relative handful of high-skilled billets in the militia as it had no service and ordnance corps units and only a few engineer units. Unemployed tradesmen were more likely to join the regulars rather than settle for the three months of initial, full-time employment offered by the militia. Men who held high-skilled jobs did not want to forego their civilian wages for militia pay during the initial training and the subsequent one-month annual encampments. During the Edwardian Era, between 20% and 25% of Regular Army recruits were of the working class upper strata; for the militia only 10% to 15%.[31]

The militia, like the army, failed to attract many recruits from the "respectable working class." The auxiliary force's political reputation was another negative factor that discouraged

---

[27] The 3/Royal Berkshire Regiment had an annual officer's dinner each spring during Derby Week. Thoyts, *History of the Royal Berkshire Militia*, 204-05.

[28] French, *Military Identities*, 209-10.

[29] Occupations of the Inhabitants of the United Kingdom, *General Report, Census of England & Wales, 1901*; Summary of Answers to Circular of Questions, *Norfolk Commission Report, Appendices*, 1904, [Cd. 2064].

[30] War Office, *General Annual Return of the British Army, 1898*, 1899, [C. 9426]. Hereafter cited as *Annual Army Report*.

[31] *Army Medical Department Report for the Year 1903*, 1905, [Cd. 2434]; Summary of Answers to Circular of Questions, *Norfolk Commission Report, Appendices*, 1904, [Cd. 2064]; Various Testimony, *Norfolk Commission Report, Minutes of Evidence*, 1904, [Cd. 2062, Cd. 2063].

enlistment. Throughout the United Kingdom, the working class opposed organized militarism within the community and viewed the militia as a force that protected the interests of the middle and upper classes.[32] In Ireland, nationalists saw the militia as an instrument of English, colonial oppression.[33] A former Under-Secretary of State for War testified before the Norfolk Commission that upon enlistment "no man gives his own address if he can help it" because "he does not like it to be known he is in the militia."[34] Though the political factor dampened enlistment enthusiasm, it was probably the financial effect of service that kept the enlisted ranks well below authorized strength: Militiamen had to serve full-time for twenty-seven days annually, a period during which they received only soldiers' wages plus a militia bonus. Additionally, militia service limited a man's employment opportunities. Many employers, particularly those with small establishments, did not hire militiamen as they would be off-the-job for one month out of every twelve.[35] Because of these political and economic factors, on Bloomsday the militia stood at 74.1% of authorized enlisted strength.

Few boys and young men joined the militia intent on long-term service or even completion of the initial, six-year enlistment term.[36] About two-thirds of recruits were youngsters who wished to try out army life before they made a multi-year commitment to the Regular Army, or who sought full-time service but were under-age, under-height, or under-weight for the regulars. Note that somewhat under half of all militia recruits transferred to full-time service within twelve months of their militia enlistment.[37] Those who joined for "the advantages of the militia" were mostly agricultural laborers in need of the

---

[32] French, *Military Identities*, 43.

[33] In *Ulysses*, the editor of the *Freeman's Journal* recalls the North Cork Militia, a regiment notorious for its heavy-handed methods to find and seize arms of the United Irishmen prior to the Rebellion of 1798. *U* (Gabler) Aeolus 7:359-63.

LTC H.H. Stewart, commanding officer 3/Royal Inniskilling Fusiliers, attributed the low level of militia enlistment to political factors. "There are political facts, although they are all more or less under the surface, which come into operation in Ireland. The National Party object tremendously to service in the Militia. You cannot prove it, but there is an under-current of opposition." *Norfolk Commission Report, Minutes of Evidence*, 1904, [Cd. 2063], at qq. 16632-33.

[34] LTC George Somerset (3rd Baron Raglan), commanding officer of the Royal Monmouth Engineers *Norfolk Commission Report, Minutes of Evidence*, 1904, [Cd. 2063], at qq. 16800-01. Raglan was a former army officer, Conservative MP, and Under-Secretary of State for War, 1900-02.

[35] Bowman and Connelly, *The Edwardian Army*, 116-17.

[36] Of the 33,205 men who enlisted in 1898, only 5,934 were on the roll five years later (14.8%). Of the 33,205 recruits in 1894, 1,587 (11.5%) were on the roll in 1904 (completed their initial term and re-enlisted). *Annual Army Report, 1898*, 1899, [C. 9426]; *Annual Army Report, 1904*, 1905, [Cd. 2268].

[37] In the 1904 reporting year, there were 35,264 enlistments, 15,648 transfers to the regular forces, 5,578 purchased releases, and 6,704 desertions indicating that up to 80% of recruits were simply "trying out" army life. *Annual Army Report, 1904*, 1905, [Cd. 2268]. Various Testimony, *Norfolk Commission Report, Minutes of Evidence*, 1904, [Cd. 2062, Cd. 2063].

extra income provided by annual training.[38] Militia training took place during the summer months between the spring hay crop harvest and the autumn general harvest. Accordingly, casual agricultural labors could spend part of that period receiving militia pay and meals.[39] Major-General Edward Spears, who began his military career with the Kildare Militia (3/Royal Dublin Fusiliers), wrote in his memoirs that the main purpose of the formation seemed to be "providing short periods of employment to a not very martial and usually unemployed section of the population."[40] The other career-oriented recruits were industrial workers who sought a temporary release from the drudgery and family obligations of civilian life. Such recruits worked in large factories, mills, and mines where the employer allowed them unpaid, time-off for military service.[41]

Some Men of 4/Royal Inniskilling Fusiliers (Tyrone Militia), 1900 [42]

The Inniskillings Museum, www.inniskillingsmuseum.com.

---

[38] Stoneman, "The Reformed British Militia," 149. *The Advantages of the Militia* was a War Office recruiting pamphlet like *The Advantages of the Army*.

[39] LTC Bernard FitzPatrick (2nd Baron Castletown), commanding officer 4/Leinster Regiment, *Norfolk Commission Report, Minutes of Evidence*, 1904, [Cd. 2063], at q. 20463.

[40] Spears, *The Picnic Basket*, 63.

[41] Stoneman, "The Reformed British Militia," 149; French, *Military Identities*, 210.

[42] The photograph is likely of a depleted company plus the battalion drummers (sergeant-drummer and seven drummers including three boy apprentices). In January 1899, the militia was understrength by 16.8%. Also, 27.8% of serving militiamen were voluntary reservists of the Regular Army who were called up in October 1899 for overseas Boer War service. As a result, in early 1900, most militia units were at half-strength. *Annual Militia Return, 1904*, 1905, [Cd. 2432]; *Appendices to the Minutes of Evidence taken before the Royal Commission on the War in South Africa*, 1903, [Cd. 1792], no. 14; *Annual Army Report, 1898*, 1899, [C. 9426].

Officers and Soldiers of the Auxiliary Forces

About one-fifth of militia recruits were the same sort of desperate men who joined the Regular Army. They were unemployed, out of money, and in immediate need of food and shelter.[43] These enlistees apparently hoped that at the end of their initial, full-time training, new employment opportunities would arise. If that turned out not to be the case, they likely transferred to the regulars or entered the workhouse.

Terms of Enlistment

A militia recruit had to be 17 years of age and generally no older than 34 (32 for the harbor mining divisions). Former soldiers of the Regular Army who had at least two years service and a character rating of "fair" or better, could enlist through age 45. The initial enlistment term was six years; re-enlistment terms were four years. Militiamen could not re-enlist if over age 45.[44]

Years' Service of Militiamen, October 1, 1904

	Recruits	Trained		Seasoned Veterans			
Years of Service:	<1	1 – 4	5 - 6	7 - 9	10 - 13	14 - 17	18 +
Percent of Men:	25.7	46.3	6.9	9.8	5.6	3.0	2.7
Percent of Men:	25.7	---53.2---		---------21.1---------			
Term of Service:	------Initial-------			2nd	3rd	4th	5th+

Few militiamen served out their first term. In 1904, only 5,934 had five to six years' service: 14.8% of the 1898 intake. About two-thirds of militiamen who completed their initial term re-enlisted for another four years. After the first re-enlistment, about half of the time-expired men re-enlisted at the end of each term. In that about three percent of militiamen were invalided out each year, it is likely that a good many veterans were denied re-enlistment on medical grounds.[45]

Under the Militia Act of 1882, a militiaman who failed to report when embodied for full-time service was guilty of desertion as defined by the Army Act. A militiaman who failed to attend militia training, annual or initial, was guilty of absence without leave (AWOL).[46] During the truant militiaman's period of annual training he was subject to military law and

---

[43] This is a rough estimate based on 65% enlisted to try out military life or to join the Regular Army; 15% enlisted for a militia career. Contemporaries acknowledged that a portion of recruits enlisted out of need for employment. Beckett, *Britain's Part-Time Soldiers*, 187; Various Testimony, *Norfolk Commission Report, Minutes of Evidence*, 1904, [Cd. 2062, Cd. 2063]. For example, LTC H.D. Fryer, commanding officer 4/Suffolk Regiment, stated that a recruit "is generally a man who is out of work and wants board and lodging, and the longer he gets that the better he likes it." Ibid., [Cd. 2063], at q. 15206.

[44] H. Jenkyns, "Constitution of the Forces" in *Manual of Military Law* edited by G.O. Morgan, Judge Advocate-General (London: HMSO, 1907), 201-04.

[45] *Annual Army Report, 1904*, 1905, [Cd. 2268].

[46] 45 & 46 Vic., c.49, §§23-25.

could be arrested by the army or militia. For the other eleven months of the year, the militiaman was subject to arrest only by the civil authorities.[47]

If the AWOL militiaman was apprehended by the police, a magistrate would notify the competent military authority of the arrest. The militiaman's commanding officer, or district officer-in-command, decided whether to receive the militiaman into military custody or allow the civil court to pursue the matter. If the AWOL militiaman was arrested by the militia or army, the same military authority decided whether to deliver the militiaman to a court of summary jurisdiction or deal with the infraction internally.[48] Conviction by a civil magistrate brought a fine of not less than £2 and not more than £25. A militiaman unable or unwilling to pay the fine would be imprisoned for at least seven days but not more than three months.[49] If the competent military authority decided to handle the breach of law within the militia, he would either bring court-martial charges or deal with the offense summarily.[50] Conviction by court-martial could carry a sentence of imprisonment not exceeding two years. Summary punishment was limited to confinement in barracks not exceeding 21 days.[51]

By Bloomsday, the militia no longer pursued legal action against members absent from training. Militia staff simply notified the civil authorities of the militiaman's absence by transmission of a completed Army Form B 124 to the police of the locality in which the militiaman last resided and the *Police Gazette*.[52] Particulars of truant militiamen appeared in that publication's Supplement D, a weekly list of absentees and deserters from the armed forces. The police ignored the absence notices as "both in London and the provinces the police authorities frankly say it is not worth their while to bother about militia absentees."[53]

Purchased Discharge

Like soldiers of the Regular Army, militiamen could purchase an early release from their military obligation. In 1904 there were 5,578 purchased discharges, representing 6.2% of the militia's beginning of year strength.[54] The purchase price for militiamen in their first enlistment was £1; for re-enlisted militiamen £1 10s. Militiamen who had completed two

---

[47] Army Act, 188, 44 & 45 Vict., §176(6).

[48] Trial of an alleged Offender under the "Militia Act, 1882."Army Circulars 1883, No. 34, Clause 7.

[49] 45 & 46 Vic., c.49, §23. If the fine were the maximum £25, the court could impose a sentence of up to three months imprisonment. Summary Jurisdiction Act (England & Wales), 1879, 42 & 43 Vict., c.49, §5.

[50] To convene a militia court-martial required War Office approval. Stoneman, "The Reformed British Militia," 156.

[51] 45 & 46 Vic., c.49, §43(4); 44 & 45 Vict., c. 58, §§36, 44.

[52] *Regulations for the Militia, 1898*, ¶408.

[53] LTC Lord Raglan, *Norfolk Commission Report, Minutes of Evidence*, 1904, [Cd. 2063], at q. 16800. Raglan complained publicly for many years about lack of police and War Office effort to apprehend AWOL militiamen. The quote is from the previously cited article, Raglan, "The Militia," *The National Review*, 250.

[54] *Annual Army Report, 1904*, 1905, [Cd. 2268].

annual trainings had to refund their most recent annual £3 non-training bounty.[55] On average, militiamen paid £1 6s. to quit the force.[56]

Initial Training

During a militiaman's first year of service, he underwent 90 days of full-time training. Nearly all basic training took place either at a depot that trained Regular Army recruits or at the militia unit's barracks.[57] Basic training provided by a militia unit was termed "preliminary drill." The instructors were the Regular Army sergeants of the militia unit's permanent staff.[58] Such training took place immediately prior to the unit's annual training so a preliminary drill recruit underwent 76 consecutive days of full-time training. At the depot, militia recruits trained alongside Regular Army recruits. They went home after completion of basic training and joined their militia unit at the next scheduled annual training. Both depot training and preliminary drill lasted 49 days.[59] Militia officers preferred that recruits receive preliminary drill. About nine-in-ten of artillery commanders and two-thirds of infantry commanders disapproved of depot training.[60]

At the annual training, novice militiamen who had completed either preliminary drill or depot training were joined into *ad hoc* companies for advanced training. An average-strength infantry battalion (618 men) would receive about 240 recruits annually; 50 straight from preliminary drill and 190 from prior depot training.[61] After completion of the training, preliminary drill recruits received the enlistment bounty of £1 10s. Depot-trained recruits received £1 as they were given 10s. upon completion of depot basic training.[62] Sometime during the first year of their militia service, recruits went on a 14-day musketry course (gunnery course for the artillery). On the musketry course, militia recruits fired the same number of rounds as regulars, 189.[63]

---

[55] *Regulations for the Militia, 1898*, ¶172.

[56] Militia Pay and Allowances, *Norfolk Commission Report, Appendices*, 1904, [Cd. 2064], no. 50.

[57] Recruits for the few medical corps and engineer units were trained at the same specialized facilities that trained regulars.

[58] In the infantry there were two sergeants of the permanent staff for each company, the company sergeant-major with the rank of colour-sergeant and the company clerk with the rank of sergeant.

[59] MG A.E. Turner, Statement Showing the Average Time Devoted to Drill and Training by Recruits and Trained Men of the Militia, *Norfolk Commission Report, Appendices*, 1904, [Cd. 2064], no. 58. Preliminary drill was somewhat different from depot training. For example, physical conditioning was not included in preliminary drill until 1896. Thoyts, *History of the Royal Berkshire Militia*, 228.

[60] Summary of Answers to Circular of Questions, *Norfolk Commission Report, Appendices*, 1904, [Cd. 2064].

[61] *Annual Army Report, 1904*, 1905, [Cd. 2268]; *Annual Militia Return, 1904*, 1905, [Cd. 2432].

[62] Army Orders, 1902, No. 191.

[63] Grierson, *The British Army*, 175.

In the post-Boer War period from 75% to 80% of militia recruits underwent basic training at the regimental depot.[64] Recruits who were in immediate need of subsistence could not wait for the annual preliminary drill so they selected depot training as regimental depots began basic training courses eight times each year.[65] Additionally, depot recruits received enlistment bounty money after 49 days; preliminary drill recruits had to wait 79 days. Many militia officers condemned depot basic training in their testimony before the Norfolk Commission. In its report, the Commission noted the charges that at depots the recruit intent on career militia service is

> "often unfairly treated at these places, that he is put on an undue proportion of fatigues, that his drill is neglected, and the he is partly teased and partly bullied into joining the line. These charges have also been strongly and categorically denied, but the weight of evidence goes to show that in many cases they are far from unfounded, and that so far as that is so, a grave abuse exits." [66]

(Note that depot training sergeants received a 4s. bonus for each militia recruit induced to join the Regular Army.[67])

The War Office required a militia recruit to spend only 390 hours in actual training (which included physical conditioning). Nearly half of each training day was allocated to fatigue duties, barracks cleaning, guard duty, and polishing uniform accoutrements and personal equipment. Like their regular counterparts, militia recruits had no required duties on Sundays other than the church parade.[68]

Enlisted Service

The War Office considered militiamen who had completed their 90 days of full-time, first-year service, trained men. Their only required service was attendance at their unit's annual assembly (the "annual training") which usually was 27 consecutive days of full-time service. Militiamen could volunteer for, or be called out for, an additional 29 days of service each year. Generally, such additional time was to assistant the permanent staff during preliminary drill. By statute, a militiaman could not be called out for training for more than

---

[64] *Annual Army Report, 1906, 1907*, [Cd. 3365].

[65] Basic training courses started monthly excepting December, January, March, and June. Grierson, *The British Army*, 172-73.

[66] *Norfolk Commission Report,* [Cd. 2061], at 29.

[67] Ibid., at 30.

[68] MG A.E. Turner, Statement Showing the Average Time Devoted to Drill and Training by Recruits and Trained Men of the Militia, *Norfolk Commission Report, Appendices*, 1904, [Cd. 2064], no. 58.

56 days annually.[69] The War Office required each militiaman to partake in 120 hours of military activity at the annual training.[70] The typical infantry training schedule was as follows:

Day	Activity for Infantry
1	Assembly, in-processing, and medical inspection.
2	Travel to training site and pitch tents.
3-25	Training Days: 17 Training Half-Days: 3 Sundays: 3
	Musketry (includes firing 77 rounds): 6 Days
	Company Drill and Other Training: 8 - 9 Days
	Battalion Drill: 3 - 6 Days
	Brigade Drill (exercise): 0 - 2 Days
	(If brigaded with other battalions during training, two days were spent on a brigade exercise, if not, those two days were devoted to battalion drill.)
26	Strike tents and travel to home barracks.
27	Disbursement of pay, out-processing, and dismissal.

The regulations, with respect to the infantry encampment, stated that "training will, as far as possible, be divided into three weeks' musketry and company training, and one week's battalion and brigade training, if possible with other arms."[71] Brigade training was uncommon and mixed-arms training very seldom took place.

For garrison artillery brigades, training was at coastal forts, practice batteries, or artillery ranges. Every three years there would be an extended training of 41 days. That training was at the unit's mobilization fort (the works it would man when embodied).[72] Many officers testified before the Norfolk Commission that the extended training was the only time militia artillery practiced with guns they would service when embodied. Usually, during the regular 27-day trainings, militiamen fired obsolete guns retained by the army for training purposes.[73] The training schedule for artillery was like that for infantry except gunnery training was

---

[69] 45 & 46 Vict., c49, §17(a)

[70] MG A.E. Turner, Statement Showing the Average Time Devoted to Drill and Training by Recruits and Trained Men of the Militia, *Norfolk Commission Report, Appendices*, 1904, [Cd. 2064], no. 58.

[71] *Regulations for the Militia, 1898*, ¶231.

[72] *Regulations for the Militia (Provisional), 1904*, ¶199.

[73] COL W.A.G. Saunders-Knox-Gore, commanding officer of the Donegal Artillery, *Norfolk Commission Report, Minutes of Evidence*, 1904, [Cd. 2063], qq. 20934-37.

FM Garnet Wolseley (1st Viscount Wolseley) testified that when he was CINC [1890-1901] "-- the guns we condemned as useless were handed over to the Militia and Volunteers" so that "…the actual guns in the hands of the Militia were obsolete…" Ibid., 1904, [Cd. 2062], qq. 1553-54.

substituted for musketry training. Gunnery training consisted of instruction, drill, and each company firing five series of 18 rounds: 45 blanks plus 45 projectiles, both low-charge explosive and slugs.[74] Two-thirds of artillery commanding officers claimed that the time allotted to gunnery training was inadequate.[75]

At nearly all annual trainings, militiamen lived "under canvas" as there were few vacant barracks at the army camps that could accommodate field training (Curragh, Colchester, Shorncliffe, Aldershot, Salisbury Plain). Small camps used by the militia included Finner Camp, Co. Donegal, Ireland and Strensall Camp, Yorkshire (North Riding), England. In some cases, militiamen were boarded in inns and rooming houses, and if the training was at the militia unit's barracks, married militiamen could spend the night at home.[76]

Training Encampment, 4/Royal Inniskilling Fusiliers, 1907

The Inniskillings Museum, www.inniskillingsmuseum.com.

Militiamen could be excused from attendance at training by their commanding officer and 5.1% were in 1904.[77] Excusals were given for illness, epidemic in the militiaman's place of abode, or if otherwise warranted.[78]

---

[74] Cavenagh-Mainwaring, *The Royal Miners*, 103; Grierson, *The British Army*, 188.

[75] Summary of Answers to Circular of Questions, *Norfolk Commission Report, Appendices*, 1904, [Cd. 2064].

[76] No more than 10% of all militiamen present at training could sleep at home. *Regulations for the Militia (Provisional), 1904*, ¶203.

[77] *Annual Militia Return, 1904*, 1905, [Cd. 2432].

[78] *Regulations for the Militia, 1883*, ¶303.

Pay and Allowances

During the annual training, militiamen received the army rate of pay plus militia bonuses termed "bounties." The training bounty of £1 10s. was paid at each training. After a militiaman attended two annual trainings, he received each winter a "non-training" bounty of £3, a retainer paid regardless of whether he attended training. Men who received the retainer but later were AWOL from training, had to refund the £3 bounty.

Table 38.
Base Pay for Militiamen
Infantry and Artillery, 1904 [79]

	Infantry			Artillery		
	Daily	27 Days	90 Days	Daily	27 Days	90 Days
Sergeant	2s. 4d.	£ 3.1	£ 10.5	3s. 2d.	£ 4.3	£ 14.3
Corporal	1s. 8d.	£ 2.2	£ 7.5	2s. 6d.	£ 3.4	£ 11.3
Lance Corporal	1s. 3d.	£ 1.7	£ 5.6	2s. 3d.	£ 3.0	£ 10.1
Private	1s.	£ 1.3	£ 4.5	1s. 2.5d.	£ 1.6	£ 5.4

Source: *Royal Warrant for the Pay, 1906.*

Militiamen who served 90 days were those who augmented the permanent staff for the preliminary drill and musketry training of new recruits. They were sergeants who served as instructors, and rank-and-file militiamen who served as cooks, sentries, officers' servants, *etc.*[80] The number of men performing such duties was limited to 25% of the recruits in training.[81] Generally, these positions were filled by volunteers. If volunteers for preliminary drill duty were lacking, the general officer commanding the army district could order militiamen to perform such additional service.[82]

During training (annual, preliminary drill, and musketry), militiamen were eligible for daily, extra-duties pay as follows:[83]

Military Police Private	2d.	Hospital Orderly	4d.
Quartermaster's Clerk	3d.	Hospital Sergeant	6d.
Transport Driver	4d.	Provost Sergeant	6d.
Military Police Sergeant	4d.	Orderly Room Clerk	6d.

---

[79] Note that in the artillery, the equivalent ranks for lance corporal and private were respectively bombardier and gunner.

[80] Grierson, *The British Army*, 187; Goodenough and Dalton, *The Army Book for the British Empire*, 368.

[81] *Regulations for the Militia, 1898*, Part I, Sec. 4.

[82] The militia statutes allowed a maximum of 56 days per year for training.

[83] *Regulations for the Militia, 1898*, ¶633.

Total annual compensation for the typical, trained militiaman (27 days of annual training), including the cost of meals provided, was as follows:

	Second Year	Subsequent Years
Pay, 1s. daily for 27 days	£1 7s.	£1 7s.
Food, 9d. daily for 25 days	18s. 9d.	18s. 9d.
Assembly/Release Meal Allowances	1s.	1s.
Messing Allowance, 3d. daily [84]	6s. 3d.	6s. 3d.
Groceries Stoppage, 3d. daily	(6s. 3d.)	(6s. 3d.)
Training Bounty	£1 10s.	£1 10s.
Non-Training Bounty	£3	
Total Compensation	£3 10s. 6d.	£6 16s. 9d.

A skilled worker in 1904 earned at least £100 annually; a minimum of £6 13s. 4d. per month.[85] Such man would incur a significant financial loss in the first two years of militia service but would "break-even" in subsequent years. For unskilled workers, particularly agricultural casual laborers, the militia paid well.

Promotion

Militiamen could not rise above the rank of sergeant as the higher NCO grades were reserved for the regulars of the permanent staff. Such regulars served as the regimental sergeant-major, the quartermaster sergeant, the company sergeant-majors, and company clerks. For the infantry, NCO positions were 5.7% of authorized enlisted strength: 2 sergeants and 4 corporals for every 100 rank-and-file militiamen (privates and lance-corporals); Promotion was somewhat easier to achieve in the artillery as NCO positions were 8.3% of authorized enlisted strength: 2 sergeants and 6 corporals for every 88 rank-and-file militiamen (gunners and bombardiers);.[86] The sergeant positions were as NCO-in-Charge of a half-company (50 men in the infantry, 44 in the artillery). Actual promotion opportunities were somewhat better as only three-fourths of the rank-and-file positions were filled (7.7% NCOs in the

---

[84] Militiamen who were under age 19 or had not completed one annual training paid a daily grocery stoppage of 3d. Army Orders 1901, No. 239. Most recruits were under age 19 during their first year of militia service. *Annual Army Report, 1904*, 1905, [Cd. 2268].

[85] £80 in Ireland. *Reports of an Enquiry by the Board of Trade into the Earnings and Hours of Labour of workpeople of the United Kingdom in 1906,* 1910, [Cd. 4844, Cd. 5086, Cd. 5196], 1912 [Cd. 6053], 1913 [Cd. 6556].

[86] War Office, *Army Estimates for the Year 1904-05*, 1904, H. C. Accounts & Papers, No. 73.

infantry, 11.1% in the artillery).[87] Also limiting promotion was the requirement that a militiaman could not be promoted unless he had at least three years remaining on his engagement.[88]

Though there were few NCO positions open to militiamen, commanding officers were unable to fill them with qualified personnel. For the artillery, 87.5% of commanding officers claimed they had difficulty in obtaining militia NCOs; for the infantry, 83.1%.[89] The men best-suited for the NCO ranks rarely sought the extra stripes. They did not want the responsibility and feared retribution in civilian life by enlisted men that they had brought up for disciplinary action.[90] Infraction of rules during training subjected militiamen to fines, forfeiture of the training bounty, and confinement to barracks. Another reason militiamen shunned promotion was they feared they could be the superior of a man who in civilian life was their supervisor, co-worker, or employer (or his son). Accordingly, commanding officers had to settle for whom they could get to fill the NCO ranks.[91] A former Inspector of Militia Artillery for Ireland testified before the Norfolk Commission that a great many NCOs in Irish artillery brigades were illiterate, though they were hard-working and capable of learning.[92]

Militiamen were eligible for the regimental staff appointments of sergeant-cook and pioneer sergeant. Note that for NCOs to receive additional pay for such appointment they had to have completed the appropriate army course; the two-month School of Cookery course at Aldershot and the three-month pioneer course at the School of Military Engineering, Chatham.[93] Few militia sergeants could afford to leave civilian employment for two or three months, so staff appointments with additional pay may have been held mostly by former regulars who had taken such courses.

Discipline

Data indicates that discipline was a greater problem in the militia than in the Regular Army. Note that to compare "misconduct" rates between militiamen and regulars one must consider the difference in time spent subject to military

---

[87] *Annual Militia Return, 1904*, 1905, [Cd. 2432].

[88] *Regulations for the Militia (Provisional), 1904*, ¶72.

[89] Summary of Answers to Circular of Questions, *Norfolk Commission Report, Appendices*, 1904, [Cd. 2064].

[90] For example: LTC W. Watts, commanding officer 3/Welsh Regiment, *Norfolk Commission Report, Minutes of Evidence*, 1904, [Cd. 2063], at qq. 18560-61.

[91] Various Testimony, *Norfolk Commission Report, Minutes of Evidence*, 1904, [Cd. 2062, Cd. 2063].

[92] COL R.W. Rainsford-Hannay, *Norfolk Commission Report, Minutes of Evidence*, 1904, [Cd. 2063], at q. 21036.

[93] Army Orders, 1902, No. 191.

law by militiamen compared to army soldiers. Militiamen were governed by military law only when they were in training, both initial and annual.[94]

Table 39.
Misconduct Cases, 1898 & 1904
Percent of Adjusted Strength

	1898		1904	
	Militia	Army	Militia	Army
Deserted	8.5	1.9	7.7	1.5
Tried by Courts-Martial	5.4	4.5	3.7	4.0
Fined, Drunkenness	20.0	12.1	15.8	9.1
Summary Punishment	134.4	100.5	NA	NA

Sources: *Annual Army Report, 1898*, 1899, [C. 9426]; *Annual Army Report, 1904*, 1905, [Cd. 2268].

"Desertion" for militiamen men was actually absence-without-leave from training. This absence rate was five times the desertion rate in the army. The court-martial trial rate was about the same for both groups while the drunkenness fine and summary punishment rates were much higher among militiamen (1.7 and 1.3 times the rate for regulars, respectively). This was due to two factors: lack of "wholesome" entertainment at militia training encampments and the holiday attitude of many seasoned militiamen. The lack of libraries, lounges, game rooms, and sporting equipment was noted by officers who testified before the Norfolk Commission as an impediment to recruitment. The only recreational activities for militiamen were off-duty, inter-company cricket and football matches.[95]

Career militiamen of the industrial working class viewed summer encampment as an escape from the drudgery and social constraints of their civil lives.[96] For many such part-time soldiers, their annual training was apparently a month-long booze-up at the regimental canteen. A militia officer told the Norfolk Commission that the greatest impediment to recruiting in the South of

---

[94] During a year, the average period of full-time service for militia units was 1.8 months (one month for seasoned militiamen, three months for recruits who on average constituted 35% of the force).

[95] Thoyts, *History of the Royal Berkshire Militia*, 226.

[96] Stoneman, "The Reformed British Militia," 149.

Ireland was the negative attitude of women towards the militia: They feared their husbands and sons would mix with the wrong sort, "drink more than is good for them," and get themselves into trouble.[97]

Pensions and Gratuities

Militia service did not give rise to a pension; however, like reservists, militiamen when embodied could qualify for a pension. Unlike reserve service, militia service did not count towards longevity pension eligibility. Accordingly, the only militiamen who qualified for such pension were those who while embodied attained 21-years' full-time service. Rarely did a militiaman qualify for a pension as embodiments in the late nineteenth-century lasted two years at most.

Militiamen disabled while embodied received disability pensions under the same conditions as regulars. If disabled, militia service reckoned towards pension eligibility.[98] Militiamen disabled during training were eligible for a temporary pension of up to six months. The maximum payment was 3s. 6d. per day.[99] Militiamen discharged as medically unfit due to injury or disease that occurred in civilian life received a gratuity of 15s.[100]

Militia Bands

A militia unit, unlike a volunteer force corps, did not have a band on its establishment; yet tradition mandated such appurtenance.[101] A regiment's band expenses in excess of the £25 annual band allowance were borne by the officers.[102] Militia bands could perform daily during the one-month training plus multiple times during the year, and bandsmen received extra-duties pay, both official and unofficial. Thus, band expense was a significant burden on militia officers.[103]

The core of a militia band was the Regular Army signaler-musicians (drummers in the infantry; buglers in the artillery) assigned to the unit's permanent staff. All such musicians,

---

[97] "… they regard the militia too as a bad school for a young man." LTC W. Cooke-Collis, commanding officer 9/King's Royal Rifles (North Cork Militia), *Norfolk Commission Report, Minutes of Evidence*, 1904, [Cd. 2063], at q. 20881.

[98] *Royal Warrant for the Pay, 1906*, Art. 1201.

[99] *Regulations for the Militia, 1898*, ¶¶710-13.

[100] Ibid., ¶545.

[101] Cavenagh-Mainwaring, *A History of the Royal Miners;* Herbert and Barlow, *Music & the British Military*; LTC C. Healy, commanding officer 3/South of Wales Borderers, *Norfolk Commission Report, Minutes of Evidence*, 1904, [Cd. 2063], at q. 18520.

[102] Army Orders, 1899, No. 94 provided for the allowance. Prior to 1899, officers bore the entire band cost.

[103] COL Kenyon-Slaney, MP for Shropshire, during Questions. 154 Parl. Deb. (4th ser.) (1906) 597.

six to ten in an infantry battalion, played instruments in addition to drums, bugles, and fifes. The sergeant-drummer, a regular, was usually bandmaster though a bandmaster could be specially recruited. The regiment recruited musician militiamen to bring the band up to concert strength (15 to 23). These bandsmen-militiamen, received extra pay (total remuneration of 13s. per day) and though they were attested militiamen, often did not undergo military training.[104] Note that prior to the Boer War, regiments augmented their bands with civilian musicians who were not on-the-strength.[105]

The Last Embodiment: The 2nd Boer War

In December 1899, the War Office discovered that the militia would be of little military value for defense of the United Kingdom. At the time, 17% of the militia's private ranks were unfilled and of the enrolled privates, 28% were also militia reservists for the Regular Army.[106] After mobilization, militia units were on average at 60% of authorized strength as all militia reservists were called up to fill out the Regular Army units slated for deployment abroad. There then was "wastage" through desertion and the invaliding out of men medically unfit for service. Accordingly, after general mobilization, nearly all militia units were at half strength with many militiamen too young, or insufficiently trained, for effective military service. In 1901, the War Office terminated enrollment in the Militia Reserve.[107] Following are descriptions of the Boer War embodiment of two militia units.

4/Princess of Wales's Own (Yorkshire Regiment)
"North York Militia" [108]

This infantry battalion of eight companies recruited in northeastern England and its barracks was at Richmond Castle, Richmond, North Riding, Yorkshire.

---

[104] LTC G.C. Twisleton-Wykeham-Fiennes, commanding officer 3/Royal Scots Fusiliers, *Norfolk Commission Report, Minutes of Evidence*, 1904, [Cd. 2063], at q. 19141. After the end of the Boer War, several imperial yeomanry bandsmen were summoned to court on charges of absence from training when they refused to attend camp after denied permission to send a substitute musician, or were told they would not receive extra pay. *Buckinghamshire Examiner*, October 9, 1903; *Volunteer Service Gazette and Military Dispatch*, October 30, 1903.

[105] Late Victorian militia regulations stated militia band uniforms would not be issued to civilian bandsmen. *Regulations for the supply of Clothing and Necessaries to Disembodied Regiments of Militia, 1878*, ¶23. Militia units advertised engagements for civilian bandsmen as late as 1899. *The Era*, November 11, 1899.

[106] Limitations on the number of militiamen that could also be in the Militia Reserve were based on the authorized number of privates. For the infantry, it was one-fourth. As that arm was only at 75% of establishment, one third of such militiamen could be Militia Reservists. For engineers, the limitation was one-fourth the artillery one-third. *Regulations for the Militia, 1883*, ¶467.

[107] Army Orders, 1901, No. 88.

[108] *Annual Militia Return, 1899*, 1900, [Cd. 84]; Turton, *The History of the North York Militia*, 150-54; *Monthly Army List, December 1904*.

In 1899, it had 11 of its authorized 23 officers and 605 of its authorized 856 militiamen. Of the enrolled militiamen, 181 (29.9%) were Militia Reservists. As with nearly all militia units, its permanent staff was complete with 2 officers, 20 sergeants, and 8 drummers.

The battalion was first embodied on May 5, 1900 and ordered to Strensall Camp. Between October 1899 and embodiment, it had acquired 30 new recruits but had lost 10 militiamen (3 absent-without-leave, 3 invalided out, 4 who joined the Regular Army). At embodiment, the battalion totaled 23 militia officers (15 newly commissioned), 444 militiamen (51.9% of authorized strength), and 30 permanent staff. All its 181 Militia Reservists had been called up to serve with the Regular Army.

In October 1900, the battalion was sent to Sheffield and in April 1901, the War Office asked it to serve abroad. Though the required 75% of enlisted men volunteered for foreign service, the War Office did not send the battalion overseas as it lacked enough men fit for such service (underage, insufficient training, temporarily unfit due to illness).[109] On June 2, 1901, the North York Militia was disembodied.

The North York Militia was embodied a second time on February 17, 1902 and sent to South Africa to relieve another militia battalion. It was disembodied seven months later.

Cornwall and Devonshire Miners' Artillery [110]

This garrison artillery brigade of four companies was based in the southwest of England with its barracks at Pendennis Castle, Falmouth, Cornwall. In 1899, it had 10 of its authorized 11 officers and 322 of its authorized 384 militiamen. Of the enrolled militiamen, 120 (37.3%) were Militia Reservists. Its permanent staff was complete with 1 officer, 12 sergeants, and 4 buglers.

The brigade was embodied on May 1, 1990, at Pendennis Castle where it was quartered under canvas on the "hornwork" of the fortress. At the time of embodiment, its strength stood at 11 officers, 207 militiamen (53.9% of authorized strength), and 17 permanent staff. All its Militia Reservists had been called up to serve with the Regular Army.

On October 5, 1990, the War Office disembodied the brigade along with most other garrison artillery militia units.

---

[109] For an infantry battalion to serve abroad as a unit it had to field at least 500 volunteers and such volunteers had to represent at least three-fourths of enrolled militiamen. Army Orders, 1899, No. 93 issued pursuant to the Reserve Forces and Militia Act, 1898, 61 & 62 Vict. c. 9; MG G. Barton, *Minutes of Evidence Vol. 2, Report of His Majesty's Commissioners Appointed to Inquire into the Military Preparations and Other Matters Connected with the War in South Africa,* 1903, [Cd. 1791], q. 16343.

[110] *Annual Militia Return, 1899,* 1900, [Cd. 84]; Cavenagh-Mainwaring, *The Royal Miners,* 95-98; *Monthly Army List, December 1904.*

The Militia Reserve Division

In 1902, Parliament authorized a reserve for the militia, the Militia Reserve Division. The Division was a pool of men available to fill-out militia units upon their embodiment. Reserve Division members could be assigned to any unit within their arm (ex: an infantry militiaman could be assigned to any militia infantry battalion but not an artillery unit).[111] The War Office formed the Militia Reserve Division in April 1903 and set its establishment at 10,000 men. By September 1904, the Division numbered 7,082 men.[112] Generally, Reserve Division men did not train, but the War Office could order three days of musketry training annually for infantry members, and fourteen days training every second year for other members.[113] Members of the Militia Reserve Division, as reservists for the militia and not the army, could not be compelled to serve abroad.

Eligible for enrollment in the Reserve Division were militiamen who had completed 10-years' service and men discharged from the Regular Army who had completed 3-years' full-time service. The War Office prohibited retired soldiers from joining. Enlistment and re-enlistment terms were four years and early discharge could be purchased for £1 10s.[114]

Members could re-enlist until they attained the age of 50 and while enlisted were paid £4 10s. annually. There were no enlistment or re-enlistment bounties, but men called up for training received a 10s. training bounty in addition to army pay. Infantry reservists were administered by the staff of the regimental district in which they resided; artillery reservists by the artillery staff of their army district; engineers by the Royal Engineer Reserve, and medical reservists by the medical staff of their army district.[115]

The Volunteers in Great Britain

Another auxiliary component of the armed forces was the Volunteer Force of which there were no units in Ireland. The "volunteers" was the "respectable" military force. Units were raised voluntarily, and their services had been offered to and accepted by the Crown. One unit extant in 1904, the Honourable Artillery Company of London, was chartered by Henry VIII in 1537.

The statutory basis for the Edwardian Era volunteer force was the Yeomanry and Volunteers in Great Britain Act, 1804 as amended by the Volunteer Act, 1863.[116] At the height of the Napoleonic Wars the volunteer force totaled 336,000 men. After the Battle of Waterloo nearly all volunteer units disbanded and by the 1850s, the force was almost extinct.

---

[111] Militia and Yeomanry Act, 1902. 2 Edw. 7, c. 39.

[112] Army Orders, 1903, No. 36; *Annual Army Report, 1904*, 1905, [Cd. 2268].

[113] 2 Edw. 7, c. 39.

[114] Army Orders, 1903, No. 36.

[115] Ibid.

[116] 44 Geo. 3, c. 54; 26 & 27, Vict. c. 65.

Fear of invasion by Emperor Louis Napoleon's France led to a revival of the force in 1859. New units were raised by colonels commissioned by county lieutenants and funded by municipalities and individuals; arms were provided by the state. Volunteers served without pay during training and the chain of command ran through the county lieutenants to the Home Office, not the War Office. If called out due to invasion or rebellion, volunteer units would be under army command and the volunteers would to be paid by the state at army rates. At the end of 1862, the force totaled 163,000.[117] As volunteers were unpaid and had to buy their uniforms, they were of a much higher socio-economic stratum than militia and army recruits. In the mid-nineteenth century "no one was in any doubt that the Volunteer Force and the Militia were recruiting from quite different sections of the community."[118]

The character of the volunteer force was altered considerably by the Regulation of the Forces Act, 1871 which placed the volunteers, like the militia, under the authority of the War Office. Henceforth, volunteer officers received Queen's commissions, all volunteers were subject to military law when assembled for training, and at all times units were in the army chain of command.[119] In 1900, the volunteer force became liable for mobilization in case of "imminent national danger or of great emergency" not just rebellion and threat of invasion.[120]

By Bloomsday, the force was no longer privately funded and manned by unpaid volunteers. While localities, or the officers, provided training grounds and drill halls, the state funded the volunteer regiments through a capitation grant for each member who had met the annual training requirement (ten or fifteen week-night or Saturday afternoon drills plus a six-day encampment).[121] Compared to the militia, the volunteer force had little contact with the Regular Army. For example, an eight-company militia infantry battalion had a permanent staff of 29 regulars, a similar-sized volunteer battalion only 10.[122]

Though volunteer service did not carry the stigma of militia and army service, and training demands were minimal, in 1904 the force was severely under-strength. Of the 343,246 volunteer force positions authorized by Parliament, only 244,537 (71.2%) were filled. The shortfall in manning was likely due to the government's enlargement of the volunteer force. In 1898, immediately prior to the Boer War, authorized volunteer strength was 80,000 lower than in 1904: 263,416. At that time there were 223,926 active volunteers, 85.0% of establishment.[123]

---

[117] Berry, *A History of the Volunteer Infantry*, 102-52, 164.

[118] Cunningham, *The Volunteer Force*, 38.

[119] 34 & 35 Vict., c. 86.

[120] Volunteer Act, 1900. 63 & 64 Vict., c. 39.

[121] Mortgage loans to commanding officers for the purchase of drill halls were available from the Board of Public Works at 3.5%. Payments were made by the regiment's officers. COL A.B. Grant, commanding officer 1st Lanarkshire Artillery Volunteers, *Norfolk Commission Report, Minutes of Evidence*, 1904, [Cd. 2062], at qq. 10728-32.

[122] *Army Estimates for the Year 1904-05*, 1904, H.C. Accounts & Papers, No. 73.

[123] War Office, *Annual Return of the Volunteer Corps for 1904*, 1905 [Cd. 2438].

Enlisted Men

In 1903, most volunteers were of the "respectable working class" and the lower-middle class. Few were the sort that enlisted in the Regular Army. In 1904, only 3,236 volunteers joined the army which represented 1.3% of beginning of year volunteer strength. For the militia that year, 16.6% of its members joined the army.[124] Following is a distribution of enlisted volunteers and militiamen by occupation in 1903:[125]

	Volunteers	Militiamen
Artisans	37%	11%
Professionals, Clerks, Shopmen, Self-employed	19	---
Agricultural Laborers	4	22

While in the mid-nineteenth century few volunteers were laborers or industrial workers, by Bloomsday the lesser-paid urban workers accounted for between 25% and 40% of the force.[126] For such men, membership in the force conferred a degree of respectability.

Service in the volunteer force was "voluntary" and except when mobilized, any member could resign on 14-days' written notice to his commanding officer.[127] Despite the ease by which a volunteer could leave the force, annual turnover was much lower than in the militia. Resignations during 1904 represented only 20.4% of beginning of year strength.[128]

Officers

While the militia had a rural orientation, the volunteer force was primarily an urban organization as reflected in the paucity of volunteers with agricultural employment. Accordingly, volunteer officers were rarely of the landed-classes or of independent means; they were employed professionals and businessmen. Following is a distribution of infantry and artillery volunteer and militia officers by occupation in 1903:[129]

	Volunteers	Militia
Independent Means	6.4%	56.3%
Professionals	28.9	5.7
Business Owners	34.3	3.3
Students	3.3	1.2

---

[124] Ibid.; *Annual Army Report, 1906*, 1907, [Cd. 3365].

[125] Summary of Answers to Circular of Questions, *Norfolk Commission Report, Appendices*, 1904, [Cd. 2064].

[126] Ibid.; Cunningham, *The Volunteer Force*, 39.

[127] 26 & 27 Vict., c. 65, §7.

[128] War Office, *Annual Return of the Volunteer Corps, 1904*, 1905, [Cd. 2438].

[129] Summary of Answers to Circular of Questions, *Norfolk Commission Report, Appendices*, 1904, [Cd. 2064]. Militia data excludes candidates for Regular Army commissions.

Officers and Soldiers of the Auxiliary Forces

The reconstituted volunteer force of the mid-nineteenth century was funded privately and commanding officers were responsible personally for the debts of their regiments.[130] In 1881, Parliament required officers and men of the volunteers to indemnify their commanding officers for repayment of such debts.[131] In practice; however, only the officers signed as obligors on regimental debt.[132] This financial liability remained in the early twentieth-century as state capitation grants did not always cover regimental expenses. As of March 31, 1903, outstanding, aggregate mortgage debt on drill halls and rifle ranges was £458,517; unsecured bank debt totaled £40,669.[133]

Officers were commissioned into a "corps" the term for any multi-company formation. Officer aspirants required nomination by the lieutenant of the county in which the selected corps was located, or its commanding officer. Required approvals were those of the officer commanding the corps' territorial district and the general officer commanding the army district. Final decision as to commissioning was by the Office of the Inspector-General for Auxiliary Forces.[134] Unlike in the militia, an officer candidate did not need approval by his unit's commanding officer. Volunteer officers, unlike militia officers, were not subject to military law year-round.

As in the militia, a commission brought with it added expense. There were uniform and mess expenses plus officers usually had to contribute to the corps' maintenance: either for the drill hall's mortgage loan payment or to make up the shortfall between corps' expenses and the annual capitation grant. For a junior officer, this was from £15 to £35 annually; in "class corps" more.[135] The "class corps" were volunteer regiments that functioned like gentlemen's clubs and their enlisted personnel were of the middle class, typically professionals and civil servants. To enroll, one needed member nominations and approval by an enlistment committee. Unlike in ordinary regiments, enlisted recruits paid initiation fees. In London there were five such corps: London Scottish, Queen's Westminster, Inns of Court, Civil Service, and Artists.[136]

---

[130] 26 & 27 Vict., c. 65.

[131] Regulation of the Forces Act, 1881. 44 & 45 Vict., c. 57.

[132] LTC R.C. MacKenzie, commanding officer, 1st Volunteer Battalion, Highland Light Infantry, *Norfolk Commission Report, Minutes of Evidence*, 1904, [Cd. 2062], at qq. 10,968-74.

[133] *Norfolk Commission Report, Appendices*, 1904, [Cd. 2064], no. 94.

[134] MAJ G.S. St. Aubyn, Assistant Military Secretary, Eastern Army District, *Norfolk Commission Report, Minutes of Evidence*, 1904, [Cd. 2063], at q. 2320.

[135] Summary of Answers to Circular of Questions, *Norfolk Commission Report, Appendices*, 1904, [Cd. 2064]; Testimony of various officers, *Norfolk Commission Report, Minutes of Evidence*, 1904, [Cd. 2062]. For example: LTC F.W. Tannett-Walker, commanding officer 3rd Volunteer Battalion, West Yorkshire Regiment, at q. 7764, LTC W.C. Horseley, commanding officer 20th Middlesex Volunteers, at q. 8149, LTC J.A. Staveley, commanding officer 2nd Volunteer Battalion, East Yorkshire Regiment, at q. 8882, LTC R.C. MacKenzie, 1st Volunteer Battalion, Highland Light Infantry, at q. 10934.

[136] Rice, *My Bohemian Days in London*, 112-19; French, *Military Identities*, 209.

According to nearly all knowledgeable contemporaries, the cost of commissioned service was the primary reason for the high number of vacancies in the commissioned ranks. A secondary reason was the low status of volunteer officers compared to militia officers. In Scotland, the part of Great Britain where volunteerism was most popular, regimental commanders were referred to as "grocer colonels."[137] As noted previously, in 1904 only 74.4% of junior officer infantry and artillery positions were filled.

Volunteer officers received little or no full-time training. In their first two years of service they had to either take a one-month course for their arm of service or pass a qualifying examination. There were no subsequent training requirements; however, promotion to captain required a 14-day course in addition to passing an examination.[138] The Norfolk Commission concluded that the majority of officers "… have neither the theoretical knowledge nor the practical skill in the handling of troops which would make them competent instructors in peace or leaders in war."[139]

Volunteer Life

Military service consisted of evening and Saturday afternoon drills of one to three hours plus an annual encampment of six to fourteen days that included musketry or gunnery training. Instruction was provided mostly by regulars of the unit's permanent staff; one instructor-sergeant per company. For the infantry, drills were at the corps' drill hall where little practical training could take place. About 40% of garrison artillery companies were located near coastal fortifications and at times drilled at the works they would man when mobilized. The remainder trained at drill halls and went through artillery drill with obsolete guns.[140]

Like with the militia, the War Office considered recruits "trained men" after they completed the required, first-year training. Corps received a capitation grant for each man that completed the minimum annual training, both drill and camp. Those training requirements follow:[141]

	Recruits, 1st Year	Trained Men
Infantry	40 drills + 6-days camp	10 drills + 6-days camp
Other	40 drills + 6-days camp	15 drills + 6-days camp

---

[137] Cunningham, *The Volunteer Force*, 59-61; Beckett, *Britain's Part-Time Soldiers*, 178. In Scotland, 4.3% of the military-aged, male population in 1899 were volunteers. Cunningham, *The Volunteer Force*, 47.

[138] *Norfolk Commission Report, Appendices*, 1904, [Cd. 2064], no. 39.

[139] *Norfolk Commission Report*, 1904, [Cd. 2061], at 9.

[140] Summary of Answers to Circular of Questions, *Norfolk Commission Report, Appendices*, 1904, [Cd. 2064].

[141] Changes in the Conditions of Volunteer Efficiency, *Norfolk Commission Report, Appendices*, 1904, [Cd. 2064], no. 99.

Old Street Drill Hall, Ashton-under-Lyne, Greater Manchester, 2014

Gerald England, Creative Commons Share-Alike License

Volunteers could participate in a great many recreational activities sponsored by their corps. There were shooting teams, plus football, cricket, and dramatics clubs. A few drill halls were equipped lavishly with baths, reading rooms, and tennis courts.[142] Available recreational facilities and equipment were an important inducement to enrollment.[143] By Bloomsday, drill halls typically had libraries, game rooms, and canteens. Many corps regularly presented theatrical productions and concerts for its men.[144] For officers, there were balls and dinners throughout the year.[145] As volunteers were local men who assembled frequently throughout the year for both military and recreational purposes, there was in the volunteer force a camaraderie that the militia lacked.

---

[142] William Brodrick, Financial Secretary for War (later Secretary of State), 342 Parl. Deb. (3d ser.) (1890) 739.

[143] Beckett, *Britain's Part-Time Soldiers*, 177.

[144] Cunningham, *The Volunteer Force*, 116-19; Beckett, *Britain's Part-Time Soldiers*, 177-78; Katie Carmichael, *Drill Halls A National Overview*, English Heritage Research Report 6-2015.

[145] Beckett, *Britain's Part-Time Soldiers*, 178.

Officers received no pay but were given a camp allowance to help defray expenses. The allowance was inadequate and two weeks at camp would cost an artillery lieutenant about £7; an infantry major about £20. The daily out-of-pocket cost for camp attendance averaged as follows:[146]

Artillery		Infantry	
Lieutenants	9s. 9d	Lts. & Captains	13s. 2d.
Others	25s. 2d	Senior Officers	27s. 2d.

Enlisted men were not paid for drills but were paid for camp attendance. Annual camp was held for fourteen days but men only had to attend for six to meet the training requirement. Extra days at camp could not be used to meet the annual drill requirement. The amount a man received as camp pay was dependent on the financial condition of his corps. An artillery private could receive as little as 7s. for two weeks at camp; an infantry quartermaster-sergeant as much as 70s. Artillerymen were paid from 6d. to 2s. per day depending on rank; infantry from 1s. to 3s. for privates, 2s. to 4s. for corporals, 4s. to 5s. for senior sergeants.[147] Unlike in the militia, colour-sergeant and quartermaster-sergeant ranks were open to enlisted men. All infantry warrant officers; however, were regulars of the volunteer force's permanent staff who served as regimental sergeant-majors.

The Imperial Yeomanry

The Imperial Yeomanry was the auxiliary cavalry and had considerable prestige. The only units in Ireland were the North of Ireland Imperial Yeomanry and the South of Ireland Imperial Yeomanry. They were commonly referred to as the North Irish Horse and the South Irish Horse. The Imperial Yeomanry began as volunteer units raised by civilians to serve in South Africa during the Boer War. In August 1901, Parliament made the Imperial Yeomanry a permanent auxiliary force when it placed the old "Home" Yeomanry (extinct in Ireland since 1834) under the laws that governed the militia.[148] The old yeomanry was a collection of mounted corps organized by aristocratic landowners, officered by the gentry, and manned mostly by prosperous tenant farmers, and employees on the demesne lands of the founding colonel. Members served without pay but were armed by the state. It was created through the Volunteer Act of 1794.[149] Henry Labouchere, Liberal MP for Northampton, expressed the view of many late-Victorian Englishmen when he commented in Parliament as follows:[150]

---

[146] Summary of Answers to Circular of Questions, *Norfolk Commission Report, Appendices*, 1904, [Cd. 2064].

[147] Ibid., no. 96.

[148] Militia and Yeomanry Act, 1901, 1 Edw. 7, c. 14.

[149] 44 Geo. 3, c. 31.

[150] 337 Parl. Deb. (3d ser.) (1889) 80.

> "I regard the Yeomanry as a relic of past ages. We might as well have so many knights in armour, having regard to the service they render the country. The primary idea of a Yeomanry is that the country gentlemen should enrol their tenants, who should come with their horses and form a force perhaps rather more ornamental than useful."

Historians characterize the Home Yeomanry as a "feudal" force and although it was established to help repel an anticipated invasion force, its only employment was in aid of the civil authority. At a time when there were no organized police forces, the yeomanry and army were used by magistrates to maintain order and repress demonstrations they considered a prelude to "insurrection." As a result, the working class viewed the yeomanry as an instrument of the British establishment. Yeomanry aid to the civil authority ended in 1856 after creation of provincial police forces.[151]

The new yeomanry had characteristics of both the volunteer force and the militia. As in the militia, rankers served for fixed enlistment terms and were subject to sanctions for absence without leave; officers were always under military law. Both commissioned and enlisted members received pay at fixed rates. As with volunteers, there was relatively little required training, yeomen could attain senior NCO rank, and there were few regulars assigned to the permanent staffs of yeomanry regiments (the adjutant, a musketry sergeant, and four sergeant-instructors).

Enlisted Men

Engagement terms were three years and yeomen were supposed to be accomplished horsemen prior to enlistment. Initial training was 20 annual drills and attendance at the 16-day annual camp. Subsequent training was 10 annual drills plus the annual camp. A drill consisted of two to four hours of training, typically on a Saturday afternoon. Regulations required yeomen provide their own horse for training, though the animal could be owned by a relative or rented. For the annual training, yeomen received a £5 horse allowance plus a daily forage allowance.[152] In 1904, 45.5% of all horses at training were rented.[153] Upon mobilization, the War Office would provide the yeomanry with army horses.

Daily pay for the summer camp was 5s. 6d. for privates, 7s. 6d for corporals, 8s. 2d. for sergeants, and 8s. 10d. for senior sergeants. For two of the annual drills (mounted training by a complete squadron) yeomen received half-pay and they received 3s. daily during the three-day musketry training, regardless of rank.[154] Yeomen were not paid for other drills of which five were required for trained men, fifteen for recruits.

Of the three auxiliary forces, the yeomanry attracted the highest class of recruit. The mounted requirements (ability to ride and provide a horse) and the low amount of annual

---

[151] Beckett, *The Amateur Military Tradition*, 72-79, 134-43; Hay, *The Yeomanry Cavalry*, 137-70.

[152] *Regulations for the Imperial Yeomanry, 1903*, ¶¶271, 273. In Ireland, the daily forage allowance was 1s. 3d. in the Belfast District, 1s. 4d. elsewhere. Army Orders 1904, No. 15.

[153] War Office: *Imperial Yeomanry Training Return, 1904*, 1905, [Cd. 2267].

[154] *Army Estimates for the Year 1904-05*, 1904, H.C. Accounts & Papers, No. 73, appx. 8.

remuneration, limited the enlisted ranks to men of the middle and lower-middle class plus the highest-paid workmen.[155] By Bloomsday, artisans of the working class accounted for 10% to 20% of the yeomanry ranks.[156] Regiments in rural areas had a large percentage of farmers and their sons in the ranks, while urban regiments were predominantly middle and lower-middle class. Unlike the militia, there were no agricultural laborers in the yeomanry. A few county and city units, such as the Northhamptonshire Yeomanry and the South Nottinghamshire Hussars were "class corps."[157]

A Sergeant of the Denbighshire Hussars, 1907

GLDF, Creative Commons Share-Alike License

The War Office had little difficulty filling the enlisted ranks of this respectable, mounted force. In 1904, the yeomanry was at 94.3% of enlisted strength.[158] The two regiments in Ireland were at 96.0% of their total enlisted strength of 888 men.

Officers

Officers were commissioned into a regiment and young gentlemen were nominated for commissions by the county lieutenant.[159] Approval of the regiment's commanding officer was required. Few young men of the officer classes had the horsemanship, inclination, and money required for a yeomanry commission. Of qualified young gentlemen eager to serve, those of the landed classes were more likely to receive offers of commissions than middle

---

[155] Bowman and Connelly, *The Edwardian Army*, 108, 136-37.

[156] Hay, *The Yeomanry Cavalry*, 82-95.

[157] Bowman and Connelly, *The Edwardian Army*, 111.

[158] War Office: *Imperial Yeomanry Training Return, 1904*, 1905, [Cd. 2267].

[159] Regulation of the Forces Act 1871, 34 & 35 Vict., c. 86, §6.

class aspirants.[160] As a result, many junior officer positions went unfilled. In 1904, while only 1.5% of the senior officer positions were unfilled, 25.1% of junior officer positions were vacant.[161] The two Irish regiments had far more unfilled positions with one-third of the 36 junior positions vacant. This may have been due to the high social position required, or perceived as such, for a commission in the Irish Horse. Both commanding officers were peers and the North Irish Horse had four other titled officers, the South Irish Horse five.[162]

As in the Regular Army and the other auxiliary forces, pay and allowances received by officers did not cover their military-related expenses. The yeomanry was the costliest auxiliary force for officers. In 1902, officers on average were out-of-pocket £60 to £100 annually. For some it was more. Entry costs were also high. The cost of an officer's uniform was about £150 and entry contributions to the regimental fund were from £5 to £50.[163]

In 1904, second lieutenants received daily pay and allowances of 16s. 8d., captains £1 3s., and majors £1 6s. The commanding officer, a lieutenant-colonel, received £1 11s. 6d. daily. Officers were paid for the 16 days of camp training and received half-pay for each of two mounted, squadron drills. Unlike yeomen, officers were not paid for annual musketry training. The major who was second-in-command of the regiment received an extra shilling per day.[164] Like enlisted men, officers received a £5 horse allowance plus a daily forage allowance.

Yeomanry Service

Drill took place at many locations as yeomanry regiments were often widely dispersed territorially with each squadron having its own barracks or small drill hall. Additionally, there were small training facilities, such as meeting halls and rifle ranges, scattered throughout a squadron's territory. To limit travel, most drills were held in small facilities and attended by a few men, 10 to 15. The two mandatory squadron drills, mounted training by all 115 men of a squadron, took place on army land, rented space, or an officer's estate. Musketry training was at the regimental or squadron barracks where the yeomen were quartered and fed for three days.

The annual summer training was at an army camp or on the colonel's estate.[165] Like in the militia, this was the only time during the year that the entire regiment assembled. The annual camp was a newsworthy event for the small towns near the campsite. For example,

---

[160] Beckett, *Britain's Part-Time Soldiers*, 189-90.

[161] *Monthly Army List, December 1904*; War Office, *Imperial Yeomanry Training Return, 1904*, 1905, [Cd. 2267].

[162] The North regiment was commanded by the Earl of Shaftesbury, the South by the Marquess of Waterford. *Monthly Army List, December 1904*.

[163] Hay, *The Yeomanry Cavalry*, 34, 39. In contrast, a militia officer's out-of-pocket uniform cost was £40 to £65, a volunteer officer's £20.

[164] *Army Estimates for the Year 1904-05*, 1904, H. C. Accounts & Papers, No. 73, appx. 8.

[165] Bowman and Connelly, *The Edwardian Army*, 138. From 1905 to the outbreak of the First World War, Winston Churchill was an officer of the Queen's Own Oxford-shire Hussars. The regiment's annual camp was frequently at his family's estate, Blenheim. Hay, *The Yeomanry Cavalry*, 63.

the *Denbighshire Free Press*, a weekly publication in North Wales, gave a day-by-day account of the Denbighshire Hussars' 1907 training at Denbigh, and listed the names of winners of the military and athletic competitions.[166] The following training schedule is based on that encampment.

Day	Activity
1	Assembly, in-processing, and medical inspection.
2	Travel to training site and pitch tents.
3-14	Training Days: 9 Training Half-Days: 1 Sundays: 2

Troop Drill and Other Training:	2.5 Days
Squadron Drill:	3 Days
Regimental Drill:	2 - 3 Days
Tactical Exercise:	0 - 1 Day
Military & Athletic Competitions:	1 Day

Day	Activity
15	Strike camp and travel to home barracks.
16	Disbursement of pay, out-processing, and dismissal.

During the year, yeomanry regiments, like volunteer units, provided recreational and social activities for its members. The officers' mess was far more active socially than in the other auxiliary forces. Balls and dinners were held throughout the year, adding greatly to the cost of yeomanry commissioned service.[167]

---

[166] *Denbighshire Free Press*, June 15, 1907.

[167] Hay, *The Yeomanry Cavalry*, 60, 104.

Chapter Bibliography

Beckett, Ian F.W., *Britain's Part-Time Soldiers*. Barnsley, UK: Pen & Sword, 2011.

Berry, Robert Potter. *A History of the Formation and Development of the Volunteer Infantry*. London: Simpkin, Marshall, Hamilton & Kent, 1903.

Bowman, Timothy and Mark Connelly. *The Edwardian Army: Recruiting, Training, and Deploying the British Army, 1902-1914*. Oxford: Oxford Univ. Press, 2012.

Cavenagh-Mainwaring, James Gordon. *A History of the Stannaries Regiment of Miners," The Royal Miners."* London: Harrison, 1913.

Cunningham, Hugh. *The Volunteer Force*. London: Croom Helm, 1975.

French, David. *Military Identities: The Regimental System, the British Army, and the British People c. 1870-2000*. New York: Oxford Univ. Press, 2005.

Goodenough, W.H. and J.C. Dalton. *The Army Book for the British Empire*. London: HMSO, 1893.

[Grierson, James Moncrieff]. *The British Army*. London: Sampson, Low, Marston, 1899.

Hay, George. *The Yeomanry Cavalry and Military Identities in Rural Britain*. Cham, Switzerland: Palgrave Macmillan, 2017.

Herbert, Trevor and Helen Barlow. *Music & the British Military in the Long Nineteenth Century*. Oxford: Oxford Univ. Press, 2013.

Perry, Nicholas. "The Irish Landed Class and the British Army, 1850-1950." *War in History* 18, no. 3 (July 2011): 304-32.

Price, Julius M. *My Bohemian Days in London*. London: Laurie, 1914.

Somerset, George (3rd Baron Raglan). "The Militia," *The National Review* 27 (1896):237-56.

Spears, Edward. *The Picnic Basket*. New York: Norton, 1967.

Stoneman, Robert James. "The Reformed British Militia, c. 1852-1908." PhD Thesis, University of Kent, 2014.

Thoyts, Emma Elizabeth. *The History of the Royal Berkshire Militia*. Reading: Hawkes, 1899.

Turton, Robert Bell. *The History of the North York Militia*. Leeds: J. Whitehead, 1907.

Appendix A
The British Army and Auxiliary Forces, 1904

The War Office and Civilian Control [1]

The Secretary of State for War was responsible to the Cabinet for all military matters, headed the War Office, and was at the apex of the army chain of command. The Army Council was the War Minister's advisory panel and consisted of the Under-Secretary of State for War, the Financial Secretary, the Chief of the General Staff, the Adjutant-General, the Quartermaster-General, and the Master-General of the Ordinance. All army regulations, orders, and directives were issued by the Army Council.

Though the head of each army command, both at home and abroad, reported to the War Minister, day-to-day contact between commands and the War Office involved the department responsible for the matter at hand. For example, Southern Command would deal with the Adjutant-General's office on personnel matters and the General Staff on training matters. The War Minister appointed and relieved generals in command of army districts, the Indian Army, and colonial garrisons.

Commanders of expeditionary forces would report to the Chief of the General Staff and not the War Minister. If threatened with invasion, the army field corps formed at home would also be subordinate to the Chief of the General Staff. The same change in command would apply to colonial garrisons engaged in military operations. The Chief of Staff was the senior army officer, and, in some respects, his position was successor to that of Commander-in-Chief.

The War Minister's personal staff was headed by a military secretary and his senior civil advisor was the Permanent Under-Secretary of the War Office. The Office of the Military Secretary had ultimate authority for the grant of Regular Army commissions. The War Office was organized into seven departments, two of which were "civil."

The department headed by the Under-Secretary of State for War was responsible for barrack construction and army chaplains. The Under-Secretary, by tradition a member of the House of Lords, stood in for the Secretary in case of illness or absence. The department headed by the Financial Secretary (a member of Commons) was responsible for contracts, accounting, audits, and preparation of the annual, Parliamentary army estimates.

The General Staff was responsible for war plans and training and would direct army operations in time of war. The Adjutant-General was the army's chief administrative officer and handled personnel matters. In addition to personnel and administrative functions, his office supervised the auxiliary forces and had final authority over new militia, volunteer, and yeomanry commissions. The Quartermaster-General was the army's chief logistics officer.

---

[1] *Minutes of Evidence taken before the Committee to inquire into War Office Organization*, 1901, [Cd. 581], at appx. 2; *Reports of the War Office (Reconstituted) Committee*, 1904, [Cd. 1932, Cd. 1968, Cd. 2002].

Appendix A

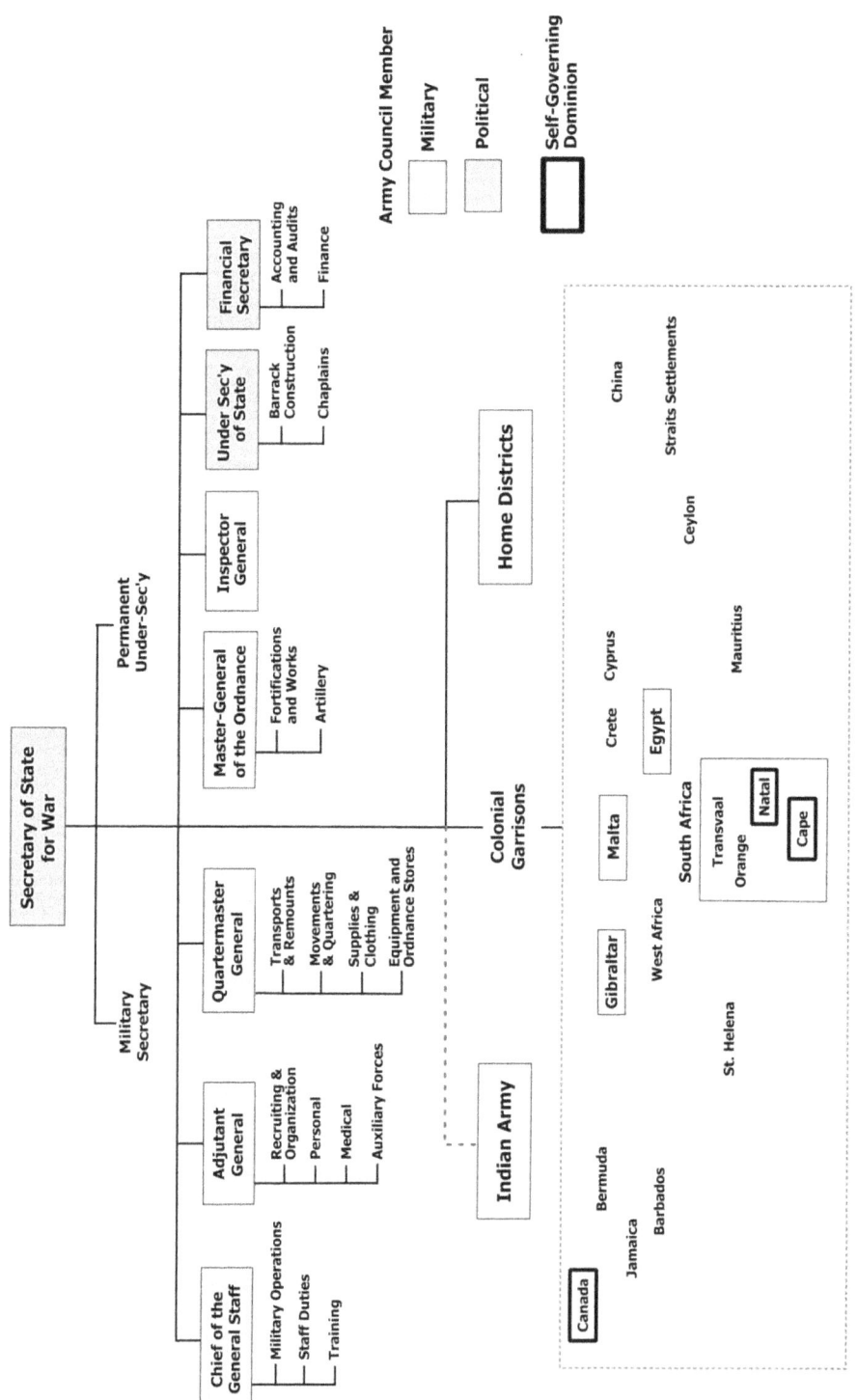

306

Appendix A

The Master-General of the Ordnance was responsible for weapons development and production, maintenance of fortifications, and supervision of the artillery. The Inspector-General's office conducted operational audits of the army to ensure that policies were implemented. It also evaluated the combat readiness of units and investigated formal complaints. Unlike the other military chiefs, the Inspector-General was not a member of the Army Council. This was to foster a degree of independence so that he would not have to evaluate programs he supported, or opposed, in Army Council deliberations.

In 1904, the War Office employed about 150 army officers, 100 enlisted personnel, and 800 civil servants (including 100 army pensioners). Its main offices were situated in Pall Mall, St. James Square, Horse Guards, and Victoria Street, London. This staff of 1,100 managed an army of about 300,000 regulars, 80,000 reservists, and 370,000 part-time auxiliaries.

<u>Statutory Army Structure and Manning</u>

British Army Strength, September 30, 1904 [2]
(All Ranks)

	Authorized	Actual	Pct.
Regular Army	294,117	282,390	96.0
Auxiliary Staff	7,122	6,862	96.4
Total Regulars	**301,239**	**289,252**	**96.0**
Army Reserve	**80,000**	**74,940**	**93.7**
Militia	126,941	89,100	70.2
Volunteer Force	341,113	254,412	74.6
Imperial Yeomanry	27,040	26,751	98.9
Total Auxiliaries	**502,216**	**370,268**	**73.7**
Militia Reserve Division	**10,000**	**7,082**	**70.8**

War Office Sources: *Annual Report on the Army for the Year Ending 30th September 1904*, 1905, [Cd. 2268]; *Imperial Yeomanry Training Report 1904*, 1905, [Cd. 2267]; *Return showing the Establishment of each Unit of Militia 1904*, 1905, [Cd. 2432]; *Annual Return of the Volunteer Corps 1904*, 1905, [Cd. 2438].

In 1904, the army accounted for about 1.8% of the United Kingdom's population: 0.7% full-time regulars, 1.1% reservists and auxiliaries. These proportions are almost the same as for 1874; 0.6% full-time regulars, 1.2% reservists and auxiliaries.[3]

Compared to the nations of Continental Europe, the United Kingdom had an exceedingly small proportion of its population under arms. That was because nearly all Continental states

---

[2] Excludes 7,538 Indian Army personnel on loan to the British Army.

[3] War Office, *General Annual Return of the British Army for the year 1884*, 1885, [C. 4570]. Hereafter cited as *Annual Army Report*.

Appendix A

had compulsory military service which facilitated the maintenance of large standing armies and reserve forces. Though in such nations all young men could be compelled to serve in the army, military service was not universal for physically and mentally fit young men. There were occupational, hardship, and "moral character" exemptions. More importantly, no European state could afford to train, arm, and equip all its militarily eligible young men. In Russia, about 30% of eligible males were conscripted, in Germany about 55%, and in France 70%.[4]

The nations that had compulsory service utilized the "Prussian Reserve System" which worked as follows: The regular army was essentially a training establishment for conscripts who served for two to four years. After conscripts completed their army service, they became reservists who underwent periodic refresher training. In time of war, these reservists would bring regular units to war-time strength, man additional units composed nearly entirely of reservists, and form a manpower pool for casualty replacement.[5] The following table shows the proportion of the population under arms in the ground forces for the United Kingdom and several other nations.[6] Note that the United States, like the United Kingdom, had wholly voluntary armed forces.

Percent of Population in the Army and its Reserve, 1902
(Population in Millions)

	Population	Percent of Population		
		Army	Reserves	Total
United Kingdom	41.5	0.7	1.1	1.8
United States	76.1	0.1	0.2	0.2
France	38.6	1.7	4.7	6.4
Germany	56.4	1.0	3.6	4.6
Austria-Hungary	45.2	0.8	3.5	4.3
Italy	32.5	0.8	2.5	3.3
Russia	141.0	0.8	1.9	2.7

Sources: *The Statesman's Year-Book 1903*; War Office, *Appendices to the Minutes of Evidence taken before the Royal Commission on the Militia and Volunteers*, 1904, [Cd. 2064], nos. 1, 3.

The United States did not have a regular army reserve. The trained, part-time soldiers of the United States were the members of the National Guard. The National Guard was

---

[4] Janet Robinson and Joe Robinson, *Handbook of Imperial Germany* (Bloomington, IN USA: Author House, 2009), 145-60; *The Statesman's Year-Book, 1903*.

[5] *Appendices to the Minutes of Evidence taken before the Royal Commission on the Militia and Volunteers*, 1904, [Cd. 2064], no. 1.

[6] Trained reserves exclude the French Territorial Army, the German *Landsturm* and untrained *Ersatz Reserve*, and the Italian Territorial Militia. Those components consisted of men who received very little training or had not served with the standing army for over twelve years.

composed of state militias organized and maintained in accordance with national statutes. The state militias were armed by the federal government and they also received federal operating subsidies. The National Guard was supervised somewhat by regular army personnel and was similar to the Volunteer Force in Great Britain. In October 1902, the U.S. Army had 63,686 men; the National Guard 122,213. The two forces together accounted for 0.24% of the U.S. population.

Corps and Departments of the Army

Regular Army Strength, September 30, 1904 [7]
(All Ranks)

	Formations	Depots	Strength	Pct.
Foot Guards	10 battalions	1	7,862	
Line Infantry	161 battalions	69	161,035	
Total Infantry			**168,897**	**59.8**
Horse & Field	188 batteries	9	31,165	
Garrison Artillery	107 companies	6	24,524	
Total Artillery			**55,689**	**19.7**
Household Cavalry	3 regiments		1,373	
Line Cavalry	28 regiments		20,317	
Total Cavalry			**21,690**	**7.7**
Colonial Corps	**30 units**	**1**	**6,599**	**2.3**
Engineers	69 companies	2	11,457	
Army Service Corps	86 companies	1	6,903	
Medical Corps	22 companies	1	4,954	
Ordnance	20 companies	1	2,562	
Pay Corps			903	
Veterinary			48	
Staff & Departmental			2,688	
Total Support			**29,515**	**10.5**
Total Army			**282,390**	**100.0**

Sources: *Annual Army Report, 1904*, 1905, [Cd. 2268]; *Monthly Army List, December 1904*.

As shown in the above table, in 1904, the army was overwhelmingly a combat organization with a "teeth-to-tail" ratio of 9 to 1. The Boer War experience resulted in an increase in support and service personnel; twenty years earlier the teeth-to-tail ratio was 16 to 1.[8]

---

[7] An additional 7,538 Indian Army personnel were on loan to the British Army and carried on the rolls of the Colonial Corps.

[8] *Annual Army Report, 1884*, 1885, [C. 4570]; *Annual Army Report, 1904*, 1905, [Cd. 2268].

Appendix A

	1904		1884
Arms (Cavalry, Infantry, Artillery, Colonial)	260,413	89.5%	94.0%
Support, Services and Staff	29,515	10.5	6.0

Infantry

The infantry, by far the largest component of the army, consisted of the elite Foot Guards (10 battalions) and the line infantry (161 battalions). The line infantry was the dumping ground for the least educated, least fit recruits. Foot Guards commissions required approval of the battalion's commanding officer and the enlisted men were of larger physique and better character than their line infantry counterparts. For example, Guardsmen had to be at least 5' 9" in height while line infantry only 5' 4". Officers and enlisted men of the Foot Guards received higher pay than their line infantry counterparts.

<u>Grenadier Guards (3 battalions)</u>: Created 1661 by King Charles II from Russell's Regiment of Foot. That regiment had been raised in England shortly after Charles returned to England from France.

<u>Coldstream Guards (3 battalions)</u>: Created 1661 by Charles II from Monck's Regiment of Foot (Cromwell's New Model Army).

<u>Scots Guards (3 battalions)</u>: Raised 1660 by the Earl of Linlithgow under commission granted by Charles II. It was part of the Scottish Military Establishment until the Union of Scotland and England in 1708.

<u>Irish Guards (1 battalion)</u>: Created in 1900 at Queen Victoria's request. It was formed by the War Office to commemorate the Irishmen who fought in the British Army during the Boer War.

Royal Artillery

The artillery had three components: Horse, Field, and Garrison. For officers, the Horse Artillery was the most prestigious branch, the Garrison Artillery the least. Artillery enlisted men, termed "gunners," were of larger physique than the typical soldier.

The artillery was organized mostly into batteries (companies for garrison artillery) which were grouped into formations termed "brigades" though they were battalion-sized. Some gunners were assigned to UK artillery districts, not batteries or companies. Such troops maintained guns in coastal fortifications that in time of war would be manned by auxiliaries; militia or volunteer force.

Cavalry

The cavalry consisted of three guards regiments termed "Household Cavalry" and 28 line regiments. Line cavalry regiments bore the title hussar, dragoon, or lancers, though in 1904 those distinctions were meaningless: All regiments were equipped, armed, and trained the same.

The Household Cavalry was at the apex of the army's social hierarchy. Commissions in these elite regiments required approval of the colonel-in-chief, an honorary position held by an aristocrat. Like the Foot Guards, its enlisted men had to meet higher standards than troopers of the line cavalry. Both officers and enlisted men received higher pay than their line cavalry counterparts.

1st Life Guards: Formed in 1661 by Charles II from his mounted bodyguard and those of the Duke of York and General George Monck (a lieutenant-general in Cromwell's New Model Army).

Royal Horse Guards: Raised in 1661 by Charles II.

2nd Life Guards: By 1788 the Life Guards regiment formed in 1661 had grown to six squadrons. That year it was split into two separate regiments. The 2nd Life Guards was formed from the 2nd Troop of Horse Guards and the 2nd Troop of Horse Grenadier Guards.

Royal Engineers

Of necessity, the Royal Engineers obtained the best educated and highest skilled recruits. By 1904, all recruits had to be qualified in a trade to serve with the Royal Engineers. Engineers were organized into companies some of which were specialized (telegraph, bridging, survey, balloon, mounted). The corps was a combat arm.

Army Service Corps

Though a support organization, the War Office classified the ASC as a combat arm because on active service its members were exposed to enemy fire. This corps was responsible for supply of the meat and bread ration and all army transport and. The ASC had many skilled, high paid positions for enlisted men. The corps was organized in companies.

Royal Army Medical Corps

All officers of the RAMC were qualified surgeons and its quartermasters were qualified dispensing chemists (pharmacists). The RAMC was organized in medical companies and hospitals. There were detachments assigned to each regiment, garrison, camp, and station.

Ordnance Department / Ordnance Corps

This was the army's logistics component that also developed, acquired, and maintained weapons. The Ordnance maintained warehouses and depots and manned munitions companies for active service. Officers were commissioned

Appendix A

into the Ordnance "Department" and enlisted men served in the Ordnance "Corps."

Pay Corps

Responsible for pay and all other cash disbursements, its officers, all with combatant commissions, were the highest paid in the army. Pay corps appointments were prized by impecunious officers.

Veterinary Corps

The smallest of army corps, it was responsible for the care of army horses. Its officers were all qualified veterinarians.

Post Office Corps

This corps consisted solely of enlisted reservists who were experienced Post Office employees.

The Colonial Corps

Infantry	Garrison Artillery	Engineer
West Indies Regiment (2 battalions + depot)	West Indies Battalion (5 companies)	West Indies Company (harbor mining)
	Royal Malta Artillery (5 companies)	
West African Regiment (1 battalion)	West African Company	Sierra Leone Company (fortress)
Chinese Regiment (1 battalion)	Hong Kong/Singapore Battalion (6 companies)	Hong Kong Company (harbor mining)
		Singapore Company (harbor mining)
	Ceylon/Mauritius Battalion (4 companies)	Ceylon Company (harbor mining)
		Mauritius Company (harbor mining)

Appendix A

The men of this corps were native to the colonies in which its units served. Except for the Royal Malta Artillery, all Colonial Corps officers were British. Officer appointments in the West Indies Regiment were permanent as the regiment was on the "home" establishment. Officers for all other colonial units were seconded from their regiments. The enlisted ranks were filled by natives.

In terms of social standing, the colonial units were at the bottom of the hierarchy. The West Indies Regiment appealed to officers without independent means as in that regiment they could live comfortably on army pay alone.

The Colonial Corps was separate and apart from the local forces of the colonies which were ultimately subordinate to the Colonial Office, not the War Office. The largest colonial forces were the West African Field Force (8 infantry battalions and 3 artillery batteries) and the King's African Rifles (5 infantry battalions). Local colonial forces included all-European part-time volunteer units, British-officered paramilitary police, and the South African Constabulary.

Rank Distribution

Rank distribution in the British Army remained practically unchanged for 30 years. Officers accounted for about 4% of personnel, NCOs and Warrant Officers 13%, and the rank-and-file 83%.[9]

	1904	1894	1884	1875
Commissioned Officers	4.2%	3.5%	3.9%	4.4%
Warrant Officers	0.5	0.4	0.3	Created 1883
Sergeants	6.7	6.2	6.7	6.9
Corporals	5.7	5.4	6.1	6.0
Buglers/Drummers	1.5	1.5	1.8	2.0
Privates	81.4	83.0	81.2	80.7

Compared to the British Army in the early twenty-first century, the Victorian army had very few officers and NCOs.[10]

	1904	2014
Officers	4%	14%
WOs & NCOs	13	27
Rank & File	83	59

---

[9] *Annual Army Report, 1904*, 1905, [Cd. 2268]; *Annual Army Report, 1894*, 1895, [C. 7885]; *Annual Army Report, 1884*, 1885, [C. 4570]. The number of corporals were not enumerated separately in reports prior to 1875. Note that the War Office created the Warrant Officer classification in 1883.

[10] Ministry of Defence, *U.K. Armed Forces Annual Personnel Report, 1 April 2014*.

Appendix A

<u>The Irish and Corps Assignment</u>

Irish soldiers were always over-represented in the unskilled and low-skilled corps and over the 20 years prior to 1904 the disparity decreased only marginally. In 1884, the Irish were disproportionate among the low-skilled soldiers by a factor of 1.09; in 1904 by 1.07.

Distribution of Enlisted Men by Corps, 1884 & 1904
Percent Irish and Army

	1884 Irish	1884 Army	1904 Irish	1904 Army
Infantry, Line	73.6	63.7	61.9	59.8
Artillery	17.7	16.9	22.4	20.4
Transport	0.9	1.5	1.6	2.4
Total, Low-Skilled	91.1	83.4	86.0	80.4
Cavalry, Line	4.8	8.4	4.6	7.5
Household Cavalry	0.4	0.7	0.2	0.8
Foot Guards	1.1	3.3	3.7	2.9
Total, Prestigious	6.2	12.4	8.5	11.7
Medical	0.8	1.2	1.3	1.5
Ordnance	0.2	0.3	0.5	0.8
Engineers	1.7	2.7	3.5	3.9
Pay	NA	NA	0.3	0.3
Total, High-Skilled	2.7	4.2	5.5	7.9

Sources: *Annual Army Report, 1884*, 1886, [C. 4570]; *Annual Army Report, 1904*, 1905, [Cd. 2268].

Irish over-representation in the least-skilled corps was likely due to the occupational and educational backgrounds of Irish recruits. For the twelve months ended September 30, 1904, 23.7% of the recruits from Ireland claimed prior employment in the least-skilled occupations compared to 15.5% of recruits from elsewhere. Those recruits were unskilled laborers, carmen, carters, and porters. The disparity in corps assignment was probably also due to the higher illiteracy rates in Ireland compared to Great Britain over the fourteen years ended in 1904. While literacy data for males five years of age or older is available from the 1891 census, such data for the 1901 census was collected only in Ireland. On Census Day in Ireland, 1901, 13.5% of Irish males could neither read nor write. For national comparison purposes, male literacy for 1904 can be measured by the percent of men married that year who couldn't sign their name to the marriage certificate.[11] Note that though for officers the artillery was viewed as a "scientific" corps, enlisted men were selected as gunners for their size and strength (heavy lifting was required to position guns and feed them ammunition). Following are the male illiteracy rates for the nations of the United Kingdom in 1891, 1901, and 1904.

---

[11] *Annual Reports of the Registrars for Ireland, England & Wales, and Scotland, 1904*, 1905 [Cd. 2673], *1905*, 1906 [Cd. 2617], *1906*, 1907 [Cd. 3200].

Appendix A

	1891	1901	1904
Ireland	13.2%	13.5%	13.4%
England & Wales	2.8	NA	1.8
Scotland	3.4	NA	1.7

While Irish recruits were disproportionately unskilled compared to recruits of other nationality, and Irish males more likely to be illiterate than British males, Irish recruits were over-represented among army applicants who were formerly somewhat educated clerks (2.8% compared to 1.7% for the rest of the UK), and were proportionately represented among former professionals and students (0.7% compared to 0.8%).[12] This raises the issue as to whether anti-Irish prejudice of recruiting officials affected the corps allocation of Irish recruits. Ethnic prejudice could have caused rejection of Irish applicants to the skilled corps, while English, Scottish, and Welsh applicants with similar qualifications as the rejected Irish, were accepted. No matter what the underlying causes, recruitment data show an Irish recruit was more likely than others to end up as cannon-fodder and should he survive the army until discharge or retirement, more likely than others to enter the ranks of the unskilled, unemployed.

The tripling of the percent Irish in the footguards within twenty years was due to the creation in 1900 of the Irish Guards Regiment (1 battalion). Previously, there were two kingdom-wide guards regiments (Coldstream and Grenadier, 3 battalions each) and a Scottish guards regiment (3 battalions).

First Recruiting Poster for the Irish Guards

---

[12] *Annual Army Report, 1904*, 1905, [Cd. 2268].

Appendix A

Recruits and Recruiting

As in earlier eras, the Victorian and Edwardian British establishments demanded a socially bifurcated army of patrician officers and plebian enlisted men. Compulsory military service was unacceptable politically, so the army relied on working class youths to fill its ranks voluntarily. For working class families, a son's enlistment was a shameful event equivalent to entry into the workhouse. Accordingly, nearly all recruits were desperate and impoverished.[13] Most were unskilled and uneducated young men whose only previous employment was as the lowest-paid laborers. Some recruits; however, were skilled, urban workers who had several years of primary schooling. It was from this group that the army obtained its clerks, soldier-tradesmen, and NCOs. What nearly all recruits had in common was the lack of both money and employment.[14]

Very few working class youngsters aspired to an army career. Those who did were either sons of soldiers, residents of orphanages and workhouses, or of the handful who sought an adventuresome life.[15]

Occupations of Recruits

In the early nineteenth century, farms and rural towns provided most of the army's enlisted personnel and the majority of recruits were formerly casual, agricultural laborers. As agricultural employment declined during the nineteenth century, so did the number of rural recruits. By the mid-nineteenth century recruits were no longer mostly agricultural laborers, but a majority still came from rural areas. In the 1890s the industrialized cities surpassed the countryside in providing soldiers for the army. In 1904, the typical recruit was the unemployed urban laborer or unskilled, factory worker.[16] Though the geographic and employment backgrounds of recruits had changed, their socio-economic status had not. Unskilled, working class young men remained a majority of new soldiers and their preponderance increased over time. The small proportion of recruits who had white-collar employment remained constant, while the proportion of skilled workmen declined. As shown in the table below, during the forty years before Bloomsday, the proportion of recruits who categorized themselves as unskilled workers or domestic servants, increased progressively from 59.2% to 67.9%. For that same period, except for 1903, the proportion of recruits with former occupations of a relatively high level of literacy and numeracy fluctuated between 6.6% and 7.6%. Skilled artisans, mechanics, and construction workers accounted for 31.5% of new soldiers in 1864 but only 22.6% in 1903.

---

[13] Alan Ramsay Skelley, *The Victorian Army at Home* (London: Croom Helm, 1977), 243-49; Edward M. Spiers, *The Army and Society* (New York: Longman, 1980), 44-52; Timothy Bowman and Mark Connelly, *The Edwardian Army* (Oxford: Oxford Univ. Press, 2012), 41-42.

[14] Medical examiners found that at the time of enlistment 90% of recruits were without work. War Office, *Report of the Health of the Army for 1909*, 1911, [Cd. 3477].

[15] Skelley, *the Victorian Army at Home*, 247-49; Spiers, *The Army and Society*, 45; Bowman and Connelly, *The Edwardian Army*, 46.

[16] Skelley, *The Victorian Army at Home*, 298.

Appendix A

Occupations of Recruits, 1860 - 1903
Percent of Total

Stated Occupation	1860	1864	1874	1884	1894	1903
Laborer, Grazier, Servant	50.3	59.2	61.9	63.3	65.1	67.9
Manufacturing Artisan	14.2	14.2	11.6	12.6	14.7	11.4
Mechanic, Construction Worker	25.0	17.3	17.6	14.5	9.9	11.2
Shopman, Clerk	9.1	6.5	5.8	6.3	6.4	4.9
Professional, Student	0.4	0.7	0.8	1.0	1.2	0.7
Boy	1.0	2.1	2.2	2.3	2.7	3.9

Sources: Army Medical Department, *Statistical, Sanitary, and Medical Reports for the Year 1860* and *1864*, 1863, [3051], 1866, [3730]; *Army Medical Department Reports for the Years 1874, 1884, 1894, and 1903*, 1876, [C. 1465], *1886*, [C. 4846], 1896, [C. 7921], 1905, [Cd. 2434].

Education and Literacy

Compared to their civilian counterparts, army recruits were poorly educated. At the turn of the twentieth century only 6% to 7% of them had a complete, primary school education; however, hardly any recruits were completely illiterate.[17] In the late-nineteenth century, Parliament expanded educational opportunities for the working class and later imposed compulsory education; first in Scotland (1872), then in England & Wales (1880), and finally Ireland (1892).[18] Presumably, mandated school attendance together with widely available, free primary schools, caused a decline in the recruit illiteracy rate from 26.1% in 1868, to 1.8% in 1904.[19]

Overall, young men who entered the Edwardian Era army were poorly educated. The percentage of recruits the army considered well-educated had changed little in the 46 years ending in 1904, having risen from 4.7% to only 6.6%.[20] In 1907, the War Office began to measure the education level of recruits according to civilian standards. From that year until the outbreak of the First World War, only one in five met or exceeded the standard for eleven-year-olds and one in ten was functionally illiterate. The balance was at the educational level required of children aged eight through ten.[21]

---

[17] *Army Medical Department Report for the Year 1904*, 1906, [Cd. 2700].

[18] Education (Scotland) Act, 1872, 35 & 36 Vict., c. 62; Elementary Education Act 1880, 43 & 44 Vict., c. 23; Education (Ireland) Act, 1892, 55 & 56 Vict., c. 42.

[19] *Army Medical Department Reports for 1869* and *1904* - 1871, [C. 336], at appx. xxx; 1906, [Cd. 2700], at 50.

[20] Army Medical Department, *Statistical, Sanitary, and Medical Reports for the Year 1860*, 1863, [3051]; *Army Medical Department Reports for 1904*, 1906, [Cd. 2700].

[21] Spiers, *The Army and Society*, 64-66. For 1909, 26% met or exceeded the standard for 10-year olds and 13% were functionally illiterate. F.D. Acland, Financial Secretary to the War Office, 1 Parl. Deb. (H.C. 5th ser.) (1909) 1670.

Appendix A

Boy Recruits

As shown in the previous table, in the last half of the nineteenth century the army increasingly relied on boys to fill its ranks (proportion increased progressively from 1.0% to 3.9%). Eighteen was the minimum age for regular enlistment; however, a few seventeen-year-olds could be "specially enlisted."[22] Boys were recruits who were at least fourteen years of age but younger than seventeen. In the 1890s, boys were enlisted as either apprentice musicians or tailors.[23] There was a need for such apprentices as for example, each infantry battalion was authorized a bandmaster, band-sergeant, band-corporal, twenty bandsmen plus a sergeant-drummer and sixteen drummers.[24] The army also required one tailor for every 200 to 250 enlisted men. By Bloomsday, boy recruits could be placed into the Royal Army Medical Corps and Army Service Corps, or apprenticed to telegraphists and artificers of the Royal Engineers.[25] Boys who had received institutional training as shoemakers or saddlers could be enlisted as apprentices of such trades.[26]

Most boys came from industrial and workhouse schools, orphanages, or the army boarding schools for soldiers' children. Those army institutions were the Royal Hibernian Military School in Dublin (410 boys) and the Duke of York's Royal Military School in London (570 boys) and the four Lawrence Military Asylums in India (about 785 boys).[27] From 1900 through 1904, the military schools provided the army with about 225 boys annually. Approximately the same number came straight from industrial schools and workhouses. As with the military schools, consent of the boy and his parents, if alive, were required for enlistment.

Many boy recruits volunteered while in orphanages; the Foundling Hospital in London alone provided about 20 to 25 apprentice bandsmen annually.[28]

---

[22] Seventeen-year-olds accounted for 0.3% of recruits in 1904 and 0.5% in 1903. *Annual Army Report, 1904*, 1905, [Cd. 2268].

[23] War Office Pamphlet, *The Advantages of the Army (1896)*, 1898, H.C. Accounts & Papers, No. 81.

[24] *The Queen's Regulations and Orders for the Army, 1899*, ¶972, hereafter cited as *King's/Queen's Regulations*.

[25] *Army Estimates, 1904-05*, appx. 2; *King's Regulations, 1908*, ¶1100.

[26] *King's Regulations, 1908*, ¶278.

[27] *Army Estimates, 1904-05*, 1904, appx. 14; *Annual Report of the Lawrence Military Asylum, Sanawar, 1897-98*; Joseph Thomas, "The Lawrence Schools," amolak.in/web/the-lawrence-schools-by-joseph-thomas/; *Website of Lawrence College, Ghora Gali*, lawrencecollege.edu.pk/.

[28] Spiers, *The Army and Society*, 45; Mansfield, *Soldiers as Workers*, 109.

The Foundling Hospital had a close relationship with army bandmasters and all boys were taught to play instruments. Each year 50% to 80% of its leaving boys became army apprentice bandsmen. Jacquelin Banerjee, "Life in the Foundling Hospital," The Victorian Web; The Foundling Museum, *Foundlings at War*, foundlingmuseum.org.uk/events/foundlings-at-war; Jim Gledhill, "Coming of Age in Uniform," *Family & Community History* 13/2 (November 2010): 114-27.

Appendix A

Nationality

As shown in the following table, enlistment rates varied among the nations of the United Kingdom. The Boer War years are indicated by bold face data cells.

Recruits per 10,000 Population, by Nation
(population interpolated)

	1865	1874	1884	1894	1897	1898	1899	1900	1901	1902	1903	1904
Ireland	3.7	5.3	7.7	7.4	8.6	9.3	**8.9**	**9.0**	**8.5**	**10.5**	7.9	8.1
Scotland	3.9	5.2	9.6	7.8	8.3	9.0	**9.9**	**12.0**	**11.5**	**11.9**	7.1	9.1
England & Wales	4.6	6.6	10.5	8.9	8.5	9.8	**9.8**	**11.7**	**10.7**	**12.4**	9.3	10.4

War Office Sources: *Annual Army Reports, 1874, 1884, 1898, and 1904* - 1876 [C. 1323]; 1886 [C. 4570]; 1899 [C. 9426]; *Annual Report of the Inspector-General of Recruiting for the Year 1899*, 1900, [Cd. 110].

Of the years shown in the above table, it was only in 1897 that Ireland's enlistment rate was higher than that of the UK's other two nations. Though Ireland had a relatively low enlistment rate, Irish-born men, which included many that lived in Great Britain, joined the army at a much higher rate than did British-born. In 1870, Irishmen were over-represented in the army by a factor of 1.6. That factor diminished steadily until 1906 when the Irish-born served in numbers proportionate to Ireland's share of the UK population.

Recruitment in Ireland

Within each nation, enlistment rates were not uniform. For example, take Ireland, where the War Office tabulated enlistments by the Irish Command's three districts. The Belfast District encompassed Ulster and parts of Leinster. The Dublin District included Connaught and the remainder of Leinster. The Cork District was primarily Munster.

Unionist Ulster was not Ireland's most fertile ground for army recruiters. The north was home to the vast majority of Ireland's Presbyterians and a disproportionate number of Methodists, Baptists, and Congregationalists. Those Protestants were far more averse to military service than Anglicans and Catholics.[29] Presbyterians were under-represented in the army by a factor of 0.6; the other Protestant denominations by a factor of 0.4.[30] Ulster was also the richest Irish province with agricultural wages markedly higher than elsewhere in Ireland, and many well-paid industrial jobs available in Belfast.[31] As shown in the following table, the political factor in Ulster that favored army enlistment was outweighed by the social and economic factors that discouraged military service. It was the Dublin Army District that

---

[29] Ibid., Census of Ireland, 1901.

[30] *Annual Army Report, 1904*, 1905, [Cd. 2268].

[31] Royal Commission on Labour, *Reports on the Agricultural Labourer (Ireland), Vol. IV*, 1893-94, [C. 6894].

Appendix A

had Ireland's highest enlistment rate, primarily because it included poverty-stricken Connaught and the extensive and densely populated slums of Dublin City.

Army Recruitment in Ireland, 1884 & 1904
Per 10,000 persons, Decennial Census

	1884		1904	
	Recruits	per 10k Pop.	Recruits	per 10k Pop.
Belfast District	1,267	7.0	1,153	7.0
Dublin District	1,381	7.7	1,525	9.8
Cork District	1,017	6.5	926	7.4

Sources: *Annual Army Reports, 1884* and *1904* - 1885, [C. 4570], 1905, [Cd. 2268].

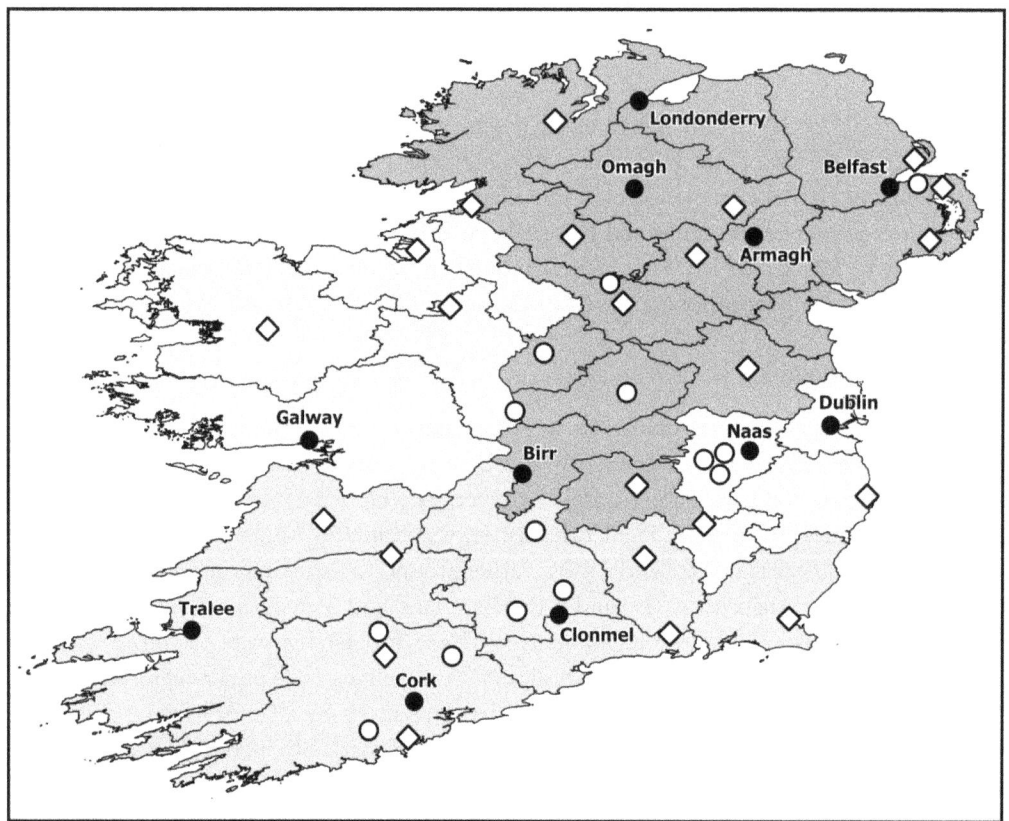

British Army Recruiting, Ireland

Appendix A

On the previous page is a map of the Irish Command Districts that shows all recruiting locations in 1904. A young man could enlist in Ireland at a training depot, a district recruiting headquarters, and any barracks with a resident recruiting sergeant, including militia barracks. On Bloomsday, Ireland had 11 dedicated recruiting offices (regimental depots and district offices), 25 locations with militia barracks, and another 14 locations with regular army units.[32]

Applicant Rejection Rates

Though the army was always in need of more soldiers, recruiters generally adhered to War Office enlistment standards and the rejection rate among army applicants was high. In most years, the army turned down about one-third of applicants. Barred from enlistment were underage boys, applicants that did not meet physique standards (height, weight, chest size), men with severe dental or vision problems, the malformed, and the diseased. Also rejected were applicants of obvious mental deficiency. In some cases, applicants were rejected on account of known bad character.

A man could be rejected for service by the recruiting sergeant, the officer in charge of the recruiting station, or the examining medical officer. In 1904, medical officers were responsible for 40.7% of rejections.[33] Additionally, each year about five percent of recruits were discharged during training as unsuitable for military service.

Occupations and Rejection Rates of Applicants and Recruit Occupations
Twelve Months ended September 30, 1904

Stated Occupation	Enlistment Applicants			Recruits	
	Applied	Rejected	Pct.	Number	Pct.
Agricultural Laborers	4,110	1,248	30.4	2,862	14.1
Other Unskilled	14,571	5,548	38.1	9,023	44.3
Domestic Servants	1,298	482	37.1	916	4.5
Skilled Workers	8,421	2,832	33.6	5,589	27.5
Clerks	1,164	449	38.6	715	3.5
Professionals & Students	270	74	27.4	196	1.0
Total, Regular Enlistment	29,834	10,633	35.6	19,201	94.9
Total, Boys	1,229	210	17.1	1,044	5.1

Source: *Army Annual Report, 1904*, 1905, [Cd. 2268].

Army Demographics

Educational Attainment of Enlisted Men (excluding Colonial Corps)

Rankers were far less educated than their civilian counterparts. The War Office measured the educational level of soldiers through examinations for Army Certificates of

---

[32] *Monthly Army List, July 1904.*

[33] *Annual Army Report, 1904*, 1905, [Cd. 2268].

Appendix A

Education. The 3rd Class Certificate was awarded for basic literacy and numeracy (Civil Standard III). Holders of the 2nd Class Certificate met the standards for eleven-year olds (Civil Standard V). The army considered holders of 1st Class Certificates as educated to the level of primary school graduates. As the attainment of the certificate required mastery of only one or two "higher" subjects, its holders were actually on par with civilians who had six or seven years' schooling (Civil Standard VI to VII).[34] Following is the distribution of education certificates among enlisted men.

Percent of Men with Certificates of Education

	1st Class	2nd Class	3rd Class	None
1904	2.8	20.3	18.7	58.2
1894	1.8	21.3	14.3	62.6
1884	0.5	17.4	16.4	65.7

Ages of Enlisted Men [35]

Age Distribution of Enlisted Men

	<19	19	20-24	25-34	35-39	40+
1904	7.0%	8.4%	48.4%	31.1%	4.0%	1.1%
1894	6.8	8.9	43.6	37.8	2.2	0.7
1884	6.8	8.6	45.1	32.0	6.1	1.4

War Office policy was that soldiers be at least age 19 for general overseas service and age 20 for service in India. Note that in the years shown above, about 15% of soldiers were not qualified for worldwide deployment due to age. The typical age group for NCOs (corporals, sergeants, colour-sergeants) was 25-34; for warrant officers and the most senior NCOs (sergeant-majors, master gunners, quartermaster-sergeants) 35-50. With the implementation of short-service enlistment in the 1870s, the number of "old sweats" (over age 35) declined from 7.5% of enlisted strength in 1884 to 2.9% in 1894 as there were far fewer career-oriented recruits in the previous 10 to 15 years. Improvement in conditions of service and increases in pay probably caused the number of long-serving soldiers to rise to 5.1% of enlisted strength by 1904.

---

[34] Council on Education (England & Wales), Day School Code 1900; *Sixth Report on Army Schools by the Director-General of Military Education*, 1896, [C. 8241]; *Army School Regulations, 1886*; [George L. Dunnett] *The First Class Army School Certificate Made Easy* (Chatham: Gale & Polden, 1888), *The Second Class Army School Certificate Made Easy* and *The Third Class Army School Certificate Made Easy* (Chatham: Gale & Polden, 1889).

[35] *Annual Army Report, 1884*, 1885, [C. 4570]; *Annual Army Report, 1894*, 1895, [C. 7885]; *Annual Army Report, 1904*, 1905, [Cd. 2268].

Appendix A

Religion, Enlisted Men (excluding Colonial Corps) [36]

	Anglican	Presbyterian	Other Protestant	Roman Catholic	Asian	Jewish	None
1904	68.1%	7.5%	6.2%	16.5%	1.2%	0.1%	0.4%
1894	68.7	7.4	6.0	17.7	----	----	0.2
1884	66.0	7.4	4.7	21.9	----	----	----

In 1884 and 1894, soldiers who professed a faith other than Anglican, Presbyterian, and Catholic were listed as "Other Protestant." In 1904, for purposes of church parade, those who did not disclose a religious preference were marched to an Anglican church on Sundays.

In 1900, the approximate distribution in the United Kingdom of civilians by religious affiliation was as follows:[37]

Anglican	Presbyterian	Other Protestant	Roman Catholic	Asian Jewish None
57%	13%	16%	13%	---------- 1% ----------

Catholics were somewhat over-represented in the army while Presbyterians and Other protestants were markedly under-represented. Low enlistment rates by non-Anglican Protestants were due to the middle class and lower-middle class status of most who embraced such faiths and the residual anti-militarism of Methodists, Baptists, Congregationalists, and Unitarians.[38]

Jews were the most under-represented of professed adherents to a faith. The data; however, may be misleading as many Jewish soldiers put down Church of England on their enlistment papers to avoid harassment. An English rabbi noted that "It is notorious that many of the Jews, as in the case of the other smaller religious bodies, prefer to 'follow the big drum,' *i.e.* attend the general Church of England parade."[39]

Overall, few soldiers were religious; most were nominal Christians who simply celebrated Christmas and Easter and behaved with decorum at mandatory church services.[40] Those who embraced their faith zealously were generally ill thought-of by their colleagues. They referred

---

[36] Ibid.

[37] Charles G. Brown, *Religion and Society in Twentieth-Century Britain* (Harlow, UK: Pearson, 2006); Robert Currie and Alan Gilbert, *Churches and Church-Goers* (Oxford: Oxford Univ. Press, 1977).

[38] Spiers, *The Army and Society*, 49; Greg Cuthbertson, "God, Empire and War: The Nonconformist Conscience and Militarism in Britain 1850-1900," *Theoria: A Journal of Social and Political Theory* 65 (October 1985): 35-48.

[39] "Jews in the Navy, Army, and Auxiliary Forces" in *The Jewish Yearbook*, edited by Isidore Harris (London: Greenberg, 1907).

[40] E.J. Hardy, *Mr. Thomas Atkins* (Toronto: Briggs, 1900), 196-241; Skelley, *The Victorian Army at Home*, 166.

Appendix A

to overtly religious soldiers as "psalm-singers" and viewed them as "more renowned for their proficiency in 'knee-drill' than in their more strictly military exercises."[41]

Nationality of Enlisted Men

Over the 30 years prior to Bloomsday, the army became less Irish and more English. By 1904, Irishmen in service reflected Ireland's proportion of the United Kingdom's population. At the outbreak of the First World War ten years later, the Irish were under-represented in the British Army. Following is a tabulation of enlisted men's nationality from 1874 through 1904.[42]

	England & Wales	Ireland	Scotland	Other
1904	76.2%	11.5%	8.2%	4.1%
1894	77.7	12.6	7.7	2.0
1884	72.2	18.6	7.9	1.3
1874	68.4	23.3	8.0	0.3

The most frequently disclosed nation of birth for the "Other" category was India. Other also included soldiers who claimed not to know their nation of birth. The following graphic shows vividly the decline and elimination of Irish over-representation in the British Army.[43]

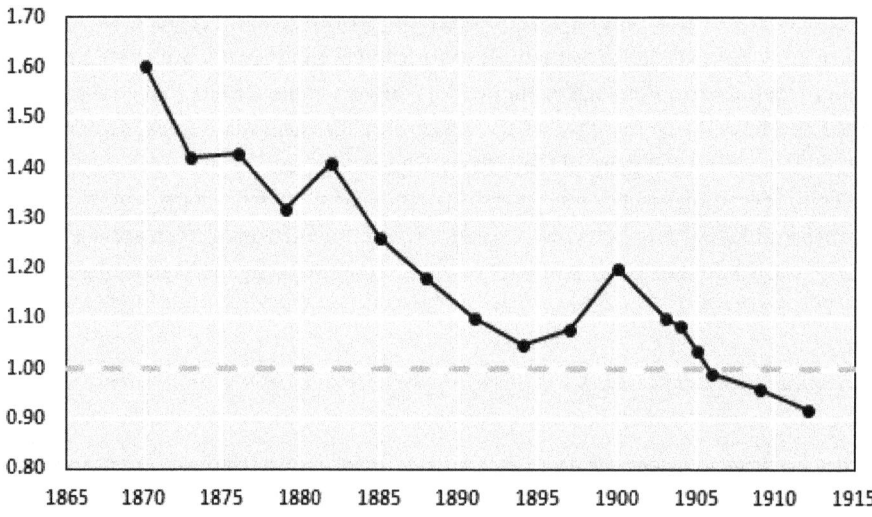

---

[41] Horace Wyndham, *The Queen's Service* (London: Heinemann, 1899), 85.

[42] *Annual Army Report, 1874*, 1876, [C. 1323]; *Annual Army Report, 1884*, 1885, [C. 4570]; *Annual Army Report, 1894,* 1895, [C. 7885]; *Annual Army Report, 1904*, 1905, [Cd. 2268].

[43] Spiers, *The Army and Society*, 50; *Annual Army Report, 1906*, 1907 [Cd. 3365].

Appendix A

Desertion, Punishment, and Rewards

During the mid-Victorian Era, soldier misconduct lessened and by 1904, discipline and desertion were not the great problems they were for the army in earlier years.

Desertion [44]

Desertions as a Percent of Beginning of Year Strength

	1884		1894		1904	
	Number	Percent	Number	Percent	Number	Percent
Deserters	4,478	2.5	3,958	1.9	3,959	1.5
Rejoined	1,568		1,649		2,073	
Net Deserters	2,910	1.6	2,309	1.1	1,886	0.7

Distribution of Deserters by Length of Service

Time-in-Service	1884	1894	1904
< 3 months	30.7%	26.1%	23.3%
3 months to 1 year	37.4	36.0	31.0
1+ years	31.9	37.9	45.7

Misconduct

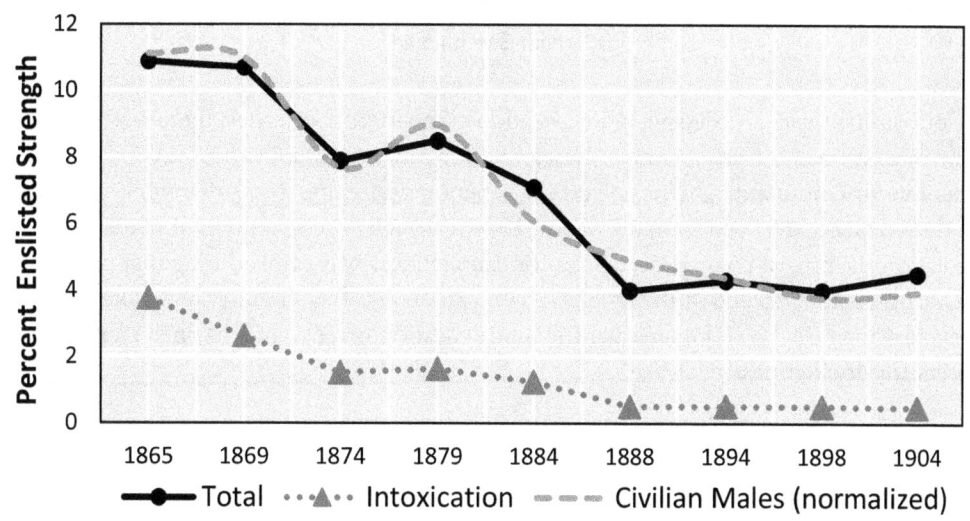

Men Tried by Courts-Martial

---

[44] *Annual Army Report, 1884*, 1885, [C. 4570]; *Annual Army Report, 1894*, 1895, [C. 7885]; *Annual Army Report, 1904*, 1905, [Cd. 2268].

Appendix A

In 1904, of courts-martial defendants convicted and punished, 89.0% were incarcerated, 10.4% were reduced in rank only, and 0.6% were fined only. The decline in court-martial cases followed the trend for trials of civilians by courts of general jurisdiction (felony cases and serious misdemeanors).[45] While overall, the court-martial rate more than halved over forty years, the rate for intoxication offenses in 1904 was less than one-fourth what it was in 1874. Intoxication offenses were drunk on duty and habitual drunkenness. The graphic on the previous page charts the decline in army crime and the corresponding decline in civilian serious crime.

Distribution of Court-Martial Offenses, Enlisted Men
(1874, 1884, 1894, 1904)

	1874		1884		1894		1904	
	Number	Pct.	Number	Pct.	Number	Pct.	Number	Pct.
Desertion	2,692	13.5	1,540	8.7	1,936	12.8	2,617	16.0
Disobedience	1,529	7.7	1,560	8.8	1,960	13.0	2,254	13.8
Theft	3,874	19.4	2,913	16.4	2,928	19.3	1,928	11.8
Insubordination	NA		1,612	9.1	2,239	14.8	1,916	11.7
Intoxication	3,834	19.2	3,102	17.5	1,589	10.5	1,763	10.8
AWOL	3,151	15.8	2,261	12.7	1,338	8.8	1,381	8.4
Other	4,850	24.4	4,747	26.8	3,140	20.8	4,523	27.5
Total	19,930	100.0	17,735	100.0	15,130	100.0	16,382	100.0

Offenses exceed the number of courts-martial as men can be tried on multiple charges. For example, drunk on duty and striking a superior officer.

Sources: See note 44.

Concomitant with the decline in intoxication court-martial offenses, there was a decrease in the rate of fines for drunkenness. Discipline for ordinary drunkenness was administered summarily by commanding officers and the sanction was a fine. Note that while army fines for drunkenness declined notably from 1874 through 1904, among the male civilian population the rate of court proceedings for drunkenness was stable during that period.[46]

During the first decade of the twentieth century, an 800-man infantry battalion, each year averaged 4 men tried by court-martial for an intoxication offense and 40 fined summarily by officers for drunkenness.

---

[45] Board of Trade: *Statistical Abstract for the United Kingdom, 1861-1876*, 1876, [C. 1573]; *Statistical Abstract for the United Kingdom, 1875-1889*, 1890, [C. 6159]; *Statistical Abstract for the United Kingdom, 1891-1905*, 1906, [Cd. 3092].

[46] George Bailey Wilson, *Alcohol and the Nation* (London: Nicholson & Watson, 1940), Tables 34, 38, 39, Persons proceeded against for drunkenness. Males for Scotland prior to 1899 estimated based on 1899-1901 date (71% male).

Appendix A

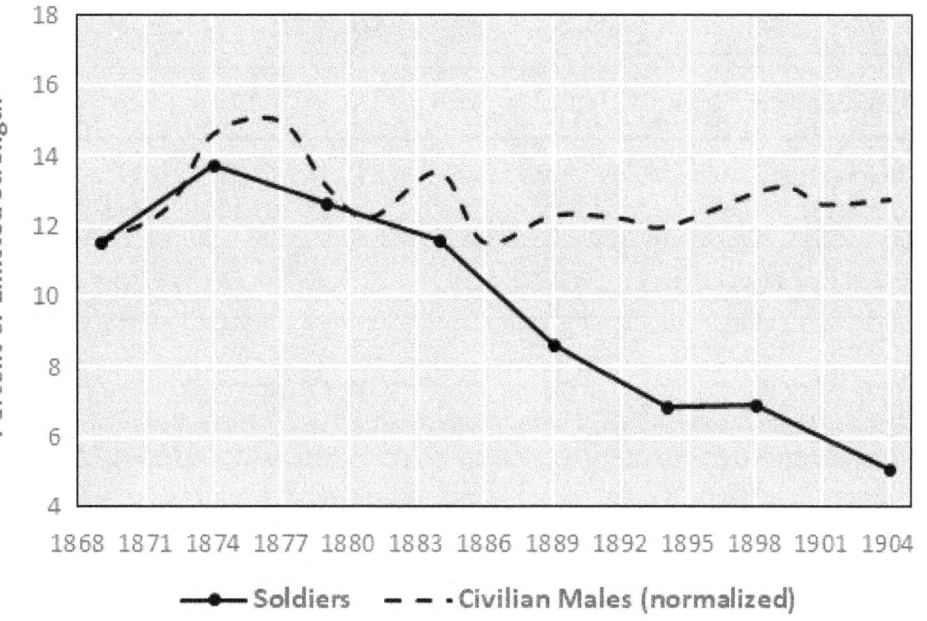

Soldiers of Good Conduct

Soldiers who served for specified intervals, and went two years without an entry in the Regimental Defaulters' Book, received Good Conduct Badges as follows:

Total Years	Badges	Interval
2 Years	1	2 Years
6 Years	2	4 Years
12 Years	3	6 Years
18 Years	4	6 Years
23 Years	5	5 Years
28 Years	6	5 Years

The Defaulters' Book was the log of all reported violations of regulations and regimental standing orders.[47] Enlisted men with eighteen or more years' service who had an exemplary discipline record received the Long Service and Good Conduct Medal.[48] The following table shows the number of enlisted men recognized officially for good conduct in 1884 and 1904.[49]

---

[47] *Royal Warrant for the Pay, Appointment, Promotion, and Non-Effective Pay of the Army, 1899*, Arts. 1088, 1095. Hereafter cited as *Royal Warrant for the Pay*.

[48] Ibid., Art. 1246.

[49] *Annual Army Reports, 1884 and 1904*, 1885, [C. 4570], 1905, [Cd. 2268].

Appendix A

	1884			1904		
	Number	Percent of Eligibles	Men	Number	Percent of Eligibles	Men
Good Conduct Medal	2,012	27.6	1.1	2,231	20.9	0.8
1 Good Conduct Badge	43,403	NA	24.0	69,421	NA	26.0
2 or More Badges	24,329	NA	13.4	47,124	NA	17.6
Good Conduct Badges	67,732	54.9	37.4	116,545	58.1	43.6

The good conduct badge award rate indicates that British Army rankers had become more well-behaved over the twenty years prior to Bloomsday. The good conduct medal award rate; however, indicates the opposite as the holders of such medals declined from 27.6% to 20.9% of long-serving soldiers. As the medal's award had the subjective standard of "exemplary" service, the decline in medal-holders may have been due to more stringent criteria.

Upon a soldier's discharge or transfer to the reserve, the character of his service was assessed by his commanding officer and noted on the soldier's record. Service was characterized as Exemplary, Very Good, Good, Fair, or Indifferent. In 1904, only 60% of men who left the army, both legally and through desertion, had character ratings of good or better.[50]

Exemplary	Very Good	Good	Indifferent and Fair	Discharged Misconduct	Deserted
2,243	11,534	11,701	9,345	3,830	3,950
5.3%	27.0%	27.5%	21.9%	9.0%	9.3%

Note that each year, about two-thirds of deserters returned to their units in the same year the commanding officer categorized their absence as desertion.

Venereal Disease in the British Army

> "disgrace to our Irish capital. an army rotten with venereal disease."
> Thoughts of Leopold Bloom in the Westland Row Post Office.[51]

Bloom's opinion of the army with respect to venereal disease was valid. In 1904, 10.8% of the home army soldiers were infected; 28.4% of the soldiers in Dublin.[52]

---

[50] *Annual Army Report, 1904*, 1905, [Cd. 2268].

[51] *U* (Gabler) Lotus Eaters 5:71-72.

[52] *Army Medical Department Report for the Year 1904*, 1906, [Cd. 2700].

Venereal Disease, Army at Home
Cases per 1,000 Average Strength

	England & Wales	Scotland	Ireland
Syphilis	34.2	36.2	36.5
Gonorrhea	52.0	55.3	60.2
Soft Chancre	18.3	19.7	21.2
All Venereal	104.5	111.2	117.9

Source: *Army Medical Department Report for the Year 1904*, 1906, [Cd. 2700].

The following table shows the venereal disease rate for the largest home garrisons. None of the camps were close to cities and the Curragh was the only one in Ireland. The venereal disease rate for the 10,621 soldiers in Greater London (which includes Woolwich), was 17.1%.

Venereal Disease Rates
Principal United Kingdom Garrisons

	Garrison Strength	Percent Infected
Dublin	4,483	28.3
London	3,829	24.3
Woolwich	5,311	13.1
Curragh Camp	3,413	13.0
Shorncliffe Camp	2,822	11.3
Aldershot Camp	21,343	7.8
Bulford Camp	3,538	6.0

Source: *Army Medical Department Report for the Year 1904*, 1906, [Cd. 2700].

In 1864, a year when one-fourth of British soldiers at home contracted syphilis or gonorrhea, Parliament passed the Contagious Diseases Act.[53] The CDA placed under regulation by medical authorities, prostitutes resident in designated localities. The act applied initially to eight towns and camp areas in England; three in Ireland (Curragh Camp, Cork City, Queenstown). The CDA was amended three times and by 1870 was effective in sixteen English and the original three Irish locales.[54] Under the CDA, police had authority to bring before a magistrate any woman they believed to be a prostitute. The magistrate could order a medical examination and if the examining physician found the woman infected, she was confined in a "lock" hospital for a maximum of nine months. Women who refused medical examination were imprisoned for one month; repeat refusants two months. The 1869 amendment authorized magistrates to order confinement for five days prior to examination.

---

[53] 27 & 28 Vict., c. 85.

[54] 29 & 30 Vict., c. 35; Contagious Diseases (Ireland) Amendment Act, 31 & 32 Vict., c. 80; 32 & 33 Vict., c. 96.

Appendix A

The CDA was opposed by religious-oriented "social purity" activists who viewed the statute as state approval of regulated prostitution. It was also opposed by libertarians who saw it as an infringement on constitutional rights, and feminists who found it degrading to women.[55] In 1883, the government suspended enforcement of the CDA and in 1886 the statute was repealed.

From 1865 through 1875, the venereal disease rate in the army declined, though this may have been a continuation of a trend that began in 1861. Additionally, from 1873 to 1879, because the War Office fined soldiers hospitalized for venereal disease, many infected men went to civilian practitioners for treatment, not army doctors, and their cases weren't in army reports.[56] The incidence of venereal disease began to rise in 1876, and by 1883 was over twice the 1875 rate.[57] To some extent, the CDA met its objective. For locations to which the act applied, syphilis incidence was about half that for locations in which the CDA was not in force. Gonorrhea incidence; however, was only about ten percent lower in the controlled locations compared to the non-controlled locations.[58]

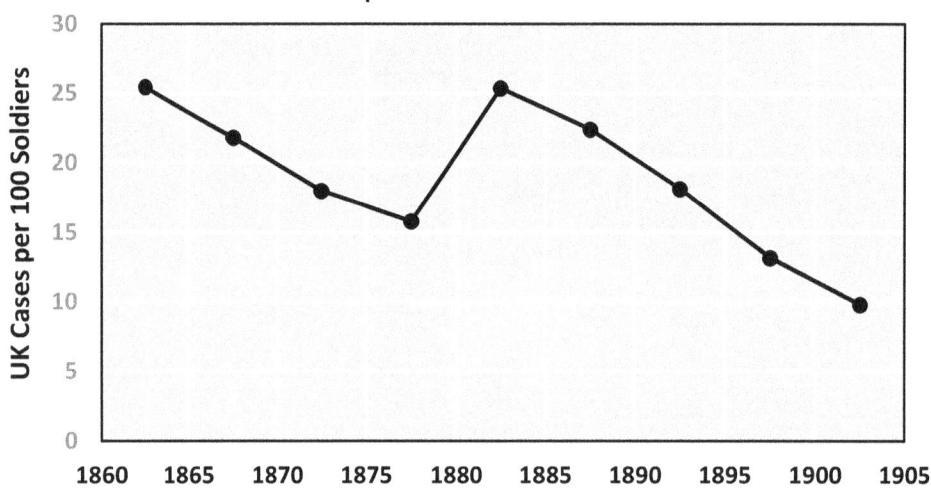

Syphilis and Gonorrhea
Cases per 100 Soldiers in the UK

---

[55] Skelley, *The Victorian Army at Home*, 53-57; Myna Trustram, *Women of the Regiment* (Cambridge: Cambridge Univ. Press, 1984), 122-26.

[56] Trustram, *Women of the Regiment*, 129-31; Testimony of William Muir, Director-General, Army Medical Department, *Report from the Select Committee on Contagious Diseases Acts*, 1878-79, H.C. Accounts & Papers, No. 323, qq. 69-71. The "fine" was disguised as an increase of the hospital stoppage from 8d. daily to 100% of pay.

[57] *Army Medical Department Reports for 1880, 1883, 1894, 1903, 1904* - 1882, [C. 3272]; 1884-85, [C.4453]; 1896 [C. 7921]; 1905, [C3d. 2434]; 1906 [Cd. 2700].

[58] Ibid. Many in the anti-CDA movement believed the War Office was under-reporting disease incidence in the controlled locality. For example: James Stansfield, MP, "On the Validity of the Annual Government Statistics of the Operation of the Contagious Diseases Acts," *Journal of the Statistical Society of London* 39, no. 3 (September 1876): 540-72.

Among European armies, the British Army had by far the highest incidence of venereal disease; however, it was less than that of the U.S. Army of the time.[59] The United Kingdom and the United States had wholly volunteer armies while the Continental states all had armies manned through compulsory service. Accordingly, the enlisted ranks of armies on the Continent reflected, in general, the socio-economic composition of the general, male population.

Venereal Disease Rates for Selected Armies (1903)

	Cases	Home Strength	Percent Infected	Type of Service
United States	7,739	42,264	13.6	Voluntary
United Kingdom	13,826	110,565	12.5	Voluntary
Italy	18,242	199,253	9.2	Compulsory
Spain	4,833	76,253	6.3	Compulsory
Austria-Hungary	17,188	291,809	5.9	Compulsory
Russia (1902)	41,631	1,058,042	3.9	Compulsory
France	13,306	489,673	2.7	Compulsory
Germany	11,393	590,859	1.9	Compulsory

Source: See note 59.

The ranks of the British and U.S. armies were filled by recruits from the lowest strata of the working classes. Additionally, in most states with conscription, the burden of full-time compulsory service fell disproportionately on the rural population. European military establishments viewed "country boys" as better military material than "city boys." Also, the autocratic, conservative governments in Europe did not want to provide too many urban workers, whom they viewed as socialists and potential revolutionaries, with military training and access to weapons.

Shortly before outbreak of the First World War, a British government commission accumulated data that indicated the incidence of syphilis among unskilled, urban workers was four times that of agricultural laborers. The data also showed that the unskilled, working class suffered the greatest infection rate of all eight social classes into which the commission divided the population. Next in disease prevalence were the upper and middle classes and semi-skilled workers. The infection rate for those classes was about three-fourths that for

---

[59] *Report of the Surgeon-General of the Army to the Secretary of War for the Fiscal Year Ending June 30, 1904* (Washington, DC: GPO, 1904); editorial note on Dr. Heinrich Schwiening's "Contributions to the Knowledge of the Spread of Venereal Diseases in European Armies and Among Young Germans Subject to Military Service," *Veröffentlichungen aus dem Gebiete des Militär-Sanitätswesen* 36 (1907), a publication of the Medical Division, Royal Prussian War Ministry. *New York Medical Journal* 85 (1907): 556-57.

Appendix A

unskilled workers.[60] One could conclude from the data that army officers were somewhat more likely to be infected by venereal disease than senior NCOs and far more likely than the least educated soldiers - former farm laborers.

An additional factor that explains the lower disease rate in conscript armies is they had much shorter enlistment terms then all-volunteer armies. Most conscripts served no more than two years with the colors before passing into the reserve; a relatively short period of celibacy. Also, as conscripts received much lower pay than voluntary soldiers, few could afford prostitutes.

The Army Abroad on Bloomsday

Deployment of British Army Units
October 1, 1904

Location	Infantry Batts.	Cavalry Regs.	Artillery Bttys.	Engineer Coys.	ASC Coys.	Ordnance Coys.	Medical Coys.
U. K.	80	16	147	47	74	14	22
India	52	10	94	1	0	0	0
Colonies	36 + 4	5	54 + 20	21 + 6	12	6	0
Total	171 + 4	31	295 + 20	69 + 6	86	20	22

For cells with two numbers, the second number is for Colonial Corps units (Malta, China, West Indies, West Africa, Ceylon, Straits Settlements).

Sources: *Monthly Army List, December 1904*; *Annual Army Report, 1904*, 1905, [Cd. 2268].

Because the Indian Army provided support services for British Army units on the subcontinent, there were almost no British support units in India (one engineer company; no ASC, Ordnance, or Medical companies). Note that almost exactly half the infantry, cavalry, and artillery formations were kept at home.

The map on the next page shows the British Empire in 1904 with its constituent parts light gray. The solid circles indicate the relative sizes of the garrisons abroad by number of men. On Bloomsday, there were no British troops in Australia and New Zealand, and a relative handful in Canada (Nova Scotia). The first of Gladstone's Liberal governments (December 1868 - February 1874), as a cost-cutting measure, had unilaterally given the self-governing colonies responsibility to provide for their own defense. In 1867, there were 46,139 British troops in the colonies; in 1874 half that number, 23,003.

In 1904, the British Army had an unusually large presence in South Africa to garrison the newly acquired colonies of the Transvaal and Orange Free State. Prior to the Boer War, these were the independent Boer states against which the British Empire fought a war of conquest. Note that the War Office and British Governments always categorized British

---

[60] Appendix I, *Final Report of the Royal Commission on Venereal Disease*, 1916, [Cd. 8190], as interpreted by Simon Szreter, "The Prevalence of Syphilis in England and Wales on the Eve of the Great War," *Social History of Medicine* 27, no. 3 (August 2014): 508-29.

Appendix A

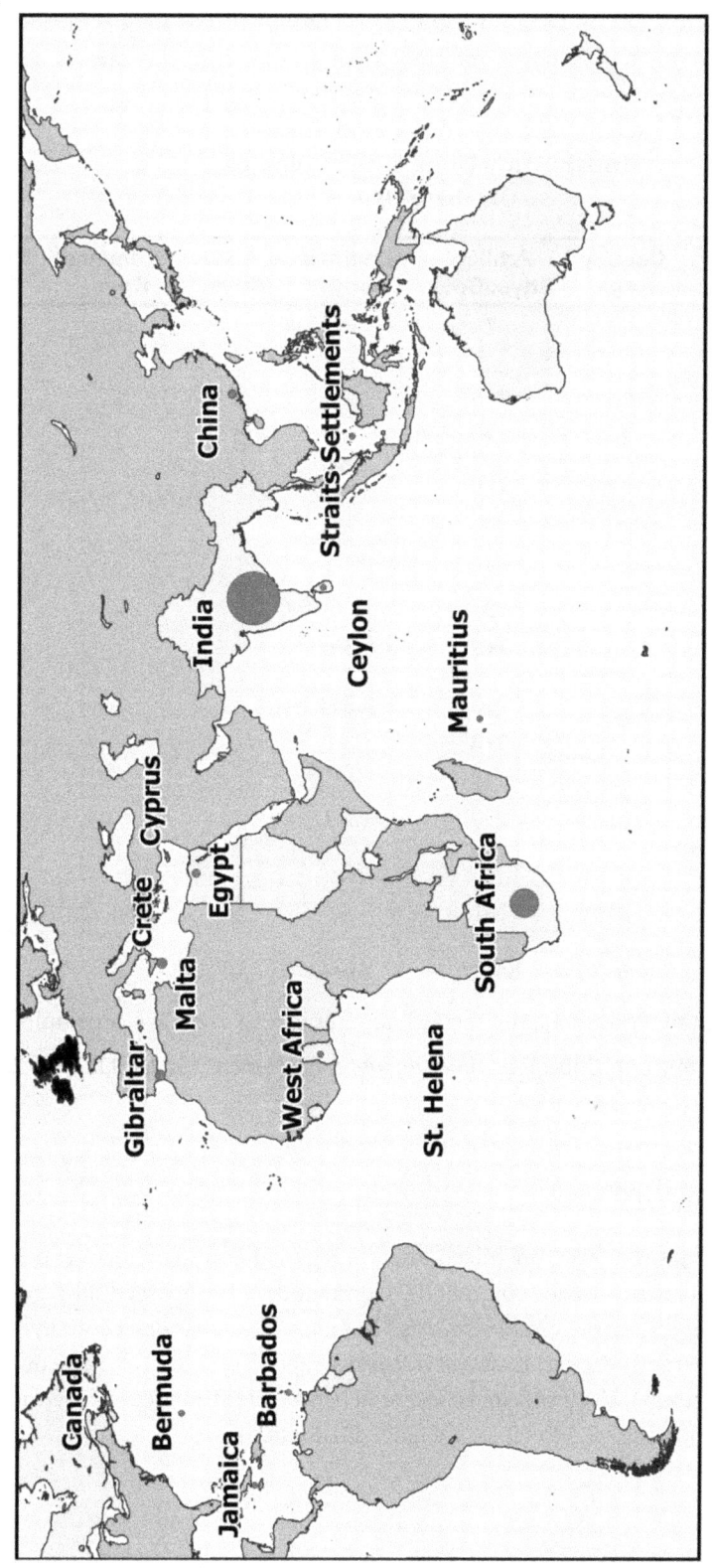

British Army Abroad, 1904

Sources: *Monthly Army List, December 1904; Annual Army Report, 1904*, 1905 [Cd. 2258].

Appendix A

possessions other than India as "the colonies" even though that category included self-governing dominions. India was never referred to as a colony; it was always something separate and apart.

British Army Units in the Colonies
October 1, 1904

Location	Infantry Batts.	Cavalry Regs.	Artillery Bttys./Coys.	Engineer Coys.	ASC Coys.	Ordnance Coys.	Medical Coys.
Gibraltar	3		7	4	1	1	
Cyprus	1						
Malta	7.5		7+5	3	1	1	
Egypt	3.5	1	2	1	1	1	
Jamaica	0+0.5		1+3	0.5+1			
Bermuda	2		3	3			
St. Helena	1		1				
Canada	1		3	3.5			
Barbados	1+0.5		1+2	0.5+1			
Ceylon	1		2+2	0.5+1			
China	1+1		3+3	1+1			
Mauritius	1		2+2	1			
Straits	1		2+3	.5+1			
West Africa	0+2		1	0+2			
South Africa	14	4	19	3	9	3	
Total	36+4	5	54+20	69+6	12	6	0

For cells with two numbers, the second number is for Colonial Corps units

Sources: *Monthly Army List, December 1904*; *Annual Army Report, 1904*, 1905, [Cd. 2268].

The Army at Home on Bloomsday

The British Army at home was organized into six regional commands, an expeditionary force, and two independent district commands. Commands encompassed all regular army units not part of departmental establishments. Regional Commands were divided into geographic areas: Districts, Coast Defence Districts, and Territorial Regiment Groups (territorial infantry districts). Each regional command had a Chief Engineer who commanded all engineering units not assigned to Coast Defence Districts.

Appendix A

Home Commands [61]

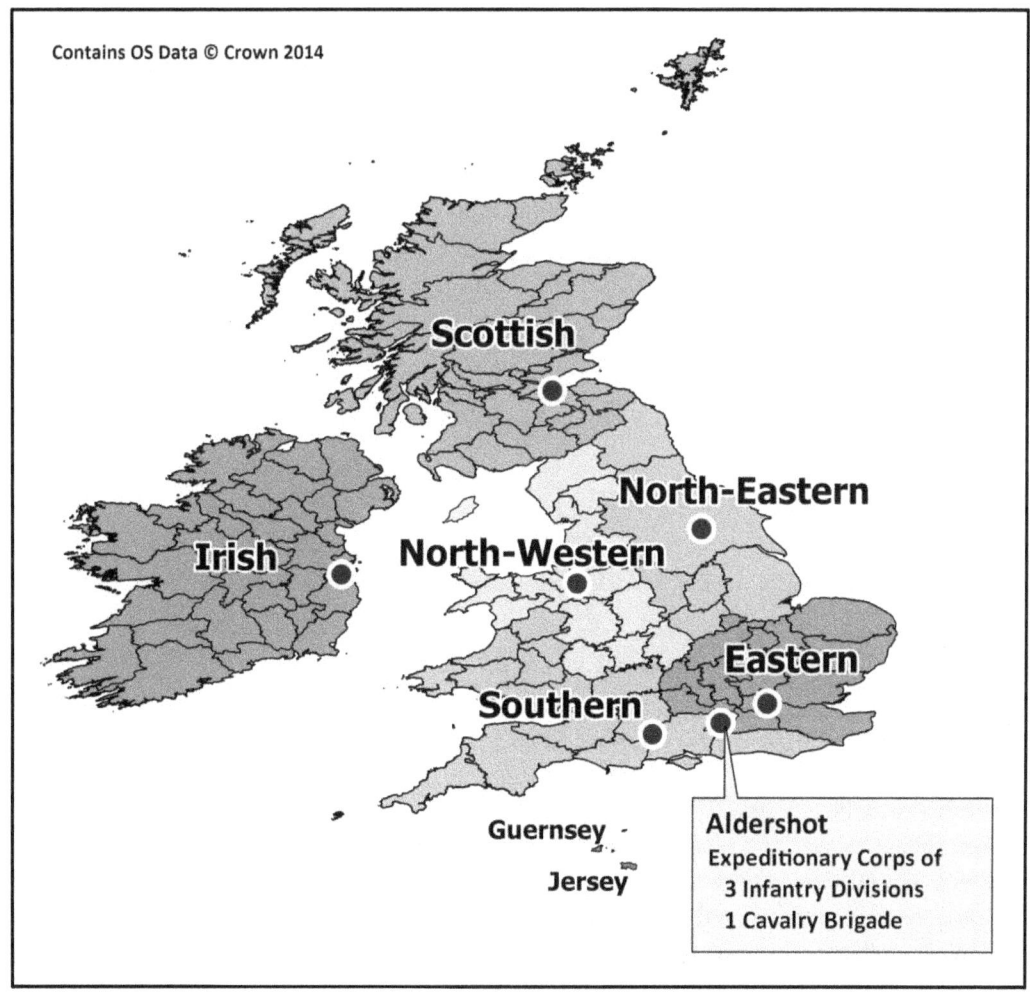

British Army Home Command Districts, 1904

Source: *Monthly Army List, July 1904.*

Aldershot (Expeditionary Force - I Corps), HQ: Aldershot Camp

1st Division	2nd Division	3rd Division	
1st, 2nd Bdes	3rd, 4th Bdes	5th, 6th Bdes	1st Cavalry Brigade
7 inf batts.	7 inf batts.	7 inf batts.	3 cavalry regs.
6 atty bttys.	6 atty bttys.	6 atty bttys.	3 atty bttys.

---

[61] *Monthly Army List, July 1904.*

Appendix A

Southern (II Corps), HQ: Salisbury
3 division HQs, 5 brigade HQs, 1 cavalry brigade HQ
2 cavalry regiments, 17 infantry battalions, 22 field artillery batteries

Irish (III Corps), HQ: Dublin
3 division HQs, 2 brigade HQs, 1 cavalry brigade HQ
3 cavalry regiments, 22 infantry battalions, 29 field artillery batteries

Eastern (IV Corps), HQ: London
1 division HQ, 1 brigade HQ
5 cavalry regiments, 16 infantry battalions, 8 field artillery batteries

North-Eastern, HQ: York
1 cavalry regiment, 2 infantry battalions, 11 field artillery batteries

Scottish, HQ: Edinburgh
1 cavalry regiment, 3 infantry battalions

North-Western, HQ: Chester
1 infantry battalion, 3 field artillery batteries

Jersey and Guernsey/Alderney - 1 infantry battalion in Jersey

Geographic Distribution of Regimental Strength

In the Edwardian and Late Victorian years about half the British Army was stationed abroad. At home, based on national population, the War Office quartered a disproportionately large part of the army in Ireland; a disproportionately small part in Scotland. Governments stated the disparity was because of the space for training available in Ireland and that nation's need for the economic benefit resulting from a large garrison. Irish nationalists were convinced that the troops stationed in Ireland were to maintain English "colonial" control. Note that from 1884 to 1904 the regimental strength of the army (men assigned to units) increased by 48.3%.[62]

1904 50.8% abroad			
England & Wales	102,099	74.3%	78.3% of UK population
Scotland	5,956	4.3	10.8%
Ireland	28,287	20.6	10.8%
Channel Islands	1,154	0.8	0.2%
Total Home:	137,496	100.0	
East Indies:	76,938		
Other:	65,268		
Total Abroad	142,206		

---

[62] *Annual Army Report, 1884*, 1885, [C. 4570]; *Annual Army Report, 1904*, 1905, [Cd. 2268].

Appendix A

```
1884   53.4% abroad
England & Wales    59,203   67.2%    74.3% of UK Population
Scotland            3,658    4.2     10.6%
Ireland            23,562   26.8     14.9%
Channel Islands     1,576    1.8      0.1%
Total Home:        87,999  100.0

East Indies:       59,375
Other:             41,283
Total Abroad      100,658
```

The Army Reserve [63]

Regular Army Reserve   Strength 1904: 74,940 men of 80,000 authorized.

Class

- Section A   2,564
- Section B   62,190
- Section D   10,186 — Replaced the old Second Class Reserve.

Corps / Department

- Infantry   52,166
- Artillery   8,895
- Cavalry   5,790
- Engineers   2,996
- All Other   5,903

Militia Reserve (old)   Strength 1904: 1,487.

Consisted of militiamen who volunteered to serve as reservists for the regular army. The War Office ended enrollment on April 1, 1901.[64] As enrollment was for a six-year term, this reserve pool was extinct in 1907.

The Auxiliary Forces, 1904

Some of the history, and most of the myths, of the auxiliary forces in Ireland were certainly known to many of the characters in *Ulysses*. Extreme nationalists, exemplified by The Citizen, viewed the part-time armed forces as instruments of British oppression. Home-

---

[63] *Annual Army Report, 1904*, 1905, [Cd. 2268].

[64] Army Orders, 1901, No. 88.

Appendix A

rulers, such as Myles Crawford, editor of *The Freeman's Journal*, were probably ambivalent (though Crown armed forces, they were local and unlike the Regular Army in Ireland, their men were all Irish). The unionists, such as Deasy, headmaster of Stephen Dedalus' school, probably admired the auxiliaries as keepers of both the existing social order and Ireland within the United Kingdom.

Auxiliary Forces, 1904
(authorized strength shaded)

	Full-Time *		Part-Time			
	Officers	Others	Officers		Others	
Militia	300	4,341	**3,431**	Pct.	**123,510**	Pct.
			2,583	75.3	91,901	74.4
Yeomanry	58	290	**1,538**	Pct.	**25,752**	Pct.
			1,232	80.1	25,502	99.0
Volunteer **	322	1,811	**11,432**	Pct.	**329,681**	Pct.
			9,247	80.9	244,632	74.2
Total	680	6,532	**16,401**	Pct.	**478,943**	Pct.
			13,062	77.7	362,035	75.6

* Regulars of the Permanent Staff   ** No Volunteer Corps in Ireland.

Sources, War Office: *Militia Return for the Training of 1904*, 1905, [Cd. 2432]; *Yeomanry Return for the Training of 1904*, 1905, [Cd. 2267]; *Return of the Volunteer Corps for 1904*, 1905, [Cd. 2438]; *Annual Army Report, 1904*, 1905, [Cd. 2268].

Leopold Bloom's attitude towards the Militia and Yeomanry, based on his anti-militarism, was surely negative.[65] His apolitical wife would likely have formed her opinion of the auxiliaries from their social aspects. Molly's opinion, if she had one of the auxiliaries, would have been based on the appearance, social conduct, and manners of the part-time officers. Additionally, as the daughter of a regular army man, she would have placed such officers several rungs down the social ladder from officers of the regular army.

Militia Reserve Division   Strength 1904: 7,082 men of 10,000 authorized.

Unlike the "Old" Militia Reserve, the Militia Reserve Division was separate and apart from the organized militia units and was a reserve for militia units, not the Regular Army. The War Office established the Division in 1903 and its initial authorized strength was 7,500.[66] During the first year of the Boer Warm the War Office ended enrollment in the Old Militia Reserve.

---

[65] Greg Winston, *Joyce and Militarism* (Gainesville, FL: Univ. of Florida, 2012), 219-21.

[66] Army Orders, 1903, No. 36.

Appendix A

Militia (organized) Strength 1904: 94,484 men of 126,941 authorized.

```
124 infantry battalions ------------------------------------------ 76,684
  1 field artillery "brigade" (3 batteries of 4 guns)}
 32 garrison artillery "brigades"                    }--------- 14,707
  2 fortress engineer "corps"                        }
 16 medical companies --------------------------------------------   793
  7 submarine (harbor) minelayer "divisions" -------------- 2,300
```

The field artillery brigade was the experimental Lancashire Royal Field Artillery with 123 regulars and 309 militiamen. Two more militia field artillery formations were approved by Parliament.

Of the 94,484 serving militia members, 1,016 were enlisted men who had volunteered in advance for foreign service with their units. Their commitment was for twelve months. The War Office classified these men as "Special Service" Militiamen.

Militia Demographics [67]

Age Distribution of Enlisted Men
Militia and Regular Army, 1904

	≤17	18	19	20-24	25+
Militia	9.6%	10.9%	9.1%	33.0%	37.4%
Army	3.0	9.2	14.2	38.8	34.8

Nearly all regulars under age 17 were boy apprentices

Age Distribution of Recruits
Militia and Regular Army, 1904

	≤18	18	19	20+
Militia	34.5%	26.9%	12.1%	26.5%
Army	4.5	45.9	18.7	31.7

Nearly all regulars under age 18 were boy apprentices.

Distribution of Militiamen by Religion, 1904

	Anglican	Presbyterian	Other Protestant	Roman Catholic	Jewish
All Militiamen	53.0%	8.2%	4.8%	33.9%	0.1%
In Ireland Only	12.8	3.5	0.4	83.3	0

---

[67] *Annual Army Report, 1904*, 1905, [Cd. 2268].

Appendix A

Militia Recruitment and Turnover, 1904 [68]

The following data are for the army's fiscal year. Militia totals given at the head of this section are as of each unit's annual training.

Strength, October 1, 1903	89,743	
Enlisted	+35,264	--- 39.3% of beginning strength
Joined Regular Forces	- 15,648	}
Discharged by Purchase	- 5,578	} -- 82.0% of year's enlistments
Discharged, Unsuitable*	- 2,145	}
Net Desertion	- 5,544	}
Died	- 425	
Invalided Out	- 2,529	
Other Losses	- 523	
Discharged, Net **	- 6,124	
Strength, October 1, 1904	86,491	(-3,252 for year)

* Released shortly after enlistment as unsuitable or discharged for misconduct.
** Term expirations less 4,590 re-enlistments.

The Imperial Yeomanry   Strength 1904: 26,734 out of 27,290 authorized.

On Bloomsday, the yeomanry consisted of 56 cavalry regiments, 2 of which were in Ireland. Many yeomen had seen active service in South Africa during the Boer War. The force was the successor to the Home Yeomanry, which after 1834 was not present in Ireland.

Regiments were organized like those of the regular cavalry except yeomanry regiments had four service squadrons, not three. In peacetime, the entire regiment was in training so there was no need for a separate training squadron. Unlike regular cavalry regiments, yeomanry regiments had medical and veterinary officers on their establishment.[69]

The Volunteer Force [70]   Strength 1904: 253,879 of 341,113 authorized.

The acts that authorized this auxiliary force applied to Great Britain only: The Yeomanry and Volunteers in Great Britain Act, 1804 as amended by the Volunteer Act, 1863.[71] There were no Volunteers in Ireland.

---

[68] *Annual Army Report, 1904*, 1905, [Cd. 2268].

[69] *Army Estimates, 1904-05*; War Office, *Yeomanry Return for the Training of 1904*, 1905, [Cd. 2267]; *Monthly Army List, December 1904*.

[70] *Annual Army Report, 1904*, 1905, [Cd. 2268].

[71] 44 Geo. 3, c. 54; 26 & 27, Vict. c. 65.

Appendix A

```
225 infantry battalions/regiments          }------- 189,012
   3 independent infantry companies        }

 68 garrison & mixed artillery "brigades" ----------------- 41,730

 22 fortress engineer battalions           }
  7 submarine (harbor) minelayer companies }
  3 telegraph companies                    }-------- 17,995
  1 railway battalion                      }
  1 electrical battalion                   }

  1 motorized transport company ------------------------- 155

 41 stretcher-bearer companies             }
  6 medical battalions                     }-------- 4,987
  2 medical companies                      }
```

Volunteer Infantry

The infantry battalions were grouped into 46 volunteer brigades, each commanded by a colonel, some of which were volunteer officers. Note that volunteer officers, unlike their militia counterparts, could attain the rank of full colonel as a brigade commander. Following was the chain-of-command for the volunteer infantry.

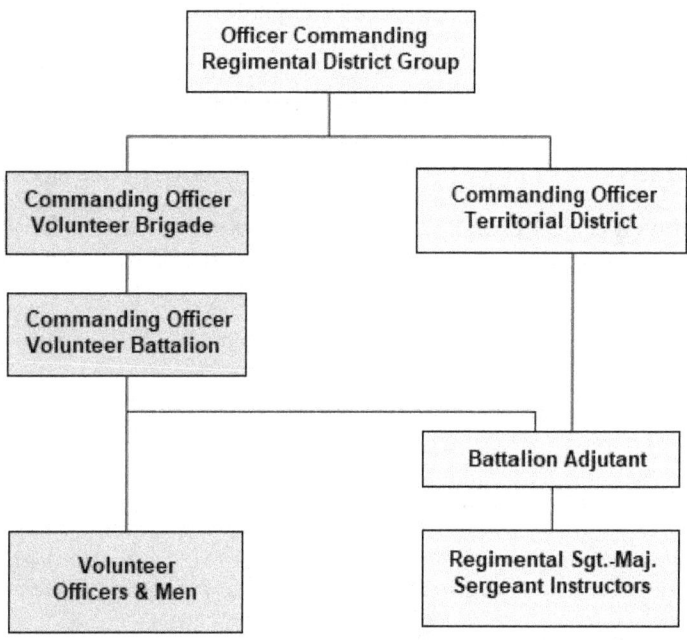

Appendix A

Volunteer Artillery [72]

The volunteer artillery consisted of 22 garrison artillery brigades and 46 mixed brigades. Mixed brigades had garrison artillery companies and one to three "position" artillery batteries. A position artillery battery could man field guns but lacked the equipment, horses, and skills to move them effectively. Upon general mobilization, the guns of position batteries would be transported to prepared positions by civilian carters. Volunteer garrison artillery batteries would man coastal fortifications.

Of the 100 position batteries, 29 were armed with heavy guns (barrel diameter of about 120mm), 71 with light guns (75mm). Only 16 of the heavy batteries and 26 of the light batteries had modern weapons. Most of the volunteer's guns were obsolete muzzle-loaders. The obsolete heavy guns had a range of 2.5 miles compared to modern guns with effective ranges of from 5.5 to 7.0 miles.

Coastal Fortifications in the United Kingdom [73]

Most of the country's coastal fortifications were in a southern arc in Great Britain that stretched from Milford Haven in South Wales to the Thames estuary. Following are the British coastal defense positions during the early-twentieth century. The most heavily fortified are denoted with bold-face type.

Ports and Naval Bases   29 RGA companies with about 5,400 men, all ranks.

<u>Major Fortifications</u>
The Clyde
Tyne & Wear
Tees & Hartlepool

Milford Haven & Pembroke
The Bristol Channel

Harwich Fortified Zone
Medway & Thames
Falmouth
Plymouth

<u>Minor Coastal Fortifications (often unarmed)</u>
Orkney, Aberdeen, Inverness, Newhaven,
Shoreham, Southampton, Swansea,
Tayside, Dartmouth, Humberside.

Belfast: Carrickfergus Castle, Kilroot Battery, and Grey Point Battery (under construction).

---

[72] War Office, *State of the Six Army Corps Commands, 1903*, [Cd. 1413]; *Army Estimates, 1904-05*; Various Testimony, *Report of the Royal Commission on the Militia and Volunteers, Minutes of Evidence*, 1904, [Cd. 2062, Cd. 2063].

[73] Paul M. Kerrigan, *Castles and Fortifications in Ireland, 1485-1945* (Cork: Collins, 1995); Ian V. Hogg, *Coastal Defences of England and Wales, 1856-1956* (London: David & Charles, 1974); *Victorian Forts and Artillery*; www.victorianforts.co.uk; Army Orders, 1906, No. 121.

Appendix A

<u>Major Fortifications</u>
Portland & Weymouth Fortified Zone
Portsmouth & Isle of Wight
Dover Fortified Zone

Fortified Naval Anchorages   4 RGA companies with about 700 men, all ranks.

These were sheltered, deep water locations to which home-based naval vessels would disperse in event of war. All were in Ireland. Scapa Flow, a sheltered body of water in the Orkney Islands, would be the Royal Navy's principal anchorage during the First World War. At the outbreak of hostilities, it was not fortified with gun emplacements.

<u>Cork Harbour</u>
3 forts plus 1 fortified battery.

<u>Lough Swilly, Co. Donegal</u>
7 fortified batteries

<u>Berehaven-Castletown, Co. Cork</u>
9 fortified batteries on Bere Island.

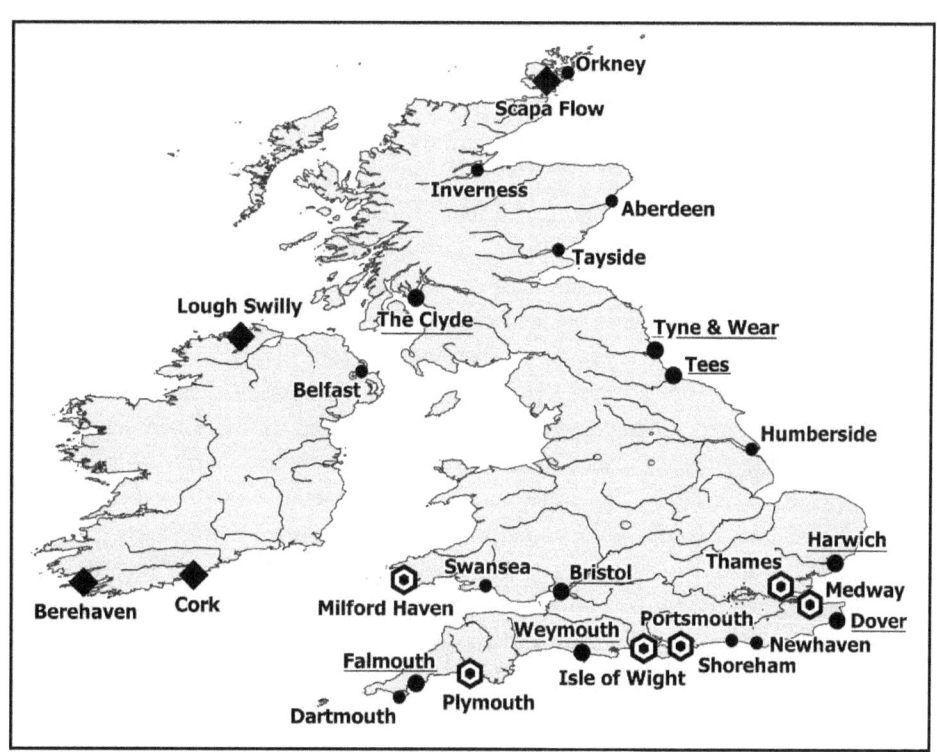

UK Coastal Fortifications, 1904

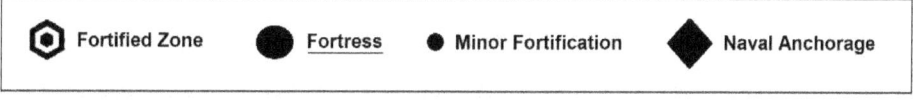

Source: Ian V. Hogg, *Coastal Defenses of England and Wales, 1856-1956* (London: David & Charles, 1974).

Appendix A

The London Defence Positions

In the 1880s, invasion scares gripped Parliament and the army leadership. They were founded on two beliefs. The first was that in a few years the navies of a hostile Continental alliance could overpower the Royal Navy and land in England an amphibious force of well over 100,000 men. The second, held by some army leaders, was that an enemy navy could "decoy" the Royal Navy away from a planned invasion site and quickly land there a large force.[74] Both views were irrational. If an enemy navy could overpower the United Kingdom's navy, there would be no need for the foreign state to invade. A blockade would quickly cause the British government to sue for peace. As for the "decoy" strategy, it would not take long for the superior Royal Navy to steam to the invasion area and cut the invader's supply line.

Cooler heads did not prevail and in 1889, Parliament appropriated funds to construct a semi-circular "mobilization" position for London. The measure introduced by the Conservative government was not to fortify London but to prepare positions for an army to occupy.[75]

> "No one connected with the War Office have for a moment suggested the fortification of London. They have proposed, and I have approved a simple scheme for mobilising the Auxiliary Forces for the defence of the capital. Some money must be necessarily expended in order to obtain sites on which to put store-houses, which must necessarily be expended in order to enable the Forces to take up their positions with the least possible delay."

By 1904, nearly the entire London Defence Position was in place. It consisted of earthwork gun emplacements from Epping northeast of the city, then south across the Thames, and on to Henley Grove southwest of the city. Also built were fourteen hardened storage depots called "mobilization centres" which contained arms, munitions, and equipment for the volunteer force units that would man the defenses. The total land acquisition and construction cost was £160,671. Under the London Defence Scheme, upon mobilization for war, pre-contracted civilian firms would dig trenches running about four miles on either side of each mobilization center while civilian carters brought in the volunteer's positional artillery. Volunteer units would assemble at the centers and form divisions. The divisions would be grouped into three static army corps each with its own supply base in London: Bishopsgate for the northeast corps, Woolwich for the southeast corps, and Nine Elms for the southwest corps. Included in the defenses was Tillbury Fort on the Thames. In all, between 100,000 and 125,000 volunteers, organized into ten divisions,

---

[74] A notable proponent of "invasion under cover of decoy" was FM Garnet Wolseley, former Adjutant-General and Commander-in-Chief of the British Army. *Minutes of Evidence taken before the Royal Commission on the Militia and Volunteers*, 1904, [Cd. 2062], at q. 1459. His view was supported by COL P.H.N. Lake, Assistant Quartermaster-General for Mobilization. Ibid., at q. 2918.

[75] Edward Stanhope, Secretary of State for War, 337 Parl. Deb. (3d ser.) (1889) 205.

Appendix A

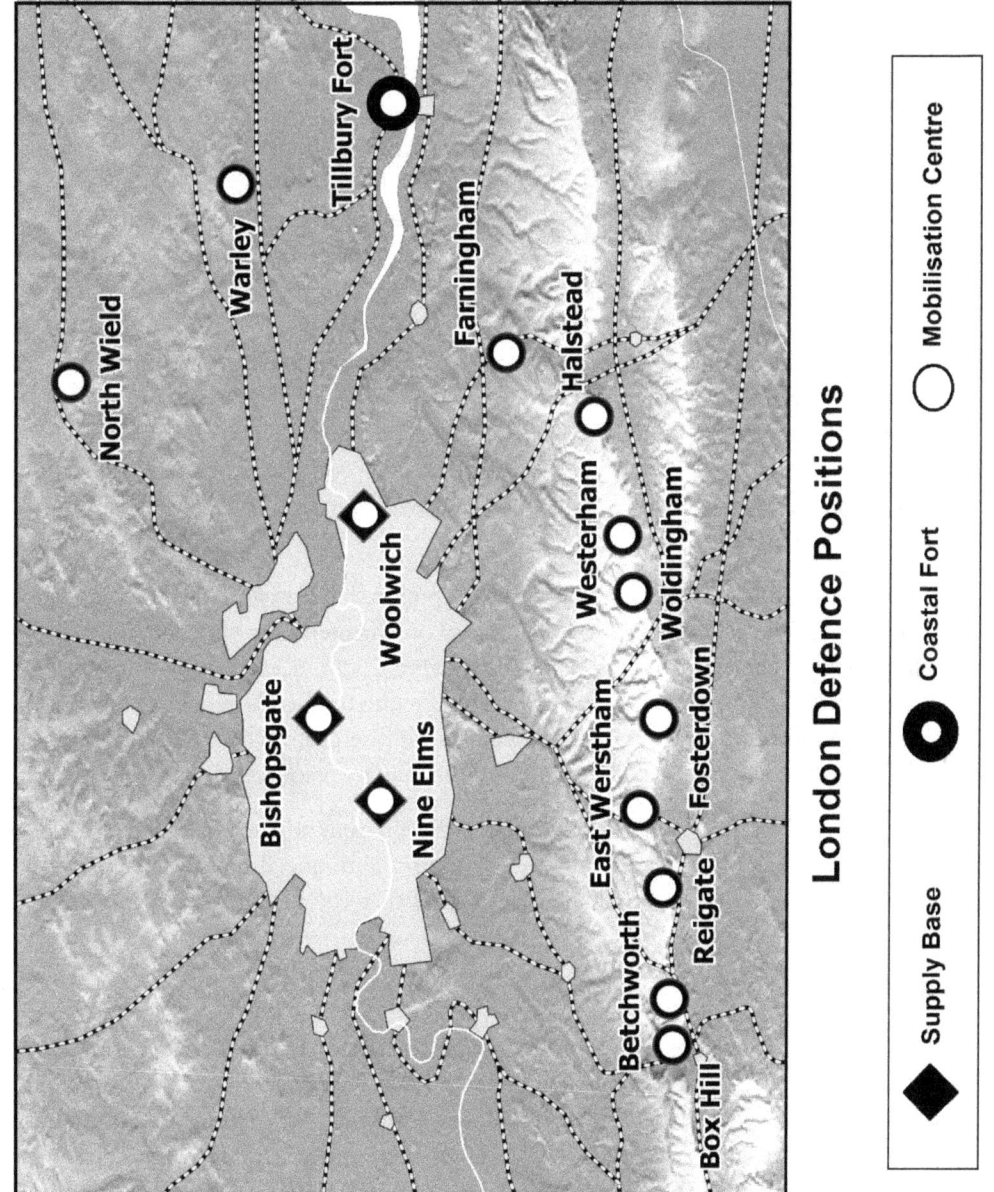

London Defence Positions

Appendix A

were slated to defend London. Should the Regular Army field force be defeated at the invasion site, it would withdraw to the defense line and reinforce the volunteers.[76]

In February 1903, Prime Minister Arthur Balfour charged the Committee of Imperial Defence to determine the likelihood of an invasion. The committee was composed of the civilian and uniformed leadership of the War Office and Admiralty, and was supervised by the prime minister. By the summer of that year, the military and naval staffs concluded that the probability of a large scale invasion was near zero. Most of the committee members accepted the findings; however, a few army members clung to their invasion beliefs, including the War Minister, William Brodrick. In October 1903, Balfour replaced Brodrick with H.O. Arnold-Foster and the other dissidents changed their position. On November 27, 1903 Balfour announced publicly his doubts as to the need for a large ground force for home defense. To appease the army, and mollify public opinion, he noted that some ground defense measures were needed. Balfour allowed that an enemy could carry out a large-scale coastal raid and that a foreign government might gamble on a suicidal invasion with a "dash on London" by up to 70,000 men.

In fact, government defense policy had changed conclusively and the burden of home defense had fallen almost entirely on the navy. Henceforth, the ground forces would be concerned primarily with imperial matters and internal security. In early 1905, Balfour made the following statement in Commons:[77]

> "The Committee of Defence are clearly of opinion that invasion of these islands in such force as to inflict a fatal blow or threaten our independence is impossible. The governing consideration which, as it seems to us, ought to determine the number of troops to be maintained depends not so much on considerations affecting home defence as upon the claim which Colonial and, still more, Indian needs may make upon our military resources."

The London Defence Positions were now officially redundant. On March 8, 1906 the new Liberal government, in power only three months, announced that the defensive works would be dismantled "root and branch and as fast as they can be made to disappear."

Reigate "Fort," A Scheduled Monument [78]

"The monument includes the main compound of Reigate Fort London mobilisation centre, situated on the southern crest of Reigate Hill. This location enjoys commanding views across the landscape to the south. The east-west aligned, elongated compound is defined by a large earthen rampart, roughly `D'-shaped in plan. A deep, unrevetted outer

---

[76] War Office, *Handbook for the London Defence Positions* (provisional) (London: HMSO, 1903); D.W. King, "The London Defence Positions and Mobilisation Centres, 1888-1907," *Journal of the Society for Army Historical Research* 55 (Winter 1977): 193-200.

[77] 142 Parl. Deb. (4th ser.) (1905) 595.

[78] Text from *Historic England*'s description of Reigate Fort, Scheduled Monument Number 1019245. From historicengland.org.uk.

ditch completely encloses the compound, creating a straight channel around 200m long at the rear, or gorge, of the installation. The ditch is in turn encircled by a hollow, designed to have been filled with barbed wire entanglements on mobilisation. In the east the hollow contains a steel palisade fence which enclosed the eastern half of the compound. Access to the interior parade is through loopholed steel gates, with flanking concrete piers, in the rampart at the eastern end of the gorge.

The entrance is reached by a causeway across the ditch. Access onto the causeway was controlled by a gate in a section of tall spiked railings, known as a Dacoit fence, parts of which remain on the outer edge of the ditch. Inside the entrance is an almost entirely subterranean magazine block, consisting of two main chambers, and covered by an earthen blast mound. To minimise the risk of explosion, the magazine passage contained a shifting lobby, where personnel changed into protective and non-spark producing clothes before entering the cartridge store. Many original features survive, including the notices labelling various components of the magazine.

...

Although not essentially designed as a fort, the mobilisation centre did possess a self-defence capability and was intended to deploy artillery in the event of a successful enemy advance. Concrete steps at each end of the casemates provided access onto the rampart, which would function as a firing step in response to enemy bombardment. Traces of a flint retaining wall survive along the inside edge of the parapet. Two additional earthworks at the western end of the parade are believed to be an additional firing platform and an earthen traverse, designed to block enfilade fire from the west.

...

The centre was sold in 1907, but was recommissioned during World War II and used by Canadian troops."

Abandoned Reigate Mobilisation Centre

© 2019 Google, Getmapping plc, Infoterra Ltd., Maxar Technologies, Geoinformation Group

Appendix B

Appendix B
Battalion and Company Establishments

Regular Army

Army formations were maintained at three levels of strength as set by Parliament in the annual army appropriation: War Strength (the highest level of manning), Foreign Service Strength, and Home Strength (the lowest level of manning). Units stationed at home were either at "high" establishment or "low" establishment. High establishment units were those designated for the standing expeditionary force, I Army Corps at Aldershot Camp. The following tables of organization are for home strength units at the high establishment. For simplification, all rank titles are infantry equivalents.[1]

Infantry Battalion

Commanding Officer	LTC	Regimental Sgt-Maj	SM
2nd in Command	MAJ	Quartermaster Sgt	QMS
Adjutant	CPT	Orderly Room Sgt	SGT
Quartermaster		Orderly Room Clerk	SGT/CPL
		Pioneer Sgt	SGT
		Musketry Sgt	SGT
		Sergeant-Cook	SGT
		Sergeant Drummer	SGT

Band
- Bandmaster
- Band-Sergeant
- Band-Corporal
- 20 Bandsmen

8 Companies

Commander	MAJ/CPT	Company Sgt-Maj	CSG
2nd in Command	LT	Company Clerk	SGT
(4 coys with an extra LT)		Staff NCO	CPL
		2 Drummers	PVT

2 Sergeants
4 Corporals
88 Privates

---

[1] The principal source for the tables of organization is the *Army Estimates*. Additional sources are: [James Moncrieff Grierson], *The British Army* (London: Sampson, Low, Marston, 1899); Nick Mansfield, *Soldiers as Workers* (Liverpool: Liverpool Univ. Press, 2016); [E. Charles Vivian], *The British Army from Within* (London: Hodder & Stoughton, 1914); *The King's Regulations and Orders for the Army, 1908*; *Monthly Army List, July 1904*.

Appendix B

Companies were divided into two half-companies with a sergeant in charge; half-companies divided into two sections with a corporal in charge. Home strength was 831 men, all ranks, foreign service strength 1,012 to 1,032, and war strength about 1,100.

About 12% to 15% of enlisted men had full-time, extra-duties jobs necessary for the battalion to function; about one service soldier for six to seven line soldiers. These soldiers, other than officer servants, plus the headquarters staff, totaled 80 to 100 men and comprised the battalion's cadre.

Servants: Appointed from establishment, 1 per officer (23)

Mess Staffs: Appointed or detailed from the establishment: Officers' Mess Sergeant, Officers' Mess Corporal, Sergeants' Mess Corporal, plus 4-5 privates per mess as waiters, kitchen help, and barmen totaling 11-12 privates.

Trades sergeants appointed from the establishment and replaced with acting sergeants (lance-sergeants). In total, they supervised 14 to 21 men and trained 2 to 4 apprentices.

Provost Sgt Supervised 5-10 regimental policemen.

Sergeant-Tailor Supervised and instructed 4 soldier-tailors and boys.

Sgt-Shoemaker Supervised and instructed 3 soldier-shoemakers and boys.

Canteen Sgt: Supervised the canteen staff of 2 to 4 men.

Tradesmen appointed the establishment (48-59). Building trades soldiers were supervised by the battalion's pioneer sergeant, cooks by the sergeant-cook, and quartermaster staff by the quartermaster sergeant.

Carpenters 3   Clerks for battalion and company orderly rooms, 10-12.
Bricklayers 2   Permanent cooks, 1 per 100 men, 8-10.
Smith   1   Quartermaster staff, 5-10.
Painter   1
Mason   1
Glazier   1
Plumber   1

Boys: Several (apprentice tailors, shoemakers, drummers, and musicians plus office boys).

Soldier-tradesmen were excused from guard and fatigue duties and received only the minimum combat training required by regulation.

When on active service, the colonel tried to limit casualties amongst the staff sergeants, quartermaster, adjutant, and 50 to 60 tradesmen. These skilled men were necessary to maintain the battalion's combat effectiveness and could not be readily replaced. Accordingly,

Appendix B

they were kept at the field depot some distance from the battlefield, beyond range of enemy artillery fire. The battalion's commander, however, could do little to limit disease casualties among the cadre.

Infantry Depot

Commanding Officer	MAJ		
Adjutant*	CPT	Depot Sgt-Maj	SM
2 Staff Officers	LT	Quartermaster Sgt*	QMS
Quartermaster*		Orderly Room Sgt	SGT
		Musketry Sergeant	SGT
		Sergeant-Tailor	SGT
		Corporal-Cook	CPL
		2 Drummers	

4 Training Companies

NCO-in-Charge   CSG
  4 Corporals
10 Privates

maximum of 80 recruits

* Militia permanent staff if a militia battalion is co-located at the depot barracks.

Servants:	Appointed from establishment, 1 per officer (5 for Depot, 1 for District CO).
Mess Staff:	Appointed from establishment were the Officers' Mess Sergeant, Sergeants' Mess Corporal, and Stewards for both messes. Mess NCOs were replaced in their companies with acting NCOs.
Canteen NCO:	Appointed from establishment and replaced by an acting NCO.
Tradesmen:	Tailors and Shoemakers, 1 or 2 of each. Cooks, 1 per 100 men. Clerks for the depot and company orderly rooms. Quartermaster's staff.
Boys:	Apprentice tailors, shoemakers, and drummers plus office boys.

Attached from the RAMC: 1 Medical Officer, 1 SGT, 2 PVTs
Attached from the ASC (Barracks Dept.): 1 SGT, 1 PVT

Appendix B

Cavalry Regiment

Commanding Officer	LTC		
2nd in Command	MAJ	Regimental Sgt-Maj	SM
Adjutant	CPT	Quartermaster Sgt	QMS
Quartermaster		Farrier-Quartermaster	CSG
Ridingmaster		Orderly Room Sgt	SGT
		Saddler Sergeant	SGT
		Sergeant-Instructor	SGT
		Sergeant-Cook	SGT
		Sergeant Master Tailor	SGT
		Sergeant-Trumpeter	SGT
		Orderly Room Clerk	CPL
		Saddletree Maker	PVT

Band

Bandmaster
Band-Sergeant
Band-Corporal
15 Bandsmen

3 Field Squadrons

Commanding Officer	MAJ		
2nd in Command	CPT	Squadron Sgt-Maj	CSG
4 Troop Officers	LT	Squadron QM-Sergeant	CSG
		Squadron Clerk	SGT
		Sergeant-Farrier	SGT
		Corporal-Shoesmith	CPL
		3 Shoesmiths	PVT
		Saddler	PVT
		2 Trumpeters	PVT

  8 Sergeants
  9 Corporals
160 Privates

Reserve Squadron

Commanding Officer	CPT		
2nd in Command	LT	Squadron Sgt-Maj	CSG
Officer	LT	Squadron QM-Sergeant	CSG
		Sergeant-Farrier	SGT
		Squadron Clerk	CPL
		2 Shoesmiths	PVT
		Saddler	PVT
		2 Trumpeters	PVT

  2 Sergeants
  2 Corporals
98 Privates (maximum)

Appendix B

Field Squadrons were divided into four troops, each with a lieutenant, two sergeants, two or three corporals, and forty privates. Like the infantry and artillery, cavalry regiments had soldier-tradesmen and others assigned to permanent extra-duties.

The "reserve" squadron trained recruits. It was within this squadron that new cavalry troopers underwent both basic and advanced training. When the regiment deployed abroad, the reserve squadron relocated to the cavalry training center at Canterbury, Kent and while there served as the regimental depot.

Field Artillery Brigade

Commanding Officer	LTC		
2nd in Command	MAJ	Regimental Sgt-Maj	SM
Adjutant	CPT	Quartermaster Sgt	QMS
Quartermaster		Orderly Room Sgt	SGT
		Sergeant-Cook	SGT
		Sergeant-Bugler	SGT

3 Batteries

Commanding Officer	MAJ		
2nd in Command	CPT	Battery Sgt-Maj	CSG
3 Section Officers	LT	Quartermaster NCO	SGT
		Sergeant-Farrier	SGT
		Battery Clerk	CPL
		3 Shoesmiths	PVT
		2 Saddlers	PVT
		1 Wheelwright	PVT
		2 Buglers	PVT
6 Sergeants			
9 Corporals			
134 Privates		**6 Artillery Guns**	

Such units were battalion-strength though styled "brigades." In the regular army they consisted of three batteries for field artillery (total 18 guns); two batteries for horse artillery (total 12 guns). Each battery had three sections, each commanded by a lieutenant who was assisted by a sergeant. Each section was divided into two sub-sections with a sergeant or corporal in charge. Sub-sections manned individual guns. War strength was about 600 men, all ranks.

Appendix B

Garrison Artillery Company

Commanding Officer	MAJ		
2nd in Command	CPT	Company Sgt-Maj	CSG
3 Section Officers	LT	Quartermaster NCO	CSG
		Gunnery Sergeant	SGT
		Company Clerk	CPL
		2 Buglers	PVT

5 Sergeants
9 Corporals
132 Privates

These units manned guns fixed in fortifications, at home and abroad. In the UK, these were at the naval anchorages, permanent naval bases, and major merchant ports. Each company was divided into sections under lieutenants and subsections under NCOs, as required by the gun complement of the garrisoned fortress.

Field Artillery Depot

Commanding Officer	MAJ		
2nd in Command	CPT	Depot Sgt-Maj	SM
Staff Officer	LT	Quartermaster Sgt	QMS
Quartermaster		Orderly Room Sgt	SGT
Ridingmaster		Sergeant-Tailor	SGT
		Sergeant-Cook	SGT
		Sergeant-Farrier	SGT
		Sergeant-Trumpeter	SGT
		Saddler	CPL
		Wheelwright	CPL
		Shoesmith	CPL
		Carriagesmith	CPL

2 Training Companies

Officer-in-Charge	LT		
		Company Sgt-Maj	CSG
		Company Clerk	CPL
3 Sergeants		Trumpeter	PVT
3 Corporals			
8 Privates			

maximum of 65 recruits

Appendix B

Service Company

Commanding Officer	CPT		
Adjutant	LT	Company Sgt-Maj	CSG
		Quartermaster Sgt	CSG
		Sergeant-Farrier	SGT
		Sergeant-Wheeler	SGT
		Sergeant-Saddler	SGT
		Farrier	CPL
		Wheelwright	CPL
		Saddler	CPL
		Company Clerk	CPL
		Shoesmith	PVT
		Trumpeter	PVT

    3 Sergeants
    4 Corporals
    45 Privates

For war service, the strength of these Army Service Corps transport companies was 106 men, all ranks.

Engineer Field Company

Commanding Officer	MAJ		
Adjutant	CPT	Company Sgt-Maj	CSG
Other Officer	LT	Quartermaster Sgt	CSG
		Company Clerk	CPL
		Shoesmith	CPL
		2 Buglers	PVT

    6 Sergeants
    13 Corporals
    156 Privates

For war service, the strength of these Engineer Field companies was somewhat over 200 men, all ranks.

Appendix B

<u>Militia</u> [2]

Militia Infantry Battalion

Commanding Officer	LTC		
2nd in Command	MAJ	Regimental Sgt-Maj*	SM
Staff Officer	MAJ	Quartermaster Sgt*	QMS
Adjutant *	CPT	Musketry Sergeant*	SGT
Quartermaster *		Sergeant-Drummer*	SGT
		Sergeant-Cook	SGT

6 to 10 Companies

Commanding Officer	CPT		
1 or 2 Other Officers	LTs	Company Sgt-Maj*	CSG
		Company Clerk*	SGT
		Drummer*	PVT
2 Sergeants			
4 Corporals			
100 Privates			

* Permanent Staff, Regular Army

Militia Medical Company

Commanding Officer (MD)	CPT		
2nd in Command (MD)	Lt	Company Sgt-Maj*	CSG
		Company Clerk*	SGT
		Bugler	PVT
4 Sergeants			
6 Corporals			
88 Privates			

* Permanent Staff, Regular Army

---

[2] *Army Estimates, 1904-05*; War Office, *Militia Return for the Training of 1904*, 1905, [Cd. 2432]; *Monthly Army List, December 1904*.

Appendix B

Militia Garrison Artillery Brigade

Commanding Officer	LTC		
2nd in Command	MAJ	Regimental Sgt-Maj*	SM
Adjutant*	CPT	Quartermaster Sgt*	QMS
Quartermaster*		Gunnery Sergeant*	SGT
		Sergeant-Bugler*	SGT

4 to 8 Companies

Commanding Officer	CPT		
1 or 2 Other Officers	LTs	Company Sgt-Maj*	CSG
		Company Clerk*	SGT
		Bugler*	PVT
2 Sergeants			
6 Corporals			
88 Privates			

* Permanent Staff, Regular Army

<u>Volunteer Force</u>

Volunteer Infantry Battalion

Commanding Officer	LTC		
2nd in Command	MAJ	Regimental Sgt-Maj*	SM
Staff Officer	MAJ	Quartermaster Sgt	QMS
Adjutant*	CPT	Armourer Sergeant	SGT
Quartermaster		Sergeant-Bugler	SGT
2 or 3 Medical Officers			
Acting Chaplain			

6 to 11 Companies

Commanding Officer	CPT		
1 or 2 Other Officers	LTs	Company Sgt-Maj	CSG
		Sergeant-Instructor*	SGT
		Company Clerk	CPL
		2 Buglers	PVT
4 Sergeants			
4 Corporals			
100 Privates			

* Permanent Staff, Regular Army

Appendix B

Volunteer Garrison Artillery Brigade

Commanding Officer   LTC
2nd in Command   MAJ
Adjutant*   CPT
Staff Officer   LT
Quartermaster
2 or 3 Medical Officers
Acting Chaplain

Regimental Sgt-Maj*   SM
Quartermaster Sgt   QMS
Orderly Room Sgt   SGT
Armourer Sergeant   SGT
Sergeant-Trumpeter   SGT

Average of 7 Sub-Units

2 to 13 Garrison Companies

Commanding Officer   CPT
1 or 2 Other Officers LT

Company Sgt-Maj   CSG
Sergeant-Instructor* SGT
Company Clerk   CPL
2 Trumpeters   PVT

3 Sergeants
5 Corporals
64 Privates

0 to 3 Position Batteries

Commanding Officer   CPT
1 or 2 Other Officers LT

Company Sgt-Maj   CSG
Sergeant Instructor* SGT
Company Clerk   CPL
2 Shoesmiths   PVT
2 Saddlers   PVT
Wheelwright   PVT
2 Trumpeters   PVT

3 Sergeants
5 Corporals
85 Gunners (PVTs)
30 Drivers (PVTs)

* Permanent Staff, Regular Army

Appendix B

Imperial Yeomanry [3]

Imperial Yeomanry Regiment

Commanding Officer	LTC		
2nd in Command	MAJ	Regimental Sgt-May	SM
Adjutant*	CPT	Quartermaster Sgt	QMS
Staff Officer	LT	Orderly Room Sgt	SGT
Quartermaster		Musketry Sergeant*	SGT
Medical Officer		Sergeant-Trumpeter	SGT
Veterinary Officer			

4 Squadrons

Commanding Officer	MAJ		
2nd in Command	CPT	Squadron Sgt-Maj	CSG
3 Other Officers	LTs	Sergeant-Instructor*	SGT
		Sergeant-Farrier	SGT
		Shoesmith	PVT
		Saddler	PVT
		2 Trumpeters	PVT
6 Sergeants			
6 Corporals			
92 Privates			

* Permanent Staff, Regular Army

Note that a six-company militia infantry battalion had a permanent staff of 24 (3.5% of total), and a four-company militia artillery brigade 17 (4.1%), while an Imperial Yeomanry regiment had only 6 regulars (1.3%). The low level of regulars reflects the force's origin as volunteer cavalry.

---

[3] *Army Estimates, 1904-05*; War Office, *Yeomanry Return for the Training of 1904*, 1905, [Cd. 2267]; *Monthly Army List, December 1904*.

Appendix C
Prestige Rankings of Army Regiments

A regiment's social status was its reputation within the British establishment. Its prestige ranking was primarily a function of its officers' family backgrounds and financial standing.[1] Regiments that had many peers, gentry, and members of socially prominent and wealthy families, were elite. Regiments officered exclusively by middle class men without private incomes and of families with relatively low social standing, were pedestrian. The general social hierarchy of the Edwardian British Army was as follows:

Household Cavalry, Foot Guards, 10th Hussars

Cavalry Regiments other than the 10th Hussars

The Rifle Brigade, The King's Royal Rifles

Highland Regiments
All Other Infantry Regiments

Royal Horse Artillery
Royal Engineers, Royal Field Artillery
Royal Garrison Artillery

Army Service Corps, Royal Army Medical Corps, Departments

At the beginning of the twentieth century nearly all European armies had a similar social hierarchy. Guards units were at the top, cavalry was above infantry, and both ranked above the technical arms. The non-combat, support branches were at the bottom. It's possible to quantify a regiment's social standing though the results won't be exact as status is a perceived quality and not a calculated attribute. For the British Army of the early twentieth century, one can use officers' names and titles to construct an index to measure regimental prestige. The more officers with hyphenated, French, or polysyllabic names, the higher the prestige score. The more officers with mono-syllabic names the lower the prestige score. Titled officers and former royal *aides-de-camp* increase the prestige score. The ranking of regiments in this appendix is from an index score calculated as follows:[2]

---

[1] David French, *Military Identities: The Regimental System, the British Army, and the British People c. 1870-2000* (New York: Oxford Univ. Press, 2005), 165-67.

[2] Weightings were set to produce scores for the Guards and the 10th Hussars much higher than those for regiments generally acknowledged as the least prestigious. The formula is applied only to captains (including rank-equivalents) as their rank was the largest stratum of the officer corps and also, they could remain in the army until age 40. Data for the Ordnance Department is from 3rd and 4th Class Ordnance Officers, all previously combatant officers.

Appendix C

Index Calculation

$$\frac{\begin{array}{l} 70 \times \text{(percent titled or ADC)} \\ + \quad 40 \times \text{(percent with posh names)} \\ + \quad 10 \\ - \quad \text{(monosyllabic names)} / \text{(posh names)} \end{array}}{\text{Prestige Score}}$$

Corps and Department Social Status
(10th Hussars included with Household Cavalry)
Names & Titles of Captains & Equivalents, December 1904

	No.	Names				Titled or Royal ADC	Pct. Titled	Index
		Hyphen	French	>=4 Syllables	Mono-Syllabic			
Household Cavalry	38	1	2	0	1	16	42.1	59.5
Foot Guards	88	3	6	6	10	26	29.5	46.3
Line Cavalry - 10th	249	18	11	0	41	16	6.4	21.7
Line Infantry	2,007	102	102	11	434	20	1.0	13.5
Royal Engineers	353	7	15	0	64	4	1.1	11.3
Artillery (Mobile)	396	10	11	2	59	3	0.8	10.9
Garrison Artillery	530	7	14	0	79	2	0.4	8.6
Pay Dept.	108	1	1	0	28	0	0.0	9.5
Ordnance Dept.	86	3	3	0	21	0	0.0	9.3
Medical Corps	328	6	12	1	81	0	0.0	8.1
Chaplains	45	1	1	0	14	0	0.0	4.8
Service Corps	175	1	4	0	35	0	0.0	4.1
Veterinarians	51	0	0	0	16	0	0.0	NA

Source: *Hart's Army List, 1905*

Note that there were no titled officers within the services and support departments and only two in the garrison artillery. Bear in mind that the index values, like temperature measurement in degrees centered on the freezing point of water, are not absolute values. To speak of a corps as "three times as prestigious" as another is a *non sequitur*.

The following two tables show the status index value for each regular cavalry and infantry regiment in the British Army. Rows for Irish regiments are shaded. All data is from *Hart's Army List, 1905*.

Appendix C

Social Status of Cavalry Regiments
Names & Titles of Captains, December 1904

	No.	Names Hyphen	French	>=4 Syllables	Mono-Syllabic	Titled or Royal ADC	Pct. Titled	Index
Household Cavalry	29	1	2	0	1	12	41.4	59.59
10th Hussars	9	0	0	0	0	4	44.4	58.89
1st Royal Dragoons	9	1	0	0	2	3	33.3	50.61
9th Lancers	11	3	0	0	2	2	18.2	40.51
2nd Dragoons	9	1	0	0	2	2	22.2	38.22
11th Hussars	12	0	0	0	4	3	25.0	36.17
7th Hussars	8	1	0	0	0	1	12.5	28.75
15th Hussars	9	1	0	0	0	1	11.1	26.67
8th Hussars	7	0	0	0	1	1	14.3	24.71
5th Lancers	10	1	0	0	2	1	10.0	24.00
5th Dragoon Guards	8	0	0	0	0	1	12.5	23.75
2nd Dragoons	8	0	0	0	1	1	12.5	22.75
16th Lancers	10	2	1	0	1	0	0	21.67
4th Hussars	8	2	0	0	1	0	0	19.50
7th Dragoon Guards	9	1	1	0	2	0	0	17.89
13th Hussars	10	0	2	0	2	0	0	17.00
17th Lancers	6	0	1	0	0	0	0	16.67
4th Dragoon Guards	7	1	0	0	0	0	0	15.71
1st Dragoon Guards	8	1	0	0	0	0	0	15.00
14th Hussars	7	1	0	0	1	0	0	14.71
3rd Dragoon Guards	13	0	2	0	4	0	0	14.15
6th Dragoons	8	0	1	0	1	0	0	14.00
8th Dragoons	7	0	1	0	2	0	0	13.71
21st Lancers	12	1	0	0	0	0	0	13.33
19th Hussars	10	1	0	0	1	0	0	13.00
20th Hussars	7	0	1	0	3	0	0	12.71
3rd Hussars	9	0	1	0	2	0	0	12.44
2nd Dragoon Guards	7	0	0	0	1	0	0	NA
12th Lancers	10	0	0	0	3	0	0	NA
18th Hussars	10	0	0	0	3	0	0	NA

Appendix C

Social Status of Infantry Regiments
Names & Titles of Captains, December 1904

	No.	Names Hyphen	French	>=4 Syllables	Mono- Syllabic	Titled or Royal ADC	Pct. Titled	Index
Foot Guards	88	3	6	6	10	26	29.5	46.30
Rifle Brigade	63	7	2	0	14	7	11.1	27.06
Cameron Highland's	24	2	1	0	4	2	8.3	23.37
Gordon Highlanders	29	4	0	0	6	2	6.9	22.10
East Surrey	25	2	4	1	4	0	0	20.63
Duke of Cornwall's	29	2	3	0	4	1	3.4	20.02
Highland Lt. Infantry	35	4	5	0	5	0	0	19.73
King's Royal Rifles	37	2	3	2	8	1	2.7	19.54
Seaforth Highlanders	28	0	3	1	3	1	3.6	19.04
Sherwood Foresters	28	1	3	0	6	1	3.6	18.44
Northhamptonshire	27	5	1	0	7	0	0	17.72
Gloucester	23	0	4	0	2	0	0	16.46
Royal West Surrey	39	3	2	0	9	1	2.6	16.45
Royal Irish	**23**	**2**	**1**	**1**	**3**	**0**	**0**	**16.21**
Warwickshire	40	3	3	1	7	0	0	16.00
West Yorkshire	30	1	2	0	7	1	3.3	15.92
The Royal Scots	31	1	2	0	7	1	3.2	15.67
Border	28	2	3	0	8	0	0	15.54
Dorsetshire	27	1	3	0	3	0	0	15.18
Shropshire Lt. Infantry	24	1	3	0	6	0	0	15.17
Royal Welch Fusiliers	21	1	0	0	4	1	4.8	15.14
Essex	26	3	1	0	5	0	0	14.90
South Lancashire	23	2	1	0	1	0	0	14.88
Somerset Lt. Infantry	30	4	1	0	9	0	0	14.87
QO West Kent	32	3	2	0	7	0	0	14.85
Black Watch	24	3	0	0	1	0	0	14.67
Oxon. & Bucks.	35	5	0	0	6	0	0	14.51
Royal Dublin Fusilr.	**29**	**0**	**3**	**1**	**5**	**0**	**0**	**14.27**
Connaught Rangers	**27**	**1**	**3**	**0**	**7**	**0**	**0**	**14.18**
Wiltshire	22	0	3	0	4	0	0	14.12
Duke of Wellington's	31	0	4	0	5	0	0	13.91
Royal Scots Fusiliers	31	2	1	1	7	0	0	13.41
Cameronians	22	2	1	0	7	0	0	13.12
S. Wales Borderers	22	0	3	0	7	0	0	13.12
Norfolk	23	2	1	0	7	0	0	12.88
Royal Fusiliers	48	2	4	0	13	0	0	12.83

Appendix C

Social Status of Infantry Regiments
Names & Titles of Captains, December 1904
(continued)

	Names					Titled or		
	No.	Hyphen	French	>=4 Syllables	Mono- Syllabic	Royal ADC	Pct. Titled	Index
Yorkshire	21	1	1	0	2	0	0	12.81
Liverpool	27	1	2	0	5	0	0	12.78
Argyll & Sutherland High.	30	2	1	0	4	0	0	12.67
Royal Sussex	22	1	1	0	2	0	0	12.64
Northumberland Fuslr.	38	3	1	0	7	0	0	12.46
Leinster	**25**	**2**	**0**	**1**	**8**	**0**	**0**	**12.13**
Durham Lt. Infantry	31	1	2	0	7	0	0	11.54
Loyal N. Lancashire	21	1	1	0	5	0	0	11.31
The Buffs	34	2	1	0	7	0	0	11.20
Scottish Borderers	23	0	2	0	5	0	0	10.98
Yorkshire Lt. Infantry	21	2	0	0	6	0	0	10.81
Bedfordshire	19	0	2	0	7	0	0	10.71
Devonshire	30	0	1	0	9	1	3.3	10.50
Royal Munster Fuslr.	**18**	**1**	**0**	**0**	**2**	**0**	**0**	**10.22**
South Staffordshire	25	1	1	0	6	0	0	10.20
Lancashire Fusiliers	49	0	2	0	4	0	0	9.63
Cheshire	27	2	0	0	8	0	0	8.96
West Indies *	35	1	0	1	7	0	0	8.79
East Lancashire	24	0	1	0	3	0	0	8.67
Manchester	42	2	0	0	7	0	0	8.40
York and Lancaster	20	0	1	0	4	0	0	8.00
Middlesex	40	1	1	0	10	0	0	7.00
Royal Irish Rifles	**26**	**1**	**0**	**0**	**5**	**0**	**0**	**6.54**
Royal Inniskillings	**24**	**0**	**1**	**0**	**6**	**0**	**0**	**5.67**
Lincolnshire	25	1	0	0	6	0	0	5.60
Worcestershire	49	0	1	1	13	0	0	5.13
North Staffordshire	24	0	1	0	7	0	0	4.67
Suffolk	24	0	1	0	7	0	0	4.67
East Yorkshire	25	0	1	0	7	0	0	4.60
Royal Berkshire	22	1	0	0	8	0	0	3.82
Leicester	24	1	0	0	8	0	0	3.67
Royal Irish Fusiliers	**28**	**1**	**0**	**0**	**10**	**0**	**0**	**1.43**
Welsh	24	0	0	0	8	0	0	NA
Hampshire	32	0	0	0	11	0	0	NA
KO Royal Lancaster	22	0	0	0	5	0	0	NA

* The West Indies Regiment was the only Colonial Corps formation that had a position in the home regimental order of precedence. It was always ranked lowest.

Appendix C

David French, in his study of the British Army's regimental system, ranked objectively the social standing of regiments in the first half of the twentieth century. His ranking was by a regiment's total number of titled officers in four specific years and the number of times it had royal patronage in those years. For such information he used the army lists for 1890, 1910, 1930, and 1950. The following table shows French's rankings. Note that he did not assign class labels and except for Class AAA and the scientific arms, regiments within a class are of equal status. Included in the following table are the prestige index scores shown in the previous tables and Irish regiments are indicated by shading.

In the following table an "X" in the variance column indicates a difference in class ranking under French's method and this book's index score by more than one class category.

Social Status of Regiments, Corps, & Departments
1890-1950 Ranking by David French

Class	Regiment, Corps, Department	Prestige Index Score	Variance
AAA	Household Cavalry	59.59	
	10th Hussars	58.89	
	Foot Guards	46.31	
	Class Average	54.93	
AA	1st Dragoon Guards	15.00	x
	1st Royal Dragoons	50.61	
	2nd Dragoons	38.22	
	5th Dragoon Guards	23.75	
	7th Hussars	28.75	
	King's Royal Rifles	19.54	
	Rifle Brigade	27.06	
	Class Average	28.99	
A	2nd Dragoon Guards	10.00	x
	3rd Dragoon Guards	14.15	x
	4th Dragoon Guards	15.71	x
	6th (Inniskilling) Dragoons	14.00	x
	7th Dragoon Guards	17.89	
	2nd Dragoons	22.75	
	8th Dragoons	13.71	x
	3rd Hussars	12.44	x
	4th Hussars	19.50	
	8th Hussars	24.71	
	14th Hussars	14.71	
	15th Hussars	26.67	
	19th Hussars	13.00	x
	20th Hussars	12.71	xx
	12th Lancers	10.00	x
	Class Average	16.13	

Appendix C

Social Status of Regiments, Corps, & Departments
1890-1950 Ranking by David French
(continued)

Class	Regiment, Corps, Department	Prestige Index Score	Variance
B+	Cameron Highlanders	23.37	x
	Gordon Highlanders	22.10	x
	Highland Light Infantry	19.73	
	Royal Welch Fusiliers	15.14	
	Seaforth Highlanders	19.04	x
	The Buffs	11.20	x
	Yorkshire Light Infantry	10.81	x
	Class Average	17.34	
B	Argyll & Sutherland Highlanders	12.67	x
	Cameronians (Scottish Rifles)	13.12	x
	Gloucester	16.46	
	Norfolk	12.88	x
	Oxford & Buckinghamshire	14.51	x
	Royal Fusiliers	12.83	x
	Royal Irish Fusiliers	**1.43**	**x**
	Royal Scots Fusiliers	13.41	x
	Scottish Borderers	10.98	x
	Somerset Light Infantry	14.87	x
	Class Average	12.32	
C	**Connaught Rangers**	**14.18**	
	Duke of Cornwall's	20.02	
	East Yorkshire	4.60	x
	Liverpool	12.78	
	Manchester	8.40	
	Northhamptonshire	17.72	
	Royal Berkshire	3.82	x
	Royal Inniskilling Fusiliers	**5.67**	**x**
	Royal West Surrey	16.45	
	South Wales Borderers	13.12	
	The Royal Scots	15.67	
	Royal Horse Artillery	10.89	
	Royal Engineers	11.28	
	Royal Field Artillery	10.89	
	Royal Garrison Artillery	8.57	
	Artillery Average	10.40	

Appendix C

Social Status of Regiments, Corps, & Departments
1890-1950 Ranking by David French
(continued)

Class	Regiment, Corps, Department	Prestige Index Score	Variance
D	Bedfordshire	10.71	
	Durham Light Infantry	11.54	
	Essex	14.9	
	Yorkshire (Green Howards)	12.81	
	Middlesex	7.00	x
	Royal Irish	**16.21**	x
	Royal Irish Rifles	**6.54**	x
	Royal Sussex	12.64	
	Sherwood Foresters	16.44	
	South Lancashire	14.68	
	West Kent	14.85	
	West Yorkshire	15.92	
	Class Average	12.77	
E	Border	15.54	x
	Cheshire	8.96	
	Devonshire	10.50	
	Dorsetshire	15.18	x
	East Lancashire	8.67	
	East Surrey	20.63	x
	Hampshire	10.00	
	King's Own Royal Lancaster	10.00	
	Lancashire Fusiliers	9.63	
	Leicester	3.67	
	Leinster	**12.13**	
	Lincolnshire	5.60	
	Loyal North Lancashire	11.31	
	North Staffordshire	4.67	
	Northumberland Fusiliers	12.46	
	Royal Dublin Fusiliers	**14.27**	x
	Royal Munster Fusiliers	**10.22**	
	Shropshire Light Infantry	15.17	x
	South Staffordshire	10.20	
	Suffolk	4.67	
	Welsh	10.00	
	West Indies	NA	
	Worcestershire	NA	
	York and Lancaster	NA	
	Class Average	10.64	

Appendix C

Social Status of Regiments, Corps, & Departments
1890-1950 Ranking by David French
(continued)

Class	Regiment, Corps, Department	Prestige Index Score	Variance
F	Army Service Corps	4.14	
	Chaplains' Department	4.78	
	Ordnance Department	9.29	
	Pay Department	9.46	
	Royal Army Medical Corps	8.05	
	Veterinary Department	NA	
	Class Average	7.14	

Appendix D

Appendix D
New Cavalry Officers' Mess at the Curragh Camp

Description of a Newly Constructed Officers' Residence and Mess
The Curragh Camp, Co. Kildare, Ireland

The Kildare Observer, July 23, 1898

"The officers' quarters and mess establishment are a fine block of buildings each 210 feet long by 86 feet deep. The main entrance is under the portico, leading to vestibule and large hall, from which the mess and ante-rooms are entered from either side. A corridor runs the entire length of the building on the ground and first floors communicating with the mess establishment and the officers' quarters which are in the wings right and left, the field officers' quarters being at the extreme end of the north wing. There is one main staircase and one staircase to each wing for access to officers' quarters, &c, on the first floor also separate staircase to field officers' quarters, and for servants at the back of left wing. The mess and ante are very fine rooms, the former being 38 feet by 22 feet [836 sq. ft.], and the latter 28 feet by 22 feet [616 sq. ft.], each being 16 feet high, and having large bay window, in the recess of which is fixed a comfortable settee. The billiard room, which is projected off the corridor at the back, is also a very fine room, 32 feet by 22 feet [704 sq. ft.], well lighted by lantern light. On each side of the room is a raised platform on which is placed comfortable seats about 16 feet long, upholstered in pig skin. At the end of the room are lavatories, water-closets, etc. At the back of the messroom and corridor is the serving room, 26 ft by 12 ft, with benches on two sides to receive dishes, &c, and lift to the kitchen above. Adjoining the serving place is the pantry, 18 ft by 15 ft, with every convenient fitting, such as sinks, cupboards, &c, and at the back of the serving space is the extensive wine celler, plate closet, mess man's quarter's passages and stairs to kitchen and servants' rooms, &c, on the first floor. The kitchen being on the first floor is a great improvement on the old plan of having it at the back of the messroom on the ground floor; it is so well cut off that there is no risk of unpleasant smell from the cooking pervading any of the rooms or corridors, as is generally the case in buildings of this class. Adjoining the kitchen, which is 27 ft by 20 ft, is the scullery, 16 ft by 15 ft, and on the landing leading to the kitchen there is a store room 11 ft by 11 ft, and a larder, 12 ft 6in by 11 ft 6 in, the walls lined with white glazed tiles.

The floors of the portico, lavatories, and water closets are paved with encaustic tiles laid in handsome pattern. The vestibules, halls, passages and corridor on the ground floor are paved De Grillo, Handret and Co's Italian marble mosaic. The screen and swing doors between the vestibule and hall are glazed with stained glass, lead lights of handsome design, which has a very pretty effect. The whole of the work appears to have been executed with materials of the best description and skilful labour …"

Appendix D

Ground Floor Plan of Cavalry Officers' Mess
Curragh Camp, 1902

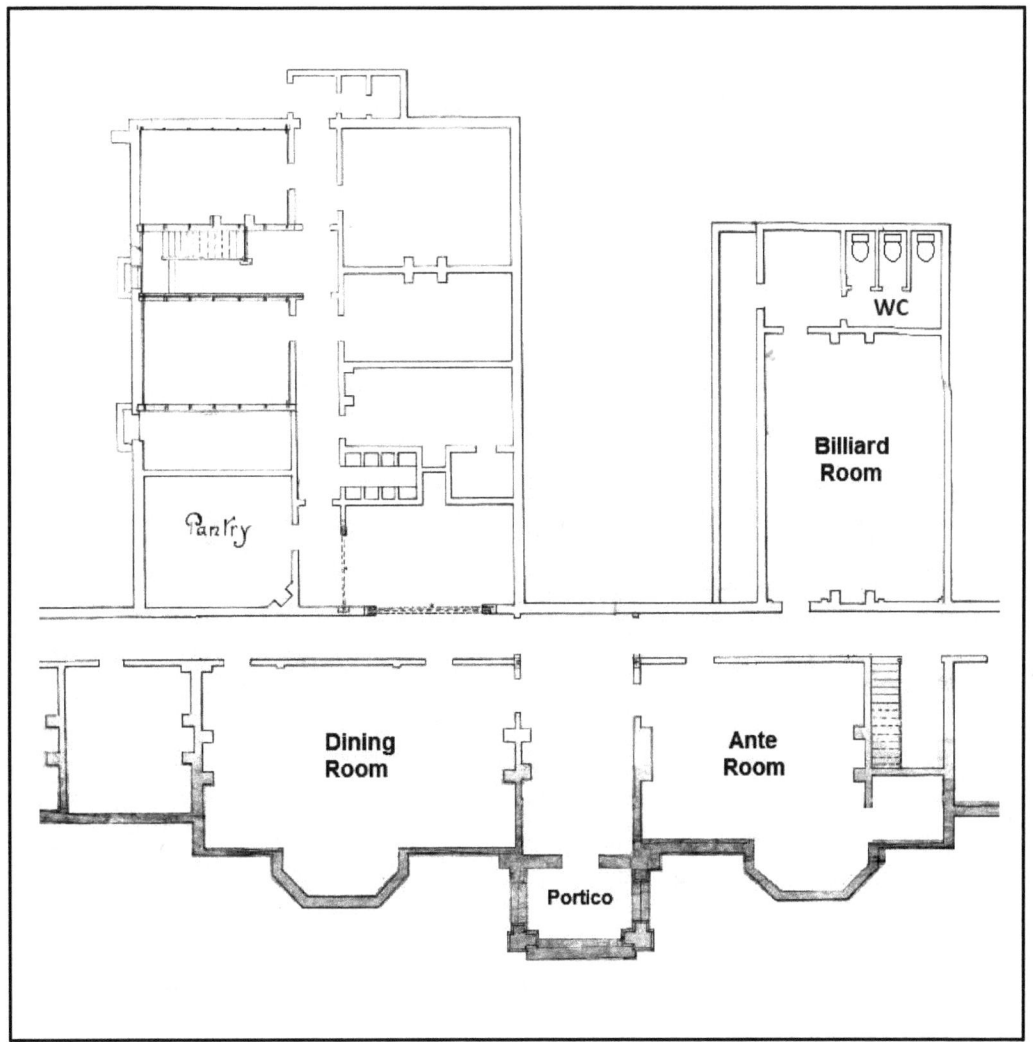

Military Archives, Defence Forces Ireland, IE/MA/MPD/ad134075-005, used with permission.

Appendix D

Upper Floor Plan of Cavalry Officers' Mess
Curragh Camp, 1902

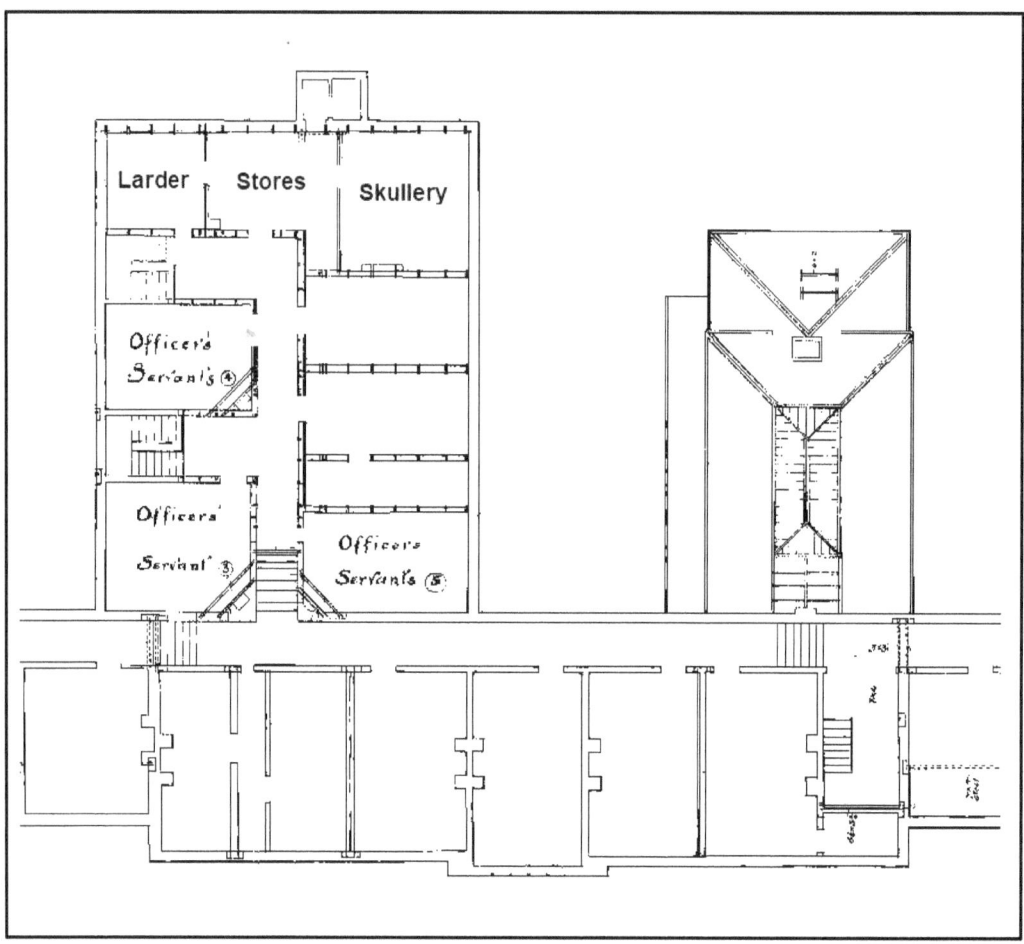

Military Archives, Defence Forces Ireland, IE/MA/MPD/ad134075-005, used with permission.

Appendix E

Appendix E
Badges of Rank

Commissioned Officers

Field Marshal		
General Officers	General	
	Lieutenant-General	
	Major General	
Senior Officers	Colonel	
	Lieutenant-Colonel	
	Major	
Junior Officers	Captain	
	First Lieutenant	
	Second Lieutenant	

Appendix E

The rank titles on the previous page are known as "combatant" ranks. There were other commissioned ranks into which only enlisted men could be promoted: Quartermaster, Ridingmaster, District Officer, Coast Battalion Officer, and Inspector of Army Schools. Such officers wore the badges of their honorary combatant rank.

Other Ranks

The 1906 *Royal Warrant for the Pay of the Army* lists for warrant officers 23 rank titles, for NCOs 34, and for the rank-and-file 19. Each rank had its own distinct badge but there was a general commonality within the six rank classes. Warrant officers, such as an infantry sergeant-major, were identified by insignia worn on the lower sleeve, which usually included a crown. NCOs wore their badges of rank on the upper sleeve and all included chevrons. Class I soldiers, all rank equivalent to the infantry quartermaster-sergeant, were identified with four chevrons worn points up. All other NCOs wore their chevrons points down. Class II soldiers, such as infantry colour-sergeants, were identified with three chevrons and a distinctive insignia. Class III soldiers, all equivalent to infantry sergeants, wore three chevrons; Class IV, corporals, two chevrons. Class V men, privates and lance corporals in the infantry, wore a single chevron (lance corporal), a distinct insignia (drummer), or no badge of rank (private). The most common rank titles for the cavalry, artillery, and engineers are listed below.

Class	Cavalry	Artillery	Engineers
WO	Sergeant-Major	Master Gunner, 2nd & 1st	Sergeant-Major
I	Quartermaster-Sgt.	Master Gunner, 3rd Class Quartermaster-Sgt	Quartermaster-Sgt.
II	Troop Sgt.-Maj	Battery Sgt-Maj.	Company Sgt.-Maj.
III	Sergeant	Sergeant	Sergeant
IV	Corporal	Corporal 2nd Corporal	Corporal
V	Lance Corporal Trooper	Bombardier Gunner	Lance Corporal Sapper

In the graphic on the following page, only differences from the infantry are shown for the cavalry, artillery, and engineers.

The parade dress tunic for the artillery and the twelve "hussar" cavalry regiments was blue; for the King's Royal Rifles and the Rifle Brigade dark green (and their lowest ranked soldiers were called "riflemen" not privates). For all other arms, it was red, but not the bright "scarlet" red of officers' uniforms.

Appendix E

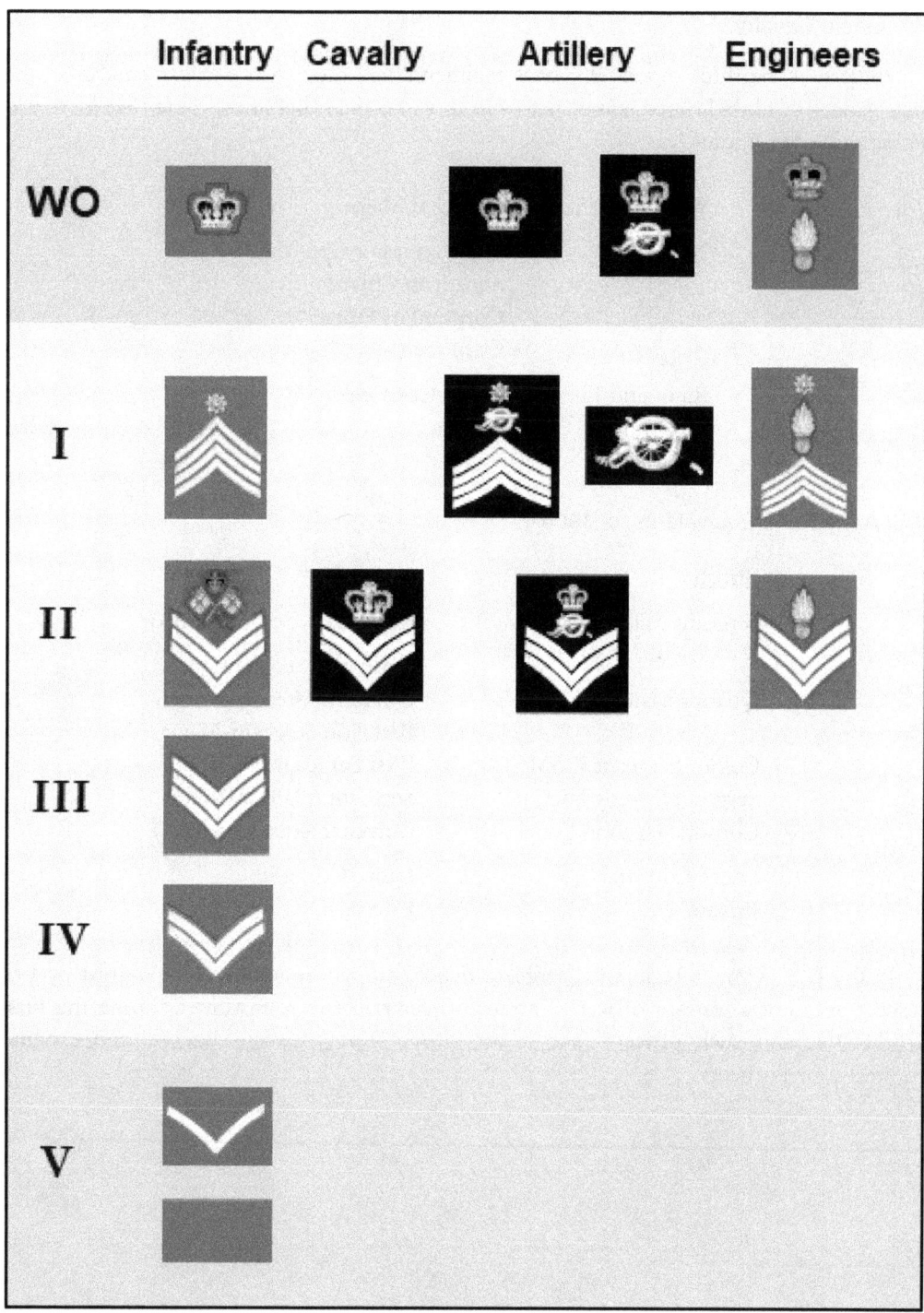

In the artillery, there were three grades of master gunner. The cannon depiction was the insignia for a master gunner third-class who was not a warrant officer but out-ranked a quartermaster-sergeant. Master gunners second and first class were warrant officers (crown for second class, crown and cannon for first class).

Appendix E

Household Cavalry

Soldiers of the three mounted guards regiments had their own peculiar rank titles and their badges of rank differed somewhat from those of the other arms. Note that there was no rank title of sergeant.

Warrant Officer	Corporal-Major
NCOs	Quartermaster Cpl.-Maj.
	Squadron Cpl.-Maj.
	Corporal of Horse
	Corporal
Rank-and-File	Lance Corporal
	Private

U.S. Army Rank Equivalents, c. 1905

British	American [1]
Sergeant-Major (RSM)	Regimental Sergeant Major
	Ordnance Sergeant
Quartermaster-Sgt.	Quartermaster-Sergeant
	Battalion Sergeant-Major
Colour-Sergeant (CSM)	First Sergeant (CSM)
Sergeant (½-coy NCO)	Sergeant (platoon NCO)
Corporal (section leader)	Corporal (squad leader)
Lance Corporal	Lance Corporal
Private	Private

In the U.S. Army, a regimental sergeant-major, though he commanded a great deal of respect, was not a warrant officer. Warrant officer ranks were instituted during the First World War and such officers (like in the Royal Navy) ranked between NCOs and commissioned officers.

---

[1] *Regulations for the Army of the United States, 1904.*

Index

1719 Jacobite Rising 65
1745 Jacobite Rebellion 66-73
1798 Insurrection in Ireland 33-37

Abbas Pasha 87
Abdi Pasha, General 90
Abdulmecid, Sultan 85, 91, 110
Aberdeen (Scotland) 67
absence-without-leave 279-80, 288, 299
Abyssinia Expedition 11, 128-30
Addiscombe (England) 119, 167, 183
Aden 121, 129
Admiralty 80, 95, 152, 157
Afghanistan 78, 121, 128
Africans 132, 134, 137, 144, 148-49
Afrikaners 132, 137, 139, 145, 149
Ahmed Bey 87
Akers-Douglas, Aretas 186
Akers-Douglas Committee 187-88, 190,
Albert, Prince of Wales 102
Aldershot Camp (England) 143, 189, 213, 214, 229, 240, 249, 285, 288
Aldershot Military District 143
Alexander II, Czar of Russia 111
Allahabad (India) 124
al-Taashi, Abdullah 130
Amsterdam (Netherlands) 27
Anglicans, 15, 32, 164, 185, 233
Anglo-Irish 16, 25, 44, 48
Anglo-Dutch Brigade 63-64
anti-recruiting movement, Ireland 150-51
Antoinette, Marie 25
Antrim, Co. 15, 25, 32, 262
Antwerp (Belgium) 26-27
apprentices
    civil 245
    army 249
Arklow, Battle of 36
Armagh, City 31
Armagh, Co. 31, 32, 45
Army Council 157
Army Enlistment Act, 1870 42, 211, 260

Army of the East, Crimean War (French) 94, 96-99, 108, 111-12, 113-14
Army Service Act, 1847 210
Army Service Corps (ASC) and predecessors 80-81, 93, 114, 129, 181, 182, 183, 187, 189, 194, 196, 199, 203, 211, 217, 229, 230
    School 184, 213
Artillery College 189
Australia 135, 142, 146, 153
Austro-Hungarian Empire and Hapsburg Dynasty 8, 25-27, 66, 95

Baden-Powell, Col. Robert 149
Bahadur Shah 124
Baji Rao II, Maratha Emperor 121
Ballynahinch, Battle of 34
Bank of England 74
Bank of Ireland 234
Basters 132
Bechuana Protectorate 137
Behar-Orissa (India) 121
Belfast 19, 20, 33, 34, 48, 49,
Belford College 185
Belgium 25–27
Bengal
    Army 120-21, 122, 123-24, 126, 169, 177
    Presidency 119, 122, 124-25, 164, 175, 177
Berehaven Naval Anchorage 43
Bermuda 252
Berwick-upon-Tweed (England) 69
Berwick-upon-Tyne (England) 70
Bill of Rights, England 65
Birmingham (England) 71, 175
Birr 177
Black Sea Fleet, Russian 90, 91, 95, 114
Blackness Castle 62
Bloemfontein (Orange Free State, South Africa) 47, 136, 143, 147

Boer War, 2nd 7, 41-43, 190, 216, 247, 251, 258, 261, 263, 272, 277, 299
   blockhouses 145
   casualties 145-49
   civilian deaths 149
   commandos 141, 143-45, 149, 150
   concentration camps 144, 149
   conventional war 138-44
   gold 136
   guerilla war 144-45
   internment 144-45
   Irish Imperial Yeomanry 47-48
   Irish opposition 149-51
   Jameson Raid 136-37
   militia embodiment 291-92
   militia in Ireland 41-47
Bombay (India) 194
   Army 119, 120, 123-24, 129
   Presidency 119, 122, 124, 164
Bompart, Commodore Jean-Baptiste-Francois 36
Bond, Oliver 33
Bristol (England) 79, 216
British Army
   aid of civil authority 53, 65, 73-75, 78-79
   bandmaster 231, 248, 291
   barracks 7, 23, 57, 59, 81, 115, 152-54, 197-98, 202, 212, 213, 223, 226-29, 231, 242, 257-58
   boy soldiers 245, 249
   brevet rank 193-94, 207
   buglers 216, 290
   canteens 152, 221, 235-39
   certificates of education 225, 234-35
   chaplains 184, 185, 208
   commissioning of officers 81, 181-86
      cost 182-83, 185
      departmental 185
      from ranks 152, 181-82, 194, 234
   courts-martial 289
   creation 54-56, 62-65
   Defaulters' Book 218
   desertion 263-64

disease and health 106, 142, 153, 168, 174, 207, 218, 229, 247, 252, 253-57
drummers 12, 230, 290, 291
enlistment terms 210-11
footguards 54-55, 62-63, 74, 155, 181, 182, 183, 187, 195, 200, 201, 204, 205, 211, 220, 252-53
general staff 157
gratuities 205, 206, 207, 255, 265
household cavalry 55, 62-63, 181, 187, 200, 204, 211, 241, 252-53
illegitimate children 251, 286
illiteracy 173, 234
impecunious officers 187, 196, 202
imperial campaigns 77-78, 128-30
India 75-78, 80, 93, 120, 126, 127, 128, 152, 168, 171, 174, 178, 246, 251, 252, 253-55
intoxication 224, 289
invalided 153, 207, 210, 255-56, 264
maneuvers 214-16
matron, nursing 197
midwifery 249
mortality rates 89, 174, 253-255
musicians 105, 290
off-the-strength families 240, 250-51
on-the-strength families 94, 153, 235, 238, 240-253
organization 80-81, 114-15, 155-58
pensioners 91, 205-06, 256-57, 259-60, 261, 266-67
quartermasters 81, 181, 182, 185, 194, 196, 199, 200, 208, 223, 224-25,
recruiters 152
recruitment 70, 81, 141, 142, 154, 157
re-enlistment 153, 155, 210, 265
reform 154-58, 206
regimental sgt.-maj. 12, 81, 199, 231, 289, 299
ridingmasters 181, 182, 194, 196
sappers 212-13, 217
schoolmaster 172, 196, 225, 234, 244, 245, 249

seniority, officer promotion 114, 189, 193, 194
sergeant-cook 229, 231, 242
service during 1815-1869 152-53
territorialization 155
trumpeters 230
veterinarians 185, 208, 213
wars of 1692-1815 65-78
wastage 141, 291
wives 94, 205, 240-42, 244, 247, 249-251
British South Africa Company (BSAC) 136-37
Brussels (Belgium) 26
Buffels River (South Africa) 131
Buller, General Redvers 138, 140-41, 143, 148
Burdett Disturbances 78
Burma 77, 78, 121, 128, 253
Bushmen (San) 131, 135
Butler, Simon 33
Buttevant Barracks 243

Calcutta (India) 119, 162-63, 164
Cameron, Capt. Charles 129, 130
Canada 77, 142, 146,
    Royal Military College 184
Canning, Charles 176, 177
Canterbury (England) 212, 214
Cape Colony and Dominion 131-36, 137, 138-39, 144, 145, 149
Cape Coloured 132, 133, 136, 137, 138, 148
Carrickfergus 19
Castlebar, Battle of 36
Catholic Confederation of Ireland 14
Catholic Defenders 31
Catholic Relief Act, 1778 74
Catholic Relief Act (Ireland) 1793 28
Catholics 8, 14-17, 28, 29, 30. 31, 32, 34, 37, 40, 41, 44, 59, 233
    attacks on, 31, 74
    disabilities of, 15, 28, 59, 173
Cavan, Co. 30
Ceylon 251, 252
Chamberlain, Joseph 137

Charlemont, Lord 21
Conservative Pary 74, 134, 135, 137, 157, 186
consols 119
Constantinople 86, 91-96, 101, 102, 105, 107, 108
Cope, Lt.-Gen. John 66-69
Cork, City of 18, 49, 175, 178
Cork, Co. 28
Cork Harbor 43
Cork Military District 169
Corn Laws 79
cornet 108
Crete 251
Crimean War
    Adrianople (Ottoman Europe) 90, 92
    Akhaltsike (Ottoman Caucasus) 91
    Alma River 99-100, 113
    Azov (Russia) 110
    Balaklava 97, 99-101, 103-04, 106, 108, 109
    Balkans 86-87, 90-92, 94-95, 104-05, 109-110
    Baltic Sea 109
    Batumi (Ottoman Tran-Caucasus) 90-91
    Bayezid (Ottoman Caucasus) 91
    Besika Bay (Ottoman Anatolia) 85, 86
    British armed forces 86, 92-95, 98, 101-07
    British financial cost 114
    Bucharest (Wallachia) 90, 92
    Bulgaria 90, 94, 96, 100-01
    Caucasus 87, 90, 95, 109
    causes 84-86
    casualties 113-114
    cholera 96, 100, 102
    Chernaya River 97, 100
    Cossacks 87, 98
    Dagestan (Russian Caucasus) 91
    Dardanelles 85, 86
    Erivan (Russian Armenia) 91
    Eupatoria 97-99, 110-111
    Fort St. Nicholas (Russian Trans-Caucasus) 90
    French forces 86, 96-99, 107, 111-13

Gallipoli (Ottoman Europe) 94
health of the allied armies 96, 100-09
Inkerman 100, 114
invasion of Crimea 95-98
Kalafa (Ottoman Bulgaria) 90
Kalamita 97, 100
Kamiesh & Kazach Bays 97, 99
Kars (Ottoman Caucasus) 91
Kerch 111-12
Kinburn Spit (Russia) 112
Malakov Bastion 112
Odessa (Russia) 91
Ottoman forces 86-87
Perecop 99, 111
Russian forces 87, 89-90
Scutari (Ottoman Anatolia) 93, 94, 95
Sevastopol 90, 91, 95-104, 107, 108, 110-13
Silistra (Ottoman Bulgaria) 92
Simferopol (Crimea) 99
Sinope (Ottoman Anatolia) 91, 95
Tiflis (Russian Georgia) 91
Treaty of Paris 114
Tunis 87
Varna (Ottoman Bulgaria) 94, 96, 100-02, 107
Wallachia 85, 90, 95
Croker, John Wilson 25
Cromwell, Oliver 14, 53, 61
Cromwell, Richard 54
Culloden (Scotland) 72-73
Cumberland, Duke of 22, 68-73
Curragh Camp 192, 214, 285

Dalhousie, Marquess of (James Brown-Ramsay) 121-22
Daniell, Louisa 240
Denbigh (Wales) 303
Derby (England) 71
Derry, Co. 9, 25
Dickeson & Co. 236-37
Dillon, John 139, 143
Disbandment Act, 1660 54-55, 64
Disraeli, Benjamin 135
Doctrine of Lapse 122
Donegal, Co. 36, 285

Dorset (England) 214
Down, Co. 15, 28, 32, 34, 262
drill halls 294, 296, 297-98, 302
drills (training sessions) 297, 299, 300, 302
Duberly, Frances (Fanny) 105
Dublin, City 7, 13, 18, 19, 20, 30, 32, 33, 34, 36, 48, 49, 53, 56, 58, 73, 107, 175, 183, 199, 202, 226, 234, 244, 245, 247, 248, 258, 266
Dublin, Co. 8, 25, 34, 262
Dublin Castle 24, 53, 57, 234
Dublin Evening Mail 41
Duke of York's Miitary School 244-46
Dumbarton Castle (Scotland) 61, 62, 69
Dunbar (Scotland) 67
Duncannon 36
Dungannon Convention 20
Dunkirk (France) 25, 55-56, 62
Durban (Natal South Africa) 138
Dutch Republic 27, 69, 70, 132
dysentery 26, 92, 108, 148

East India Company, Armies of 84, 118, 161, 165, 168-69, 171, 175-77, 183
  Commissariat 172
  enlisted men 170-74
  European regiments 9, 119, 127-28, 161, 163, 178-79
  health 174
  Military College 119, 167, 183
  officers 167-70
  officers in civil positions 166
  Ordnance Department 172
  Public Works Department 166, 172
  Recruit Depot 120, 171, 178
  Recruitment Act, 1799 167, 175-76
  "White Mutiny" 175-78
East India Company, British (EIC) 77, 80, 118-19, 128, 152, 161-62
  civil college 119
  civil administration 163-65
  clergy 164
  Control Board 80, 119, 127, 167
  Court of Directors 119
  Presidencies 122, 124, 163-65

Resident Officers 121
  statutes regulating 166-67
East India Company, Dutch (VOC) 131-32
East India Company, French 77
Easter Rising 73
Edinburgh 61, 67, 69, 70
Edinburgh Castle 62, 67, 69, 72, 74
Edinburgh, University of 66
Egypt 11, 87, 91, 96, 100, 111, 128, 129, 130, 251, 261
Elcho, Lord (David Wemyss) 70
Elgin, Earl of (James Bruce) 178
Elizabeth I, Queen of England 161
Emmet, Thomas Adis 33
Enclosure Acts 74
enlisted men, life of British Army
  adult education 234-35
  barrack housing 226-29
  busy-work 172, 231, 232
  butchers 80, 229
  childrens' education 244-46
  church parade 152, 233, 257
  civilian clothing 258
  cookhouse 220, 228, 229, 230, 238, 242
  disability pensions 255-57
  discharge gratuities 265
  discharge by purchase 264
  discipline 53, 152, 154, 172, 199, 212, 289, 245, 257, 289
  drill 152, 212, 214, 215, 230, 231
  drunkenness 224, 289
  early 19th Century 152-53
  employment of dependents 249
  extra duties 219-221
  family housing 242-44
  family medical care 242, 244
  family separation allowance 249-51
  fatigues 152, 201, 212, 230, 231, 232, 234
  flogging 53, 152, 154
  food 153, 154, 222, 229-30, 238, 240, 242
  furlough 257-258, 265
  garrison life 226-234
  good conduct awards 154, 218-19
  groceries 222-23, 230, 235, 236, 238, 251
  haircuts 171, 221, 224
  laundry 152, 171, 224, 249
  leave 258
  libraries 154, 220, 224, 229, 239-40
  lodgings 76, 285, 288
  marriage 240-42,
  mortality 253-55
  musketry 152, 212, 214, 231
  overseas service 249-53
  passes 219, 257-58
  pay 216-24
  promotion 224-26
  rations 152, 222, 224, 229-30, 242, 251, 265
  regimental institutes 235-39, 258
  retirement and pensions 265-67
  retirement homes ("hospitals") 266-67
  short-service 42, 155, 211, 225, 234, 260, 265
  soldiers' homes 240
  stoppages 29, 171, 221, 222-24, 249, 250-51
  survivor benefits 246-48
  training 211-16
ensign 29, 38
enteric fever (see typhoid fever)
epidemics 147, 285
Esher Commission 157-58
Essex (England) 120, 171, 214, 216
Eton 185

Falkirk Muir (Scotland) 72
Falmouth (England) 292
fencibles 32
Fenians 40
fensible persons 60-61
Fermoy 178
Filder, James 94
Finner Camp 285
First World War 7, 8, 262
Fish River (South Africa) 131
Fitzgerald, Lord Edward 33-34

Flanders 66, 68-69, 70, 71, 72, 75
Fleetwood, General Charles 54
Flight of the Earls 14
Foreign Enlistment Act 137
Fort Augustus (Scotland) 66, 72, 73
Fort George (Scotland) 66, 72
Fort William (Scotland) 72
Fort William (India) 162
Fowke, Brig. Thomas 67
France 19, 20, 24, 25-28, 29, 31, 36, 66, 71, 75, 76, 85, 91, 92, 96, 132, 137, 158, 294
Francis II, Hapsburg Emperor 25
Franciscans 84-85, 114
Freeman's Journal 9

gentlemen 53, 58, 118, 162, 167, 168, 170, 181, 182, 185, 186, 190, 195, 201, 203, 204, 205, 206, 239, 271, 272, 273, 296, 299, 301, 302
gentry 13, 20, 30, 31, 64, 181, 186, 201, 271, 299
George II, King of Great Britain 68, 70
George III, King of Great Britain 22-23, 26
George-Louis, Prince of Hesse-d'Armstadt 76
Gibraltar 12, 76, 86, 130, 192, 251, 252, 255
Gladstone, William 135
Glasgow (Scotland) 67, 72, 74, 145, 175
Glenfinnan (Scotland) 66
Gogarty, Oliver St. John 49
Gonne, Maud 258
Gorchakov, Prince Mikhail 85, 86, 90, 111, 112
Gordon, Lord George 74
Gordon Riots 74
Government of India Act, 1833 118, 127, 167
Government of India Act, 1858 127, 175
Graham, James 95
Gratton, Henry 21
Grenville, William 26-27
Griqua 132, 135
Guest, Lt.-Gen. Joshua 67

Gunpowder Act, 1793 25
Guinness Son & Co. 247
gymnasia 187, 214, 231

Haldane, Richard 154, 158, 270
Hamilton, John 72
Hampshire (England) 213, 214
Handasyde, Lt.-Gen. Roger 69, 70
Hanoverian soldiers 162
Harrow 185
Hawley, Gen. Henry 72
Henry VIII, King of England 56, 293
Highlanders 60, 66, 67, 69, 73
Highlands (Scotland) 62, 65, 66, 67, 69, 70, 72, 73
Hill, Arthur 28
Hindu 121, 122
Hobart, Robert 28
Home Office 93, 115, 294
home-rule 150
Hong Kong 78, 252
Hottentots (Khoi-Khoi) 131-32
Huguenot 131
Humbert, Gen. Jean Joseph 36-37

Indian Army 127, 128, 158, 161, 176-79, 183, 186, 187, 191, 246, 253
    batta 191
    native ranks 166
Insurrection Act, 1796 33
Invalid Companies and Battalions 66, 69
invasion fears
    French 19, 24, 27, 31, 71, 294
    Spanish 15
Inverness (Scotland) 43, 66, 67, 72
Irish Brigades, South Africa 150
Irish Military Department and Establishment 19, 23-24, 56-59
Irish Parliamentary (Nationalist) Party 139, 150
Isle of Wight (England) 245, 275
Ismail Pasha, General 90

Jacobites 14, 65, 66, 73
James II, King of England 14, 58-59, 67
Jameson, Leander 136-37

Jerusalem (Palestine) 84, 129
Jews 87, 233
Johannesburg (Transvaal, South Africa) 136-37, 143
Joseph, Hapsburg Emperor Franz 95
Joyce, James 7-8, 9, 49, 84, 161, 174, 219, 258
Joyce, John 149
Joyce, Stanislaus 7

Kaffirs 132
Kent (England) 171, 212, 214
Kerr, S. Parnell 7
Kerry, Co. 30, 32
Kettle, Thomas 7
Khedive of Egypt 129
Khoi-Khoi 131-32
Khomoutov, General 98
Khyber Pass (India) 119
Kildare, Co. 33, 34, 214
Kilkenny, Co. 30, 32, 34, 44, 273
Killala 36
Kilmainham Gaol 34
Kimberley (Cape Colony, South Africa) 137, 139, 143, 147
King's County 13, 32
Kingstown 8
Kipling, Rudyard 128
Kitchener, Field Marshal Herbert 130, 144-45, 149
Kornilov, Admiral 91, 98, 99
Kronstadt (Russia) 89
Kruger, Paul 136-38

Langdale, Thomas 74
Lawrence Military Asylums 246
Leicester, Earl of (Robert Dudley) 63
Leinster 32, 34
Levy Acts of Scotland 60
Liberal Party 135, 154, 157, 185, 270, 299
Lichfield (England) 71
Lille (France) 25
Limerick, City of 18, 28, 49
Limerick, Co. 30
Lindley (Orange Free State, South Africa) 48

Lipton, Ltd. 236-38
Lord Mar's Revolt 65
London, University of 184
London Globe 41
Londonderry 9, 49
Longford, Co. 36,
Longford, Earl of (Thomas Pakenham) 48
Longwy (France) 26
Lord Mars' Revolt 65
Loudon, Earl of (John Campbell) 67, 70, 72
Lough Swilly 36, 43
Louis XIV, King of France 75
Louis XV, King of France 71
Louis XVI, King of France 26, 27
Louth, Co. 46
Louth Militia "Mutiny" 44
lower-middle class 13, 181, 195, 197, 295, 301
Lowlands (Scotland) 62, 66, 69
Lucknow (India) 124
Luddites 78
Lumsden, John 247-48, 250, 251

MacDonald, Ranald 66
MacNevin, William 33
McNeill, John 107-08
Madras (India) 162, 163
    Army 119, 120, 123
    Presidency 119, 120, 124, 164, 177
Mafeking (Cape Colony, South Africa) 137, 139, 143, 147, 149
Magdala (Abyssinia) 129-30, 141
Mahdi 130
malaria 148
Malaya 121, 252
Malta 43, 86, 93, 255
Manchester (England) 70, 79, 175, 300
    University 184
Manoeuvres of 1898 214-16
Maratha Empire 77, 121
marines 76
Marlborough Barracks 226
Marlborough College 185

Mary I, Queen of England 13
Massawa (Egypt) 129
Master-General of the Ordnance 92, 182
Maubuege (France) 25
Mauritius 251, 252
Mayo, Co. 28, 36, 150
Meath, Co. 34
Medway, River (England) 275
Menshikov, Gen. Prince Alexander 98-99
mercenaries 59, 61, 63, 66, 69, 71, 77, 87, 89, 90, 162
merchants 27, 123, 161, 162, 163, 170, 173
Methodists 185, 240
middle class 13, 18, 20, 64, 106, 152, 167, 181, 186, 206, 272, 295, 296, 301, 302
Military Orphan Society 169
Military Service Act, 1916 158
militia 271-91
militia, Irish
    history of 14-19, 25-31, 38-41
    Insurrection of 1798 36, 37
    Boer War 41-47
Militia Act, 1882 270, 280
Militia Act (Ireland), 1716 14-15, 16, 17
Militia Act (Ireland), 1793 28-29, 30
Militia (Ireland) Act, 1802 38
Militia (Ireland) Act, 1854 39
Militia and Yeomanry Acts, 1901 & 1902 270
militia ballot 18, 29-30, 38, 154
militia bands 290-91
Militia Reserve 42, 139, 270, 291
Militia Reserve Division 293
militia service
    enlisted men 277-90
    officers 271-77
Milner, Alfred 137-38, 145
Mir Jafar, General 163
Moldavia 85-86, 95
Monaghan, Co. 30
militia ballot 18, 29-30, 38, 154
militia bands 290-91
Militia Reserve 42, 139, 270, 291

Militia Reserve Division 293
militia service
    enlisted men 277-90
    officers 271-77
Milner, Alfred 137-38, 145
Mir Jafar, General 163
Moldavia 85-86, 95
Monaghan, Co. 30
Monck, Gen. George (Duke of Albemarle) 54-56, 58
Mountjoy Convict Prison 234
Mughal Empire 121, 124, 162
Munster 32, 34
Murray, John 66, 69
Murray, Lord George 69
Musa Pasha, General 92
Muslims 86, 87, 122, 125, 129, 130
Mutiny Act 65

Nagpore (India) 122
Nakhimov, Admiral 91
Nana Sahib 121, 124
Napier, Lt.-Gen. Robert 129-30
Napoleon III, Emperor of France 84, 85, 92, 93, 96, 110, 294
Napoleon Bonaparte 65, 76, 152
Napoleonic Wars 38, 76-77, 93, 102, 114, 132, 146, 254, 293
Natal (South Africa) 11, 77, 133-36, 138, 139, 141, 143, 144, 145, 148, 149
nationalists (Irish) 7, 34, 41, 44, 65, 149, 150, 151, 278
Native States 121-22, 123, 124
needlework 244
Netherlands 14, 25, 27, 132
New Model Army 53-55, 57, 58, 61-63
New Zealand 77, 78, 128, 142, 146, 153
Newcastle, Duke of (Henry Pelham-Clinton) 95, 105
Newcastle-upon-Tyne (England) 70-71
Newtownbarry "Massacre" 41
Nicholas I, Czar of Russia 85-86, 90, 95, 111
Nightingale, Florence 105
Nine Years War 14

Norfolk Commission 157, 270, 271, 283, 284, 288, 289, 297
nurses 105, 108, 197
O'Connor, Arthur 33
O'Dougherty Rebellion 14
O'Sullivan, Captain John William 66
oaths 15, 28, 33, 176
    Anglican Oath of Allegiance 15
officers, life of British Army
    garrison life 199-200
    half-pay 168-69, 192, 206
    housing 197-99
    leave 200-201
    marriage 204-05
    mess 201-03
    musketry 188, 200, 212
    pay 190-93
    promotion 193-96
    ranks, types 193
    retirement and pensions 205-08
    social life 200-205
    staff appointments 192-93
    training 186-91
Omar Pasha (Mihaylo Latas), Gen. 87, 111
Omdurman (Sudan) 130, 141
opium 77, 78, 118, 119, 128, 167
Orange Free State 47, 48, 133, 135, 137, 138, 145
Orange River (South Africa) 133, 135, 139
Orange River Colony 144
Order of Orange 31, 31
Ordnance College 189
Ordnance Corps and predecessors 12, 53, 56, 59, 80, 81, 93, 115, 155, 181, 182, 183, 185, 191, 195-96, 208, 213, 218, 223, 253
orphanages 244-45, 247
Osman III, Sultan 84
Ottoman Empire 84-85, 86-88, 90, 95, 103
Oudh (India) 121-22, 124

Palmerston, Viscount (Henry John Temple) 106, 108

Panmure, Baron (Fox Maule-Ramsay) 106-08
Parkhurst (England) 178
parliamentary reform protests 79
Parr, Maj.-Gen. Henry Hallam 93
Paskevitch, General Count 90, 92
Patriotic Fund 246-47
Pay Department 185, 195-96, 213
Paymasters 105, 182
Paymaster-General 56
Peep O'Day Boys 31
Pelham, Thomas 31
Pembroke (England) 178
Persia 121
Perth, Duke of (James Drummond) 69, 70
Perth (Scotland) 67
Peshawar (India) 124
Peshwa of Maratha Empire 121
Peterloo "Massacre" 79, 300
Petropaulovsk (Russia) 90
Piedmont and Sardinia, Kingdom of 74, 100, 113
Pitsani (Bechuana Protectorate, South Africa) 137
Plassey (India) 163
Plymouth (England) 275
Police Gazette 281
Poor Law Act, 1834 79
Portobello Barracks 225
Portugal 61, 63, 162
Poyning's Law 21
Presbyterians 15, 16, 17, 20, 21, 32, 33, 34, 164, 185
Prestonpans (Scotland) 69, 71
Pretoria (Transvaal, South Africa) 47, 133, 135, 136, 143, 145, 147
Pretoria Convention, 1881 136
Pretorius, Andries 135
prostitution 152, 171, 219
Protestant Ascendancy 15, 16, 32
Protestants 8, 14-20, 29, 31, 32, 33, 34, 39, 41, 44, 59, 64, 74, 150
Prussia 25, 95, 152

Quartermaster-General 101, 115

Queen Alexandra's Nursing Service 197
Queen's County 13, 30, 45
radicals 20, 41, 78
Raghoji III Bosale, Raja of Nagpur 122
Raglan, General Lord (Fitzroy Somerset) 92, 95-97, 99, 101, 104, 108
Rajputs 121-22
Redmond, John 150
reserve, British Army 42, 139, 153, 155, 211, 225, 257, 259-63, 270
Reserve Forces Acts, 1859 & 1867 260
Reserve Forces Act, 1882 261
Reserve Forces and Militia Act, 1898 270
Reynolds, Thomas 33
Rhodes, Cecil 136-37, 139
Richmond (England) 291
Richmond Barracks 202, 226
Riot Act, 1715 74
Roberts, Field Marshal Frederick 141, 143, 148
Roebuck, John Arthur 106
Rooke, Admiral George 76
Royal Army Medical Corps (RAMC)
  and predecessors 93, 115, 147, 183, 185, 191, 213, 218, 220, 224, 241, 256, 271
Royal Artillery 12, 43, 56, 66, 67, 69, 81, 115, 127, 178, 181-89, 191, 194-96, 199, 200, 202, 205-06, 210-12, 213, 214, 217, 221, 224, 231, 253, 271-73, 275, 282, 284, 286, 287, 290, 292, 295, 297, 299
Royal Barracks 226
Royal Chelsea Hospital 81, 107, 266-67
Royal Engineers 12, 81, 115, 129, 153, 168, 181-89, 191, 194, 195, 196, 202, 205-06, 211, 212-13, 214, 217-18, 222, 224, 271, 277, 293
Royal Hibernian Military School 244-45
Royal Kilmainham Hospital 81, 107, 266-67
Royal Military Academy (Woolwich) 181-85, 187, 188, 191, 273
Royal Military College (Sandhurst) 181-87, 273
Royal Navy 36, 43, 54, 66, 72, 80, 139, 152, 157

Royal Patriotic Fund 294-95
Royal University 184
rupee 168, 171, 191
Russell, William Howard 102, 107
Ruthven Barracks (Scotland) 69, 70, 72, 73
Sacramental Test Act 15, 21
Saint-Arnaud, Gen. Leroy de 92, 96, 97, 99
Saint Helena 127, 152, 252
Salisbury Plain (England) 214-16, 285
Salvation Army 240
San 131, 135
Sandes, Elise 240
Sardinia and Piedmont 25, 100, 113
SASFA 250-51
Savoy Dynasty 100
School of Cookery 229, 288
School of Gunnery 188, 189, 275
School of Gymnastics 189
School of Military Engineering 188, 189, 212, 213, 288
School of Musketry 188, 275
School of Signaling 189
Schroeder, Christian 22
Scottish Army 60-61
Scottish Military Establishment 61-62
Secretary-at-War 80, 81, 108, 115
Secretary of State for India 127
Secretary of State for War 80, 81, 95, 106, 108, 115, 157, 229
separatists, Irish 8, 150
serfs, Russian 87, 89
Seven Years War 23
Seventh Day Adventists 233
Shamyl, Ali 91
Sheerness (England) 275
Sheffield (England) 43, 44, 45, 292
Shoeburyness (England) 188, 189
Shorapur (India) 124
Shorncliffe Camp (England) 214, 285
Sikhs 78, 125
Simpson, Lt.-Gen. James 108
Singapore 252
Sirah-ud Daula, Nawab 162, 163

soldier-tradesmen 264-76
 barber 221, 224
 bandsman 230, 290-91
 barman 201
 canteen assistant 221, 236, 249
 canteen steward 236
 clerk 220, 199, 219, 230
 cook 229
 drummer, *etc.* 216, 230, 290
 farrier 216, 230
 harness-maker 230
 master-tailor 249
 mess steward 201
 pioneer sergeant 288
 saddler 216
 servant 193, 197, 199, 219, 220-21, 234
 shoemaker 221, 224
 tailor 221, 224, 249
Staff College 155, 189-90
Stair, Earl of (General John Dalrymple) 71
Steyn, Martinus 138
Stirling Castle (Scotland) 61, 62, 66, 67, 69, 72
Straits Settlements 251, 252
Strathallan, Viscount (William Drummond) 67, 71, 72
Strensall Camp 285, 292
Stuart, "Bonnie" Charles Edward 66-69
slave-apprentices 132, 133
slavery 131, 132, 133
slums 254
smallpox 131
Smith, Harry 135
smuggling 75
Soldiers' and Sailors' Family Association 250-51
South Africa 41-47, 131-40, 153, 251, 292
Southampton (England) 245
Spain 55, 75, 76
"Spitalfields" Act, 1773 74
Spears, Maj.-Gen. Edward 279
Spoo, Robert 8
Spraage, Lt.-Col. Basil 48

Stuart, James Francis (Pretender) 66, 69, 71
Stuarts (Scottish Royal Family) 14, 65, 160
Sudan Expedition 128, 130
Swiss mercenaries 26, 69, 70, 162

Tandy, James Napper 33
Tangiers 61, 62, 63
Tantia Tope 121
Territorial and Reserve Forces Act, 1907 270
Tewodros II, Emperor of Abyssinia 129-30
Tipperary, Co. 30, 43, 273
tithes 30, 41
Tory Island, Battle of 36-37
Townley, Francis 72
tradesmen 13, 18, 131, 175, 231, 266
trainings, militia 42, 273, 282, 284, 285, 286
Tralee 240
Transvaal Colony 135-36, 144
Transvaal Republic 47, 133, 135, 136-38, 145
Treasury Department 80, 93, 94, 102, 108, 256
Treaty of Tournai 71
Trekboers 131–32
Trieste (Hapsburg Empire) 7, 8
Tugela River (South Africa) 133
Tulloch, Col. Alexander Murray 108-09
Tuileries Palace 26
Tweedy, Major Brian 8, 9, 84, 130, 154, 161, 167, 174, 178, 181, 210
typhoid fever 142, 144, 147-48
Tyrconnell, Earl of (Richard Talbot) 14, 59

Uitlanders 136-38
Ulster 14, 15, 16, 17, 19, 20, 31, 32, 33, 34, 36, 41, 48
Ulysses 7-9, 13, 53, 130, 161, 219, 229, 258
unemployment 77, 79, 152, 172, 174, 210, 263, 279, 280

Union of England and Scotland 62
Union of Great Britain and Ireland 37, 38, 59
unionists (Irish) 8, 13, 137, 150
United Irishmen 32-34, 65
United States and predecessors 16, 17, 19, 20, 23, 24, 77
unskilled workers 13, 18, 173, 175, 210, 216, 248, 266, 277, 278, 287, 295
Vaal River (South Africa) 133, 135
Valmy (France) 26
venereal disease 218
Verdun (France) 26
Vernon, Admiral Edward 71
Veterinary Corps and predecessors 185, 208, 213
Vice-Regal Lodge 234
Victor Emanuel II, King of Piedmont and Sardinia 100
Victoria, Queen 129, 130
Vinegar Hill, Battle of 36
Volunteers, Great Britain 93, 273, 293-99
Volunteers, Ireland 9, 13, 19-21, 24-25
Voortrekkers 133-35

Wade, Field Marshal George 70, 71, 72
Wajid Ali, Nawab of Oudh 122, 124
War Office Act, 1870 155
War of the Spanish Succession 75-76
Ward, John 185
Warley (England) 120, 171, 178
Waterloo, Battle of 65, 92
Wellington Barracks 198-99, 226
Wellington College 185
West Africa 131, 132, 252, 254
West Indies 242, 2534
Westmoreland (England) 70
Wexford, Co. 36, 37, 41, 273
Wexford Rising 34, 36, 65
Whig Party 74, 134
Wicklow, Co. 34, 36
William & Mary, King and Queen 14, 59, 65
Williamite Wars 14, 75
Wiltshire (England) 214

Witwatersrand (Transvaal, South Africa) 136
Wolfe Tone, Theobold 33, 36, 37
Woolwich (England) 181, 275
workhouse 79, 152, 210, 247, 257, 280
Wyndham, George 45-46

Xhosa 78, 128, 131, 132, 135

Yelverton, Barry 20
Yeomanry, English 31, 94, 170, 300
Yeomanry, Irish 31-32, 34, 37, 41
Yorkshire (England) 78, 285, 292

Zula (Egypt) 129, 130
Zulu 128, 135

Regiments

Regular Army **(Irish in bold.)**
   4th Regiment of Foot 76, 129
   8th Light Dragoons (King's Royal
      Irish Hussars) 105
   19th-21st Hussars 214
   19th-21st Light Dragoons 214
   31st Regiment of Foot 76
   32nd Regiment of Foot 76
   Cameron Highlanders 43
   Coldstream Guards 55
   Connaught Rangers 43, 141
   Durham Light Infantry 178
   Horse Guards 55, 62, 63
   King's Own Yorkshire Light
      Infantry 178
   King's Royal Rifles 155
   Leinster Regiment 179
   Life Guards 55, 56, 62, 63
   Manchester Regiment 203
   Rifle Brigade 155
   Royal Dublin Fusiliers 9, 12, 84, 150,
      161, 174, 178, 279
   Royal Inniskilling Fusiliers 179, 276,
      279, 285
   Royal Irish Fusiliers 150, 200
   Royal Irish Regiment 43, 273
   Royal Irish Rifles 43, 44, 45, 47, 140
   Royal Munster Fusiliers 46, 161, 178,
      214
   Royal Sussex Regiment 179
   Yorkshire Regiment 291

Volunteer Corps
   Artists 296
   Civil Service 296
   Inns of Court 296
   London Scottish 296
   Queen's Westminster 296

Irish Militia
   Antrim 40
   Armagh 40
   Donegal 36, 38
   Dublin City 28, 38
   Kerry 32, 37
   Kilkenny 32, 37, 273
   Longford 37
   Louth 44-46
   North Cork 9, 36-37, 40
   Roscommon 40

Imperial Yeomanry **(Irish in bold.)**
   13th Imperial Yeomanry 48-49
   Denbighshire Hussars 301, 303
   North of Ireland Imperial
      Yeomanry 49, 299
   Northhamptonshire Yeomanry 301
   South Nottinghamshire Hussars 301
   South of Ireland Imperial
      Yeomanry 49, 299

www.ingramcontent.com/pod-product-compliance
Lightning Source LLC
Chambersburg PA
CBHW081202170426
43197CB00018B/2894